T0134883

Slawomir Koziel and Xin-She Yang (Eds.)

Computational Optimization, Methods and Algorithms

Studies in Computational Intelligence, Volume 356

Editor-in-Chief
Prof. Janusz Kacprzyk
Systems Research Institute
Polish Academy of Sciences
ul. Newelska 6
01-447 Warsaw
Poland
E-mail: kacprzyk@ibspan.waw.pl

Further volumes of this series can be found on our homepage: springer.com

Vol. 333. Fedja Hadzic, Henry Tan, and Tharam S. Dillon
Mining of Data with Complex Structures, 2011
ISBN 978-3-642-17556-5

Vol. 334. Álvaro Herrero and Emilio Corchado (Eds.)
Mobile Hybrid Intrusion Detection, 2011
ISBN 978-3-642-18298-3

Vol. 335. Radomir S. Stankovic and Radomir S. Stankovic
From Boolean Logic to Switching Circuits and Automata, 2011
ISBN 978-3-642-11681-0

Vol. 336. Paolo Remagnino, Dorothy N. Monekosso, and Lakhmi C. Jain (Eds.)
Innovations in Defence Support Systems – 3, 2011
ISBN 978-3-642-18277-8

Vol. 337. Sheryl Brahnam and Lakhmi C. Jain (Eds.)
Advanced Computational Intelligence Paradigms in Healthcare 6, 2011
ISBN 978-3-642-17823-8

Vol. 338. Lakhmi C. Jain, Eugene V. Aidman, and Canicious Abeynayake (Eds.)
Innovations in Defence Support Systems – 2, 2011
ISBN 978-3-642-17763-7

Vol. 339. Halina Kwasnicka, Lakhmi C. Jain (Eds.)
Innovations in Intelligent Image Analysis, 2010
ISBN 978-3-642-17933-4

Vol. 340. Heinrich Hussmann, Gerrit Meixner, and Detlef Zuehlke (Eds.)
Model-Driven Development of Advanced User Interfaces, 2011
ISBN 978-3-642-14561-2

Vol. 341. Stéphane Doncieux, Nicolas Bredeche, and Jean-Baptiste Mouret(Eds.)
New Horizons in Evolutionary Robotics, 2011
ISBN 978-3-642-18271-6

Vol. 342. Federico Montesino Pouzols, Diego R. Lopez, and Angel Barriga Barros
Mining and Control of Network Traffic by Computational Intelligence, 2011
ISBN 978-3-642-18083-5

Vol. 343. Kurosh Madani, António Dourado Correia, Agostinho Rosa, and Joaquim Filipe (Eds.)
Computational Intelligence, 2011
ISBN 978-3-642-20205-6

Vol. 344. Atilla Elçi, Mamadou Tadiou Koné, and Mehmet A. Orgun (Eds.)
Semantic Agent Systems, 2011
ISBN 978-3-642-18307-2

Vol. 345. Shi Yu, Léon-Charles Tranchevent, Bart De Moor, and Yves Moreau
Kernel-based Data Fusion for Machine Learning, 2011
ISBN 978-3-642-19405-4

Vol. 346. Weisi Lin, Dacheng Tao, Janusz Kacprzyk, Zhu Li, Ebroul Izquierdo, and Haohong Wang (Eds.)
Multimedia Analysis, Processing and Communications, 2011
ISBN 978-3-642-19550-1

Vol. 347. Sven Helmer, Alexandra Poulovassilis, and Fatos Xhafa
Reasoning in Event-Based Distributed Systems, 2011
ISBN 978-3-642-19723-9

Vol. 348. Beniamino Murgante, Giuseppe Borruso, and Alessandra Lapucci (Eds.)
Geocomputation, Sustainability and Environmental Planning, 2011
ISBN 978-3-642-19732-1

Vol. 349. Vitor R. Carvalho
Modeling Intention in Email, 2011
ISBN 978-3-642-19955-4

Vol. 350. Thanasis Daradoumis, Santi Caballé, Angel A. Juan, and Fatos Xhafa (Eds.)
Technology-Enhanced Systems and Tools for Collaborative Learning Scaffolding, 2011
ISBN 978-3-642-19813-7

Vol. 351. Ngoc Thanh Nguyen, Bogdan Trawiński, and Jason J. Jung (Eds.)
New Challenges for Intelligent Information and Database Systems, 2011
ISBN 978-3-642-19952-3

Vol. 352. Nik Bessis and Fatos Xhafa (Eds.)
Next Generation Data Technologies for Collective Computational Intelligence, 2011
ISBN 978-3-642-20343-5

Vol. 353. Igor Aizenberg
Complex-Valued Neural Networks with Multi-Valued Neurons, 2011
ISBN 978-3-642-20352-7

Vol. 354. Ljupco Kocarev and Shiguo Lian (Eds.)
Chaos-Based Cryptography, 2011
ISBN 978-3-642-20541-5

Vol. 355. Yan Meng and Yaochu Jin (Eds.)
Bio-Inspired Self-Organizing Robotic Systems, 2011
ISBN 978-3-642-20759-4

Vol. 356. Slawomir Koziel and Xin-She Yang (Eds.)
Computational Optimization, Methods and Algorithms, 2011
ISBN 978-3-642-20858-4

Slawomir Koziel and Xin-She Yang (Eds.)

Computational Optimization, Methods and Algorithms

Springer

Dr. Slawomir Koziel
Reykjavik University
School of Science and Engineering
Engineering Optimization & Modeling Center
Menntavegur 1
101 Reykjavik
Iceland
E-mail: Koziel@hr.is

Dr. Xin-She Yang
Mathematics and Scientific Computing
National Physical Laboratory
Teddington TW11 0LW
UK
E-mail: xin-she.yang@npl.co.uk

ISBN 978-3-662-52004-8 ISBN 978-3-642-20859-1(eBook)

DOI 10.1007/978-3-642-20859-1

Studies in Computational Intelligence ISSN 1860-949X

Softcover re-print of the Hardcover 1st edition 2011

Typeset & Cover Design: Scientific Publishing Services Pvt. Ltd., Chennai, India.

Printed on acid-free paper

9 8 7 6 5 4 3 2 1

springer.com

Preface

Computational modelling is becoming the third paradigm of modern sciences, as predicted by the Nobel Prize winner Ken Wilson in 1980s at Cornell University. This so-called third paradigm complements theory and experiment to problem solving. In fact, a substantial amount of research activities in engineering, science and industry today involves mathematical modelling, data analysis, computer simulations, and optimization. The main variations of such activities among different disciplines are the type of problem of interest and the degree as well as extent of the modelling activities. This is especially true in the subjects ranging from engineering design to industry.

Computational optimization is an important paradigm itself with a wide range of applications. In almost all applications in engineering and industry, we almost always try to optimize something - whether to minimize the cost and energy consumption, or to maximize the profit, output, performance and efficiency. In reality, resources, time and money are always limited; consequently, optimization is far more important. The optimal use of available resources of any sort requires a paradigm shift in scientific thinking, which is because most real-world applications have far more complicated factors and parameters as well as constraints to affect the system behaviour. Subsequently, it is not always possible to find the optimal solutions. In practice, we have to settle for suboptimal solutions or even feasible ones that are satisfactory, robust, and practically achievable in a reasonable time scale.

This search for optimality is complicated further by the fact that uncertainty almost always presents in the real-world systems. For example, materials properties always have a certain degree of inhomogeneity. The available materials which are not up to the standards of the design will affect the chosen design significantly. Therefore, we seek not only the optimal design but also robust design in engineering and industry. Another complication to optimization is that most problems are nonlinear and often NP-hard. That is, the solution time for finding optimal solutions is exponential in terms of problem size. In fact, many engineering applications are NP-hard indeed. Thus, the challenge is to find a workable method to tackle the problem and to search for optimal solutions, though such optimality is not always achievable.

Contemporary engineering design is heavily based on computer simulations. This introduces additional difficulties to optimization. Growing demand for accuracy and ever-increasing complexity of structures and systems results in the simulation process being more and more time consuming. Even with an efficient optimization algorithm, the evaluations of the objective functions are often time-consuming. In many engineering fields, the evaluation of a single design can take as long as several hours up to several days or even weeks. On the other hand, simulation-based objective functions are inherently noisy, which makes the optimization process even more difficult. Still, simulation-driven design becomes a must for a growing number of areas, which creates a need for robust and efficient optimization methodologies that can yield satisfactory designs even at the presence of analytically intractable objectives and limited computational resources.

In most engineering design and industrial applications, the objective cannot be expressed in explicit analytical form, as the dependence of the objective on design variables is complex and implicit. This black-box type of optimization often requires a numerical, often computationally expensive, simulator such as computational fluid dynamics and finite element analysis. Furthermore, almost all optimization algorithms are iterative, and require numerous function evaluations. Therefore, any technique that improves the efficiency of simulators or reduces the function evaluation count is crucially important. Surrogate-based and knowledge-based optimization uses certain approximations to the objective so as to reduce the cost of objective evaluations. The approximations are often local, while the quality of approximations is evolving as the iterations proceed. Applications of optimization in engineering and industry are diverse. The contents are quite representative and cover all major topics of computational optimization and modelling.

This book is contributed from worldwide experts who are working in these exciting areas, and each chapter is practically self-contained. This book strives to review and discuss the latest developments concerning optimization and modelling with a focus on methods and algorithms of computational optimization, and also covers relevant applications in science, engineering and industry.

We would like to thank our editors, Drs Thomas Ditzinger and Holger Schaepe, and staff at Springer for their help and professionalism. Last but not least, we thank our families for their help and support.

Slawomir Koziel
Xin-She Yang
2011

List of Contributors

Editors

Slawomir Koziel
Engineering Optimization & Modeling Center, School of Science and Engineering, Reykjavik University, Menntavegur 1, 101 Reykjavik, Iceland (koziel@ru.is)

Xin-She Yang
Mathematics and Scientific Computing, National Physical Laboratory, Teddington, Middlesex TW11 0LW, UK (xin-she.yang@npl.co.uk)

Contributors

Carlos A. Coello Coello
CINVESTAV-IPN, Departamento de Computación, Av. Instituto Politécnico Nacional No. 2508, Col. San Pedro Zacatenco, Delegación Gustavo A. Madero, México, D.F. C.P. 07360. MEXICO (ccoello@cs.cinvestav.mx)

David Echeverría Ciaurri
Department of Energy Resources Engineering, Stanford University, Stanford, CA 94305, USA (echeverr@stanford.edu)

Kathleen R. Fowler
Clarkson University, Department of Math & Computer Science, P.O. Box 5815, Postdam, NY 13699-5815, USA (kfowler@clarkson.edu)

Amir Hossein Gandomi
Department of Civil Engineering, University of Akron, Akron, OH, USA (a.h.gandomi@gmail.com)

Genetha Anne Gray
Department of Quantitative Modeling & Analysis, Sandia National Laboratories, P.O. Box 969, MS 9159, Livermore, CA 94551-0969, USA (gagray@sandia.gov)

Christian A. Hochmuth
Manufacturing Coordination and Technology, Bosch Rexroth AG, 97816, Lohr am Main, Germany (christian.hochmuth@boschrexroth.de)

Ming-Fu Hsu
Department of International Business Studies, National Chi Nan University, Taiwan, ROC (s97212903@ncnu.edu.tw)

Ivan Jeliazkov
Department of Economics, University of California, Irvine, 3151 Social Science Plaza, Irvine CA 92697-5100, U.S.A. (ivan@uci.edu)

Jörg Lässig
Institute of Computational Science, University of Lugano, Via Giuseppe Buffi 13, 6906 Lugano, Switzerland (joerg.laessig@usi.ch)

Slawomir Koziel
Engineering Optimization & Modeling Center, School of Science and Engineering, Reykjavik University, Menntavegur 1, 101 Reykjavik, Iceland (koziel@ru.is)

Oliver Kramer
UC Berkeley, CA 94704, USA, (okramer@icsi.berkeley.edu)

Leifur Leifsson
Engineering Optimization & Modeling Center, School of Science and Engineering, Reykjavik University, Menntavegur 1, 101 Reykjavik, Iceland (leifurth@ru.is)

Alicia Lloro
Department of Economics, University of California, Irvine, 3151 Social Science Plaza, Irvine CA 92697-5100, U.S.A. (alloro@uci.edu)

Alfredo Arias-Montaño
CINVESTAV-IPN, Departamento de Computación, Av. Instituto Politécnico Nacional No. 2508, Col. San Pedro Zacatenco, Delegación Gustavo A. Madero, México, D.F. C.P. 07360. MEXICO (aarias@computacion.cs.cinvestav.mx)

Efrén Mezura-Montes
Laboratorio Nacional de Informática Avanzada (LANIA A.C.), Rébsamen 80, Centro, Xalapa, Veracruz, 91000, MEXICO (emezura@lania.mx)

Stanislav Ogurtsov
Engineering Optimization & Modeling Center, School of Science and Engineering, Reykjavik University, Menntavegur 1, 101 Reykjavik, Iceland (stanislav@ru.is)

Ping-Feng Pai
Department of Information Management, National Chi Nan University, Taiwan, ROC (paipf@ncnu.edu.tw)

Stefanie Thiem
Institute of Physics, Chemnitz University of Technology, 09107 Chemnitz, Germany, (stefanie.thiem@cs.tu-chemnitz.de)

Xin-She Yang
Mathematics and Scientific Computing, National Physical Laboratory, Teddington, Middlesex TW11 0LW, UK (xin-she.yang@npl.co.uk)

Table of Contents

Chapter 1
Computational Optimization: An Overview

Xin-She Yang and Slawomir Koziel

Abstract. Computational optimization is ubiquitous in many applications in engineering and industry. In this chapter, we briefly introduce computational optimization, the optimization algorithms commonly used in practice, and the choice of an algorithm for a given problem. We introduce and analyze the main components of a typical optimization process, and discuss the challenges we may have to overcome in order to obtain optimal solutions correctly and efficiently. We also highlight some of the state-of-the-art developments in optimization and its diverse applications.

1.1 Introduction

Optimization is everywhere, from airline scheduling to finance and from the Internet routing to engineering design. Optimization is an important paradigm itself with a wide range of applications. In almost all applications in engineering and industry, we are always trying to optimize something – whether to minimize the cost and energy consumption, or to maximize the profit, output, performance and efficiency. In reality, resources, time and money are always limited; consequently, optimization is far more important in practice [1, 7, 27, 29]. The optimal use of available resources of any sort requires a paradigm shift in scientific thinking, this is because most real-world applications have far more complicated factors and parameters to affect how the system behaves. The integrated components of such an optimization process are the computational modelling and search algorithms.

Xin-She Yang
Mathematics and Scientific Computing,
National Physical Laboratory, Teddington, Middlesex TW11 0LW, UK
e-mail: xin-she.yang@npl.co.uk

Slawomir Koziel
Engineering Optimization & Modeling Center,
School of Science and Engineering, Reykjavik University,
Menntavegur 1, 101 Reykjavik, Iceland
e-mail: koziel@ru.is

S. Koziel & X.-S. Yang (Eds.): Comput. Optimization, Methods and Algorithms, SCI 356, pp. 1–11.
springerlink.com

Computational modelling is becoming the third paradigm of modern sciences, as predicted by the Nobel Prize winner Ken Wilson in 1980s at Cornell University. This so-called third paradigm complements theory and experiment to problem solving. It is no exaggeration to say almost all research activities in engineering, science and industry today involve a certain amount of modelling, data analysis, computer simulations, and optimization. The main variations of such activities among different disciplines are the type of problem of interest and the degree and extent of the modelling activities. This is especially true in the subjects ranging from engineering design to oil industry and from climate changes to economics.

Search algorithms are the tools and techniques of achieving optimality of the problem of interest. This search for optimality is complicated further by the fact that uncertainty almost always presents in the real-world systems. For example, materials properties such as Young's modulus and strength always have a certain degree of inhomogeneous variations. The available materials which are not up to the standards of the design will affect the chosen design significantly. Therefore, we seek not only the optimal design but also robust design in engineering and industry. Optimal solutions, which are not robust enough, are not practical in reality. Suboptimal solutions or good robust solutions are often the choice in such cases.

Contemporary engineering design is heavily based on computer simulations. This introduces additional difficulties to optimization. Growing demand for accuracy and ever-increasing complexity of structures and systems results in the simulation process being more and more time consuming. In many engineering fields, the evaluation of a single design can take as long as several days or even weeks. On the other hand, simulation-based objective functions are inherently noisy, which makes the optimization process even more difficult. Still, simulation-driven design becomes a must for a growing number of areas, which creates a need for robust and efficient optimization methodologies that can yield satisfactory designs even at the presence of analytically intractable objectives and limited computational resources.

1.2 Computational Optimization

Optimization problems can be formulated in many ways. For example, the commonly used method of least-squares is a special case of maximum-likelihood formulations. By far the most widely formulation is to write a nonlinear optimization problem as

$$\text{minimize } f_i(x), \quad (i = 1, 2, ..., M), \tag{1.1}$$

subject to the constraints

$$h_j(x), \quad (j = 1, 2, ..., J), \tag{1.2}$$

$$g_k(x) \leq 0, \quad (k = 1, 2, ..., K), \tag{1.3}$$

where f_i, h_j and g_k are in general nonlinear functions. Here the design vector $x = (x_1, x_2, ..., x_n)$ can be continuous, discrete or mixed in n-dimensional space. The functions f_i are called objective or cost functions, and when $M > 1$, the optimization is multiobjective or multicriteria [21]. It is possible to combine different objectives into a single objective, and we will focus on the single-objective optimization problems in most part of this book. It is worth pointing out that here we write the problem as a minimization problem, it can also be written as a maximization by simply replacing $f_i(x)$ by $-f_i(x)$.

In a special case when $K = 0$, we have only equality constraints, and the optimization becomes an equality-constrained problem. As an equality $h(x) = 0$ can be written as two inequalities: $h(x) \leq 0$ and $-h(x) \leq 0$, some formulations in the optimization literature use constraints with inequalities only. However, in this book, we will explicitly write out equality constraints in most cases.

When all functions are nonlinear, we are dealing with nonlinear constrained problems. In some special cases when f_i, h_j, g_k are linear, the problem becomes linear, and we can use the widely linear programming techniques such as the simplex method. When some design variables can only take discrete values (often integers), while other variables are real continuous, the problem is of mixed type, which is often difficult to solve, especially for large-scale optimization problems.

A very special class of optimization is the convex optimization [2], which has guaranteed global optimality. Any optimal solution is also the global optimum, and most importantly, there are efficient algorithms of polynomial time to solve such problems [3]. These efficient algorithms such the interior-point methods [12] are widely used and have been implemented in many software packages.

On the other hand, some of the functions such as f_i are integral, while others such as h_j are differential equations, the problem becomes an optimal control problem, and special techniques are required to achieve optimality.

For most applications in this book, we will mainly deal with nonlinear constrained global optimization problems with a single objective. In one chapter by Coello Coello, multiobjective optimization will be discussed in detail. Optimal control and other cases will briefly be discussed in the relevant context in this book.

1.3 Optimization Procedure

In essence, an optimization process consists of three components: model, optimizer and simulator (see Fig. 1.1).

The mathematical or numerical model is the representation of the physical problem using mathematical equations which can be converted into a numerical model and can then be solved numerically. This is the first crucial step in any modelling and optimization. If there is any discrepancy between the intended mathematical model and the actual model in use, we may solve the wrong mathematical model or deal with a different or even wrong problem. Any mathematical model at this stage should be double-checked and validated. Once we are confident that the mathematical model is indeed correct or right set of approximations in most cases, we can proceed to convert it into the right numerical model so that it can be solved numerically and efficiently. Again it is important to ensure the right numerical schemes for dicretization are used; otherwise, we may solve a different problem numerically. At this stage, we should not only ensure that numerical model is right, but also ensure that the model can be solved as fast as possible.

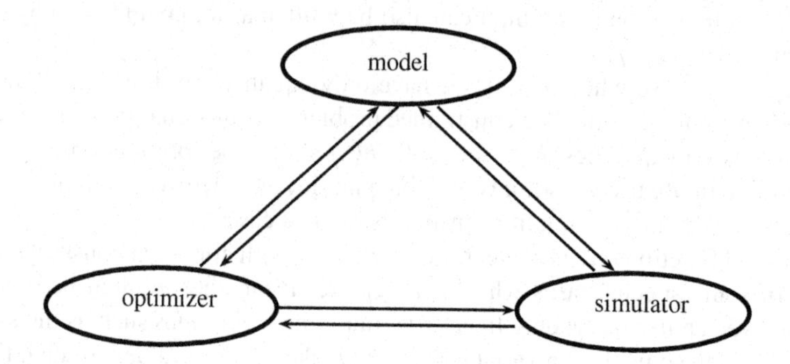

Fig. 1.1 A typical optimization process

Another important step is to use the right algorithm or optimizer so that an optimal set of combination of design variables can be found. An important capability of optimization is to generate or search for new solutions from a known solution (often a random guess or a known solution from experience), which will lead to the convergence of the search process. The ultimate aim of this search process is to find solutions which converge at the global optimum, though this is usually very difficult.

In term of computing time and cost, the most important step is the use of an efficient evaluator or simulator. In most applications, once a correct model representation is made and implemented, an optimization process often involves the evaluation of objective function (such as the aerodynamical efficiency of an airfoil) many times, often thousands and even millions of configurations. Such evaluations often involve the use of extensive computational tools such as a computational fluid dynamics simulator or a finite element solver. This is the step that is most time-consuming, often taking 50% to 90% of the overall computing time.

1.4 Optimizer

1.4.1 Optimization Algorithms

An efficient optimizer is very important to ensure the optimal solutions are reachable. The essence of an optimizer is a search or optimization algorithm implemented correctly so as to carry out the desired search (though not necessarily efficiently). It can be integrated and linked with other modelling components. There are many optimization algorithms in the literature and no single algorithm is suitable for all problems, as dictated by the No Free Lunch Theorems [24].

Optimization algorithms can be classified in many ways, depending on the focus or the characteristics we are trying to compare. Algorithms can be classified as gradient-based (or derivative-based methods) and gradient-free (or derivative-free methods). The classic method of steepest descent and Gauss-Newton methods are gradient-based, as they use the derivative information in the algorithm, while the Nelder-Mead downhill simplex method [18] is a derivative-free method because it only uses the values of the objective, not any derivatives.

From a different point of view, algorithms can be classified as trajectory-based or population-based. A trajectory-based algorithm typically uses a single agent or solution point which will trace out a path as the iterations and optimization process continue. Hill-climbing is trajectory-based, and it links the starting point with the final point via a piecewise zigzag path. Another important example is the simulated annealing [13] which is a widely used metaheuristic algorithm. On the other hand, population-based algorithms such as particle swarm optimization use multiple agents which will interact and trace out multiple paths [11]. Another classic example is the genetic algorithms [8, 10].

Algorithms can also be classified as deterministic or stochastic. If an algorithm works in a mechanically deterministic manner without any random nature, it is called deterministic. For such an algorithm, it will reach the same final solution if we start with the same initial point. Hill-climbing and downhill simplex are good examples of deterministic algorithms. On the other hand, if there is some randomness in the algorithm, the algorithm will usually reach a different point every time we run the algorithm, even though we start with the same initial point. Genetic algorithms and hill-climbing with a random restart are good examples of stochastic algorithms.

Analyzing the stochastic algorithms in more detail, we can single out the type of randomness that a particular algorithm is employing. For example, the simplest and yet often very efficient method is to introduce a random starting point for a deterministic algorithm. The well-known hill-climbing with random restart is a good example. This simple strategy is both efficient in most cases and easy to implement in practice. A more elaborate way to introduce randomness to an algorithm is to use randomness inside different components of an algorithm, and in this case, we often call such algorithm heuristic or more often metaheuristic [23, 26]. A very good example is the popular genetic algorithms which use randomness for crossover and mutation in terms of a crossover probability and a mutation rate. Here, heuristic means to search by trial and error, while metaheuristic is a higher level of heuristics. However, modern literature tends to refer all new stochastic algorithms as metaheuristic. In this book, we will use metaheuristic to mean either. It is worth pointing out that metaheuristic algorithms form a hot research topics and new algorithms appear almost yearly [25, 28].

Memory use can be important to some algorithms. Therefore, optimization algorithms can also be classified as memoryless or history-based. Most algorithms do not use memory explicitly, and only the current best or current state is recorded and all the search history may be discarded. In this sense, such algorithms can thus be considered as memoryless. Genetic algorithms, particle swarm optimization and cuckoo search all fit into this category. It is worth pointing out that we should not confuse the use of memory with the simple record of the current state and the elitism or selection of the fittest. On the other hand, some algorithms indeed use memory/history explicitly. In the Tabu search [9], tabu lists are used to record the move history and recently visited solutions will not be tried again in the near future, and it encourages to explore completely different new solutions, which may save computing effort significantly.

Another type of the algorithm is the so-called mixed-type or hybrid, which uses some combination of deterministic and randomness, or combines one algorithm with another so as to design more efficient algorithms. For example, genetic algorithms can be hybridized with many algorithms such as particle swarm optimization; more specifically, may involve the use of generic operators to modify some components of another algorithm.

From the mobility point of view, algorithms can be classified as local or global. Local search algorithms typically converge towards a local optimum, not necessarily (often not) the global optimum, and such algorithms are often deterministic and have no ability of escaping local optima. Simple hill-climbing is an example. On the other hand, we always try to find the global optimum for a given problem, and if this global optimality is robust, it is often the best, though it is not always possible to find such global optimality. For global optimization, local search algorithms are not suitable. We have to use a global search algorithm. Modern metaheuristic algorithms in most cases are intended for global optimization, though not always successful or efficiently. A simple strategy such as hill-climbing with random restart may change a local search algorithm into a global search. In essence, randomization is an efficient component for global search algorithms. A detailed review of optimization algorithms will be provided later in the chapter on optimization algorithms by Yang.

Straightforward optimization of a given objective function is not always practical. Particularly, if the objective function comes from a computer simulation, it may be computationally expensive, noisy or non-differentiable. In such cases, so-called surrogate-based optimization algorithms may be useful where the direct optimization of the function of interest is replaced by iterative updating and re-optimization of its model - a surrogate [5]. The surrogate model is typically constructed from the sampled data of the original objective function, however, it is supposed to be cheap, smooth, easy to optimize and yet reasonably accurate so that it can produce a good prediction of the function's optimum. Multi-fidelity or variable-fidelity optimization is a special case of the surrogate-based optimization where the surrogate is constructed from the low-fidelity model (or models) of the system of interest [15]. Using variable-fidelity optimization is particularly useful is the reduction of the computational cost of the optimization process is of primary importance.

Whatever the classification of an algorithm is, we have to make the right choice to use an algorithm correctly and sometime a proper combination of algorithms may achieve better results.

1.4.2 Choice of Algorithms

From the optimization point of view, the choice of the right optimizer or algorithm for a given problem is crucially important. The algorithm chosen for an optimization task will largely depend on the type of the problem, the nature of an algorithm, the desired quality of solutions, the available computing resource, time limit, availability of the algorithm implementation, and the expertise of the decision-makers [27].

The nature of an algorithm often determines if it is suitable for a particular type of problem. For example, gradient-based algorithms such as hill-climbing are not suitable for an optimization problem whose objective is discontinuous. Conversely, the type of problem we are trying to solve also determines the algorithms we possibly choose. If the objective function of an optimization problem at hand is highly nonlinear and multimodal, classic algorithms such as hill-climbing and downhill simplex are not suitable, as they are local search algorithms. In this case, global optimizers such as particle swarm optimization and cuckoo search are most suitable [27, 28].

Obviously, the choice is also affected by the desired solution quality and computing resource. As in most applications, computing resources are limited, we have to obtain good solutions (not necessary the best) in a reasonable and practical time. Therefore, we have to balance the resource and solution quality. We cannot achieve solutions with guaranteed quality, though we strive to obtain the quality solutions as best as we possibly can. If time is the main constraint, we can use some greedy methods, or hill-climbing with a few random restarts.

Sometimes, even with the best possible intention, the availability of an algorithms and the expertise of the decision-makers are the ultimate defining factors for choosing an algorithm. Even some algorithms are better, we may not have that algorithm implemented in our system or we do not have such access, which limits our choice. For example, Newton's method, hill-climbing, Nelder-Mead downhill simplex, trust-region methods [3], interior-point methods [19] are implemented in many software packages, which may also increase their popularity in applications.

Even we may have such access, but we may have not the experience in using the algorithms properly and efficiently, in this case we may be more comfortable and more confident in using other algorithms we have already used before. Our experience may be more valuable in selecting the most appropriate and practical solutions than merely using the best possible algorithms.

In practice, even with the best possible algorithms and well-crafted implementation, we may still do not get the desired solutions. This is the nature of nonlinear global optimization, as most of such problems are NP-hard, and no efficient (in the polynomial sense) exist for a given problem. Thus the challenge of the research in computational optimization and applications is to find the right algorithms most suitable for a given problem so as to obtain the good solutions, hopefully also the global best solutions, in a reasonable timescale with a limited amount of resources. We aim to do it efficiently in an optimal way.

1.5 Simulator

To solve an optimization problem, the most computationally extensive part is probably the evaluation of the design objective to see if a proposed solution is feasible and/or if it is optimal. Typically, we have to carry out these evaluations many times, often thousands and even millions of times [25, 27]. Things become even more challenging computationally, when each evaluation task takes a long time via some black-box simulators. If this simulator is a finite element or CFD solver, the running time of each evaluation can take from a few minutes to a few hours or even weeks. Therefore, any approach to save computational time either by reducing the number of evaluations or by increasing the simulator's efficiency will save time and money.

1.5.1 Numerical Solvers

In general, a simulator can be a simple function subroutines, a multiphysics solver, or some external black-box evaluators.

The simplest simulator is probably the direct calculation of an objective function with explicit formulas, this is true for standard test functions (e.g, Rosenbrock's function), simple design problems (e.g., pressure vessel design), and many problems in linear programming [4]. This class of optimization with explicit objectives and constraints may form the majority of optimization problems dealt with in most textbooks and optimization courses.

In engineering and industrial applications, the objectives are often implicit and can only be evaluated through a numerical simulator, often black-box type. For example, in the design of an airfoil, the aerodynamic performance can only be evaluated either numerically or experimentally. Experiments are too expensive in most cases, and thus the only sensible tool is a finite-volume-based CFD solver, which can be called for a given setting of design parameters. In structural engineering, a design of a structure and building is often evaluated by certain design codes, then by a finite element software package, which often takes days or even weeks to run. The evaluation of a proposed solution in real-world applcations is often multidisciplinary, it could involve stress-strain analysis, heat transfer, diffusion, electromagnetic waves, electrical-chemistry, and others. These phenomena are often coupled, which makes the simulations a daunting task, if not impossible. Even so, more and more optimization and design requires such types of evaluations, and the good news is that computing speed is increasing and many efficient numerical methods are becoming routine.

In some rare cases, the optimization objective cannot be written explicitly, and cannot be evaluated using any simulation tools. The only possibility is to use some external means to carry out such evaluations. This often requires experiments, or trial-or-error, or by certain combination of numerical tools, experiment and human expertise. This scenario may imply our lack of understanding of the system/mechanisms, or we may not formulate the problem properly. Sometimes, certain reformulations can often provide better solutions to the problem. For example, many design problems can be simulated by using neural networks and support vector machines. In this case, we know certain objectives of the design, but the relationship between the parameter setting and the system performance/output is not only implicit, but also dynamically changing based on iterative learning/training. Fuzzy system is another example, and in this case, special techniques and methods are used, which is essentially forms a different subject.

In this book, we will mainly focus on the cases in which the objective can be evaluated either using explicit formulas or using black-box numerical tools/solvers. Some case studies of optimization using neural networks will be provided as well.

1.5.2 Simulation Efficiency

In terms of computational effort, an efficient simulator is paramount in controlling the overall efficiency of any computational optimization. If the objectives can be evaluated using explicit functions or formulas, the main barrier is the choice and use of an efficient optimizer. In most cases, the evaluation via a numerical solver such as FE/CFD package is very expansive. This is the bottleneck of the whole optimization process. Therefore, various methods and approximations are designed either to reduce the number of such expensive evaluations or to use some approximation (though more often a good combination of both).

The main way to reduce the number of objective evaluations is to use an efficient algorithm, so that only a small number of such evaluations are needed. In most cases, this is not possible. We have to use some approximation techniques to estimate the objectives, or to construct an approximation model to predict the solver's outputs without actual using the solver. Another way is to replace the original objective function by its lower-fidelity model, e.g., obtained from a computer simulation based on coarsely-discretized structure of interest. The low-fidelity model is faster but not as accurate as the original one, and therefore it has to be corrected. Special techniques have to be applied to use an approximation or corrected low-fidelity model in the optimization process so that the optimal design can be obtained at a low computational cost. All of this falls into the category of surrogate-based optimization[20, 14, 15, 16, 17].

Surrogate models are approximate techniques to construct response surface models, or metamodels [22]. The main idea is to approximate or mimic the system behaviour so as to carry out evaluations cheaply and efficiently, still with accuracy comparable to the actual system. Widely used techniques include polynomial response surface or regression, radial basis functions, ordinary Kriging, artificial neural networks, support vector machines, response correction, space mapping and others. The data used to create the models comes from the sampling of the design space and evaluating the system at selected locations. Surrogate models can be used as predictive tools in the search for the optimal design of the system of interest. This can be realized by iterative re-optimization of the surrogate (exploitation), filling the gaps between sample points to improve glocal accuracy of the model (exploration of the design space) or a mixture of both [5]. The new data is used to update the surrogate. A detailed review of surrogate-modeling techniques and surrogate-base optimization methods will be given by Koziel et al. later.

1.6 Latest Developments

Computational optimization has been always a major research topic in engineering design and industrial applications. New optimization algorithms, numerical methods, approximation techniques and models, and applications are routinely emerging. Loosely speaking, the state-of-the-art developments can put into three areas: new algorithms, new models, and new applications.

Optimization algorithms are constantly being improved. Classic algorithms such as derivative-free methods and pattern search are improved and applied in new applications both successfully and efficiently.

Evolutionary algorithms and metaheuristics are widely used, and there are many successful examples which will be introduced in great detail later in this book. Sometimes, complete new algorithms appear and are designed for global optimization. Hybridization of different algorithms are also very popular. New algorithms such as particle swarm optimization [11], harmony search [6] and cuckoo search [28] are becoming powerful and popular.

As we can see later, this book summarize the latest development of these algorithms in the context of optimization and applications.

Many studies have focused on the methods and techniques of constructing appropriate surrogate models of the high-fidelity simulation data. Surrogate modeling methodologies as well as surrogate-based optimization techniques have improved significantly. The developments of various aspects of surrogate-based optimization, including the design of experiments schemes, methods of constructing and validating the surrogate models, as well as optimization algorithms exploiting surrogate models, both function-approximation and physically-based will be summarized in this book.

New applications are diverse and state-of-the-art developments are summarized, including optimization and applications in network, oil industry, microwave engineering, aerospace engineering, neural networks, environmental modelling, scheduling, structural engineering, classification, economics, and multi-objective optimization problems.

References

1. Arora, J.: Introduction to Optimum Design. McGraw-Hill, New York (1989)
2. Boyd, S.P., Vandenberghe, L.: Convex Optimization. Cambridge University Press, Cambridge (2004)
3. Conn, A.R., Gould, N.I.M., Toint, P.L.: Trust-region methods. SIAM & MPS (2000)
4. Dantzig, G.B.: Linear Programming and Extensions. Princeton University Press, Princeton (1963)

5. Forrester, A.I.J., Keane, A.J.: Recent advances in surrogate-based optimization. Prog. Aerospace Sciences 45(1-3), 50–79 (2009)
6. Geem, Z.W., Kim, J.H., Loganathan, G.V.: A new heuristic optimization: Harmony search. Simulation 76, 60–68 (2001)
7. Gill, P.E., Murray, W., Wright, M.H.: Practical optimization. Academic Press Inc., London (1981)
8. Goldberg, D.E.: Genetic Algorithms in Search, Optimization and Machine Learning. Addison-Wesley, Reading (1989)
9. Glover, F., Laguna, M.: Tabu Search. Kluwer Academic Publishers, USA (1997)
10. Holland, J.: Adaptation in Natural and Artificial Systems. University of Michigan Press, Ann Anbor (1975)
11. Kennedy, J., Eberhart, R.C.: Particle swarm optimization. In: Proc. of IEEE International Conference on Neural Networks, Piscataway, NJ, pp. 1942–1948 (1995)
12. Karmarkar, N.: A new polynomial-time algorithm for linear programming. Combinatorica 4(4), 373–395 (1984)
13. Kirkpatrick, S., Gelatt, C.D., Vecchi, M.P.: Optimization by simulated annealing. Science 220(4598), 671–680 (1983)
14. Koziel, S., Bandler, J.W., Madsen, K.: Quality assessment of coarse models and surrogates for space mapping optimization. Optimization and Engineering 9(4), 375–391 (2008)
15. Koziel, S., Cheng, Q.S., Bandler, J.W.: Space mapping. IEEE Microwave Magazine 9(6), 105–122 (2008)
16. Koziel, S., Bandler, J.W., Madsen, K.: Space mapping with adaptive response correction for microwave design optimization. IEEE Trans. Microwave Theory Tech. 57(2), 478–486 (2009)
17. Koziel, S., Yang, X.S.: Computational Optimization and Applications in Engineering and Industry. Springer, Germany (2011)
18. Nelder, J.A., Mead, R.: A simplex method for function optimization. Computer Journal 7, 308–313 (1965)
19. Nesterov, Y., Nemirovskii, A.: Interior-Point Polynomial Methods in Convex Programming. Society for Industrial and Applied Mathematics (1994)
20. Queipo, N.V., Haftka, R.T., Shyy, W., Goel, T., Vaidynathan, R., Tucker, P.K.: Surrogate-based analysis and optimization. Progress in Aerospace Sciences 41(1), 1–28 (2005)
21. Sawaragi, Y., Nakayama, H., Tanino, T.: Theory of Multiobjective Optimisation. Academic Press, London (1985)
22. Simpson, T.W., Peplinski, J., Allen, J.K.: Metamodels for computer-based engienering design: survey and recommendations. Engienering with Computers 17, 129–150 (2001)
23. Talbi, E.G.: Metaheuristics: From Design to Implementation. John Wiley & Sons, Chichester (2009)
24. Wolpert, D.H., Macready, W.G.: No free lunch theorems for optimization. IEEE Trans. Evolutionary Computation 1, 67–82 (1997)
25. Yang, X.S.: Introduction to Computational Mathematics. World Scientific Publishing, Singapore (2008)
26. Yang, X.S.: Nature-Inspired Metaheuristic Algoirthms. Luniver Press, UK (2008)
27. Yang, X.S.: Engineering Optimization: An Introduction with Metaheuristic Applications. John Wiley & Sons, Chichester (2010)
28. Yang, X.S., Deb, S.: Engineering optimization by cuckoo search. Int. J. Math. Modelling Num. Optimisation 1(4), 330–343 (2010)
29. Yang, X.S., Koziel, S.: Computational optimization, modelling and simulation – a paradigm shift. Procedia Computer Science 1(1), 1291–1294 (2010)

Chapter 2
Optimization Algorithms

Xin-She Yang

Abstract. The right choice of an optimization algorithm can be crucially important in finding the right solutions for a given optimization problem. There exist a diverse range of algorithms for optimization, including gradient-based algorithms, derivative-free algorithms and metaheuristics. Modern metaheuristic algorithms are often nature-inspired, and they are suitable for global optimization. In this chapter, we will briefly introduce optimization algorithms such as hill-climbing, trust-region method, simulated annealing, differential evolution, particle swarm optimization, harmony search, firefly algorithm and cuckoo search.

2.1 Introduction

Algorithms for optimization are more diverse than the types of optimization, though the right choice of algorithms is an important issue, as we discussed in the first chapter where we have provided an overview. There are a wide range of optimization algorithms, and a detailed description of each can take up the whole book of more than several hundred pages. Therefore, in this chapter, we will introduce a few important algorithms selected from a wide range of optimization algorithms [4, 27, 31], with a focus on the metaheuristic algorithms developed after the 1990s. This selection does not mean that the algorithms not described here are not popular. In fact, they may be equally widely used. Whenever an algorithm is used in this book, we will try to provide enough details so that readers can see how they are implemented; alternatively, in some cases, enough citations and links will be provided so that interested readers can pursue further research using these references as a good start.

Xin-She Yang
Mathematics and Scientific Computing,
National Physical Laboratory,
Teddington, Middlesex TW11 0LW, UK
e-mail: xin-she.yang@npl.co.uk

S. Koziel & X.-S. Yang (Eds.): Comput. Optimization, Methods and Algorithms, SCI 356, pp. 13–31.
springerlink.com

2.2 Derivative-Based Algorithms

Derivative-based or gradient-based algorithms use the information of derivatives. They are very efficient as local search algorithms, but may have the disadvantage of being trapped in a local optimum if the problem of interest is not convex. It is required that the objective function is sufficiently smooth so that its first (and often second) derivatives exist. Discontinuity in objective functions may render such methods unsuitable. One of the classical examples is the Newton's method, while a modern example is the method of conjugate gradient. Gradient-base methods are widely used in many applications and discrete modelling [3, 20].

2.2.1 *Newton's Method and Hill-Climbing*

One of the most widely used algorithms is Newton's method, which is a root-finding algorithm as well as a gradient-based optimization algorithm [10]. For a given function $f(x)$, its Tayler expansions

$$f(x) = f(x_n) + (\nabla f(x_n))^T \Delta x + \frac{1}{2}\Delta x^T \nabla^2 f(x_n)\Delta x + ..., \tag{2.1}$$

in terms of $\Delta x = x - x_n$ about a fixed point x_n leads to the following iterative formula

$$x = x_n - H^{-1}\nabla f(x_n), \tag{2.2}$$

where $H^{-1}(x^{(n)})$ is the inverse of the symmetric Hessian matrix $H = \nabla^2 f(x_n)$, which is defined as

$$H(x) \equiv \nabla^2 f(x) \equiv \begin{pmatrix} \frac{\partial^2 f}{\partial x_1^2} & \cdots & \frac{\partial^2 f}{\partial x_1 \partial x_n} \\ \vdots & & \vdots \\ \frac{\partial^2 f}{\partial x_n \partial x_1} & \cdots & \frac{\partial f^2}{\partial x_n^2} \end{pmatrix}. \tag{2.3}$$

Starting from an initial guess vector $x^{(0)}$, the iterative Newton's formula for the nth iteration becomes

$$x^{(n+1)} = x^{(n)} - H^{-1}(x^{(n)})\nabla f(x^{(n)}). \tag{2.4}$$

In order to speed up the convergence, we can use a smaller step size $\alpha \in (0,1]$ and we have the modified Newton's method

$$x^{(n+1)} = x^{(n)} - \alpha H^{-1}(x^{(n)})\nabla f(x^{(n)}). \tag{2.5}$$

It is often time-consuming to calculate the Hessian matrix using second derivatives. In this case, a simple and yet efficient alternative is to use an identity matrix I to approximate H so that $H^{-1} = I$, which leads to the quasi-Newton method

$$x^{(n+1)} = x^{(n)} - \alpha I \nabla f(x^{(n)}). \tag{2.6}$$

In essence, this is the steepest descent method.

For a maximization problem, the steepest descent becomes a hill-climbing. That is, the aim is to climb up to the highest peak or to find the highest possible value of an objective $f(x)$ from the current point $x^{(n)}$. From the Taylor expansion of $f(x)$ about $x^{(n)}$, we have

$$f(x^{(n+1)}) = f(x^{(n)} + \Delta s) \approx f(x^{(n)} + (\nabla f(x^{(n)}))^T \Delta s, \tag{2.7}$$

where $\Delta s = x^{(n+1)} - x^{(n)}$ is the increment vector. Since we are trying to find a better (higher) approximation to the objective function, it requires that

$$f(x^{(n)} + \Delta s) - f(x^{(n)}) = (\nabla f)^T \Delta s > 0. \tag{2.8}$$

From vector analysis, we know that the inner product $u^T v$ of two vectors u and v is the largest when they are parallel. Therefore, we have

$$\Delta s = \alpha \nabla f(x^{(n)}), \tag{2.9}$$

where $\alpha > 0$ is the step size. In the case of minimization, the direction Δs is along the steepest descent in the negative gradient direction.

It is worth pointing out that the choice of the step size α is very important. A very small step size means slow movement towards the local optimum, while a large step may overshoot and subsequently makes it move far away from the local optimum. Therefore, the step size $\alpha = \alpha^{(n)}$ should be different at each iteration and should be chosen so as to maximize or minimize the objective function, depending on the context of the problem.

2.2.2 Conjugate Gradient Method

The conjugate gradient method is one of most widely used algorithms and it belongs to a wider class of the so-called Krylov subspace iteration methods. The conjugate gradient method was pioneered by Magnus Hestenes, Eduard Stiefel and Cornelius Lanczos in the 1950s [13]. In essence, the conjugate gradient method solves the following linear system

$$Au = b, \tag{2.10}$$

where A is often a symmetric positive definite matrix. This system is equivalent to minimizing the following function $f(u)$

$$f(u) = \frac{1}{2} u^T A u - b^T u + v, \tag{2.11}$$

where v is a vector constant and can be taken to be zero. We can easily see that $\nabla f(u) = 0$ leads to $Au = b$. In theory, these iterative methods are closely related to the Krylov subspace $\mathcal{K}_n$ spanned by A and b as defined by

$$\mathcal{K}_n(A, b) = \{Ib, Ab, A^2 b, ..., A^{n-1} b\}, \tag{2.12}$$

where $A^0 = I$.

If we use an iterative procedure to obtain the approximate solution u_n to $Au = b$ at nth iteration, the residual is given by

$$r_n = b - Au_n, \tag{2.13}$$

which is essentially the negative gradient $\nabla f(u_n)$.

The search direction vector in the conjugate gradient method can subsequently be determined by

$$d_{n+1} = r_n - \frac{d_n^T A r_n}{d_n^T A d_n} d_n. \tag{2.14}$$

The solution often starts with an initial guess u_0 at $n = 0$, and proceeds iteratively. The above steps can compactly be written as

$$u_{n+1} = u_n + \alpha_n d_n, \quad r_{n+1} = r_n - \alpha_n A d_n, \tag{2.15}$$

and

$$d_{n+1} = r_{n+1} + \beta_n d_n, \tag{2.16}$$

where

$$\alpha_n = \frac{r_n^T r_n}{d_n^T A d_n}, \quad \beta_n = \frac{r_{n+1}^T r_{n+1}}{r_n^T r_n}. \tag{2.17}$$

Iterations stop when a prescribed accuracy is reached. In the case when A is not symmetric, we can use other algorithms such as the generalized minimal residual (GMRES) algorithm developed by Y. Saad and M. H. Schultz in 1986.

2.3 Derivative-Free Algorithms

Algorithms using derivatives are efficient, but may pose certain strict requirements on the objective functions. In case of discontinuity exists in objective functions, derivative-free algorithms may be more efficient and natural. Hooke-Jeeves pattern search is among one of the earliest, which forms the basis of many modern variants of pattern search. Nelder-Mead downhill simplex method [19] is another good example of derivative-free algorithms. Furthermore, the widely used trust-region method use some form of approximation to the objective function in a local region, and many surrogate-based models have strong similarities to the pattern search method.

2.3.1 *Pattern Search*

Many search algorithms such as the steepest descent method experience slow convergence near the local minimum. They are also memoryless because the past information during the search is not used to produce accelerated moves in the future. The only information they use is the current location $x^{(n)}$, gradient and value of the objective itself at step n. If the past information such as the steps at $n-1$ and n is properly used to generate a new move at step $n+1$, it may speed up the convergence. The Hooke-Jeeves pattern search method is one of such methods that incorporate the past history of iterations in producing a new search direction.

The Hooke-Jeeves pattern search method consists of two moves: exploratory move and pattern move. The exploratory moves explore the local behaviour and information of the objective function so as to identify any potential sloping valleys if they exist. For any given step size (each coordinate direction can have a different increment) $\Delta_i (i = 1, 2, ..., p)$, exploration movement performs from an initial starting point along each coordinate direction by increasing or decreasing $\pm\Delta_i$, if the new value of the objective function does not increase (for a minimization problem), that is $f(x_i^{(n)}) \leq f(x_i^{(n-1)})$, the exploratory move is considered as successful. If it is not successful, then a step is tried in the opposite direction, and the result is updated only if it is successful. When all the d coordinates have been explored, the resulting point forms a base point $x^{(n)}$.

The pattern move intends to move the current base $x^{(n)}$ along the base line $(x^{(n)} - x^{(n-1)})$ from the previous (historical) base point to the current base point. The move is carried out by the following formula

$$x^{(n+1)} = x^{(n)} + [x^{(n)} - x^{(n-1)}]. \tag{2.18}$$

Then $x^{(n+1)}$ forms a new temporary base point for further new exploratory moves. If the pattern move produces improvement (lower value of $f(x)$), the new base point $x^{(n+1)}$ is successfully updated. If the pattern move does not lead to any improvement or a lower value of the objective function, then the pattern move is discarded and a new search starts from $x^{(n)}$, and the new search moves should use a smaller step size by reducing increments D_i/γ where $\gamma > 1$ is the step reduction factor. Iterations continue until the prescribed tolerance ε is met.

2.3.2 *Trust-Region Method*

The so-called trust-region method is among the most widely used optimization algorithms, and its fundamental ideas have developed over many years with many seminal papers by a dozen of pioneers. A good history review of the trust-region methods can be found [5, 6]. Then, in 1970, Powell proved the global convergence for the trust-region method [22].

In the trust-region algorithm, a fundamental step is to approximate the nonlinear objective function by using truncated Taylor expansions, often in a quadratic form in a so-called trust region which is the shape of the trust region is a hyperellipsoid.

The approximation to the objective function in the trust region will make it simpler to find the next trial solution x_{k+1} from the current solution x_k. Then, we intend to find x_{k+1} with a sufficient decrease in the objective function. How good the approximation ϕ_k is to the actual objective $f(x)$ can be measured by the ratio of the achieved decrease to the predicted decrease

$$\gamma_k = \frac{f(x_k) - f(x_{k+1})}{\phi_k(x_k) - \phi_k(x_{k+1})}. \tag{2.19}$$

If this ratio is close to unity, we have a good approximation and then should move the trust region to x_{k+1}. The trust-region should move and update iteratively until the (global) optimality is found or until a fixed number of iterations is reached.

There are many other methods, and one of the most powerful and widely used is the polynomial-time efficient algorithm, called the interior-point method [16], and many variants have been developed since 1984.

All these above algorithms are deterministic, as they have no random components. Thus, they usually have some disadvantages in dealing with highly nonlinear, multimodal, global optimization problems. In fact, some randomization is useful and necessary in algorithms, and metaheuristic algorithms are such powerful techniques.

2.4 Metaheuristic Algorithms

Metaheuristic algorithms are often nature-inspired, and they are now among the most widely used algorithms for optimization. They have many advantages over conventional algorithms, as discussed in the first chapter for introduction and overview. There are a few recent books which are solely dedicated to metaheuristic algorithms [27, 29, 30]. Metaheuristic algorithms are very diverse, including genetic algorithms, simulated annealing, differential evolution, ant and bee algorithms, particle swarm optimization, harmony search, firefly algorithm, cuckoo search and others. Here we will introduce some of these algorithms briefly.

2.4.1 Simulated Annealling

Simulated annealing developed by Kirkpatrick et al. in 1983 is among the first metaheuristic algorithms, and it has been applied in almost every area of optimization [17]. Unlike the gradient-based methods and other deterministic search methods, the main advantage of simulated annealing is its ability to avoid being trapped in local minima. The basic idea of the simulated annealing algorithm is to use random search in terms of a Markov chain, which not only accepts changes that improve the objective function, but also keeps some changes that are not ideal.

In a minimization problem, for example, any better moves or changes that decrease the value of the objective function f will be accepted; however, some changes that increase f will also be accepted with a probability p. This probability p, also called the transition probability, is determined by

$$p = \exp[-\frac{\Delta E}{k_B T}], \tag{2.20}$$

where k_B is the Boltzmann's constant, and T is the temperature for controlling the annealing process. ΔE is the change of the energy level. This transition probability is based on the Boltzmann distribution in statistical mechanics.

The simplest way to link ΔE with the change of the objective function Δf is to use

$$\Delta E = \gamma \Delta f, \tag{2.21}$$

where γ is a real constant. For simplicity without losing generality, we can use $k_B = 1$ and $\gamma = 1$. Thus, the probability p simply becomes

$$p(\Delta f, T) = e^{-\Delta f / T}. \tag{2.22}$$

Whether or not a change is accepted, a random number r is often used as a threshold. Thus, if $p > r$, or

$$p = e^{-\Delta f / T} > r, \tag{2.23}$$

the move is accepted.

Here the choice of the right initial temperature is crucially important. For a given change Δf, if T is too high ($T \to \infty$), then $p \to 1$, which means almost all the changes will be accepted. If T is too low ($T \to 0$), then any $\Delta f > 0$ (worse solution) will rarely be accepted as $p \to 0$, and thus the diversity of the solution is limited, but any improvement Δf will almost always be accepted. In fact, the special case $T \to 0$ corresponds to the classical hill-climbing because only better solutions are accepted, and the system is essentially climbing up or descending along a hill. Therefore, if T is too high, the system is at a high energy state on the topological landscape, and the minima are not easily reached. If T is too low, the system may be trapped in a local minimum (not necessarily the global minimum), and there is not enough energy for the system to jump out the local minimum to explore other minima including the global minimum. So a proper initial temperature should be calculated.

Another important issue is how to control the annealing or cooling process so that the system cools down gradually from a higher temperature to ultimately freeze to a global minimum state. There are many ways of controlling the cooling rate or the decrease of the temperature. geometric cooling schedules are often widely used, which essentially decrease the temperature by a cooling factor $0 < \alpha < 1$ so that T is replaced by αT or

$$T(t) = T_0 \alpha^t, \quad t = 1, 2, ..., t_f, \tag{2.24}$$

where t_f is the maximum number of iterations. The advantage of this method is that $T \to 0$ when $t \to \infty$, and thus there is no need to specify the maximum number of iterations if a tolerance or accuracy is prescribed. Simulated annealling has been applied in a wide range of optimization problems [17, 20].

2.4.2 Genetic Algorithms and Differential Evolution

Simulated annealing is a trajectory-based algorithm, as it only uses a single agent. Other algorithms such as genetic algorithms use multiple agents or a population to carry out the search, which may have some advantage due to its potential parallelism.

Genetic algorithms are a classic of algorithms based on the abstraction of Darwin's evolution of biological systems, pioneered by J. Holland and his collaborators in the 1960s and 1970s [14]. Holland was the first to use genetic operators such as the crossover and recombination, mutation, and selection in the study of adaptive and artificial systems. Genetic algorithms have two main advantages over traditional algorithms: the ability of dealing with complex problems and parallelism. Whether the objective function is stationary or transient, linear or nonlinear, continuous or discontinuous, it can be dealt with by genetic algorithms. Multiple genes can be suitable for parallel implementation.

Three main components or genetic operators in genetic algorithms are: crossover, mutation, and selection of the fittest. Each solution is encoded in a string (often binary or decimal), called a chromosome. The crossover of two parent strings produce offsprings (new solutions) by swapping part or genes of the chromosomes. Crossover has a higher probability, typically 0.8 to 0.95. On the other hand, mutation is carried out by flipping some digits of a string, which generates new solutions. This mutation probability is typically low, from 0.001 to 0.05. New solutions generated in each generation will be evaluated by their fitness which is linked to the objective function of the optimization problem. The new solutions are selected according to their fitness – selection of the fittest. Sometimes, in order to make sure that the best solutions remain in the population, the best solutions are passed onto the next generation without much change, this is called elitism.

Genetic algorithms have been applied to almost all area of optimization, design and applications. There are hundreds of good books and thousand of research articles. There are many variants and hybridization with other algorithms, and interested readers can refer to more advanced literature such as [12, 14].

Differential evolution (DE) was developed by R. Storn and K. Price by their nominal papers in 1996 and 1997 [25, 26]. It is a vector-based evolutionary algorithm, and can be considered as a further development to genetic algorithms. It is a stochastic search algorithm with self-organizing tendency and does not use the information of derivatives. Thus, it is a population-based, derivative-free method.

As in genetic algorithms, design parameters in a d-dimensional search space are represented as vectors, and various genetic operators are operated over their bits of strings. However, unlikely genetic algorithms, differential evolution carries out operations over each component (or each dimension of the solution). Almost everything is done in terms of vectors. For example, in genetic algorithms, mutation is carried out at one site or multiple sites of a chromosome, while in differential evolution, a difference vector of two randomly-chosen population vectors is used to perturb an existing vector. Such vectorized mutation can be viewed as a self-organizing search, directed towards an optimality.

For a d-dimensional optimization problem with d parameters, a population of n solution vectors are initially generated, we have x_i where $i = 1, 2, ..., n$. For each solution x_i at any generation t, we use the conventional notation as

$$x_i^t = (x_{1,i}^t, x_{2,i}^t, ..., x_{d,i}^t), \tag{2.25}$$

which consists of d-components in the d-dimensional space. This vector can be considered as the chromosomes or genomes.

Differential evolution consists of three main steps: mutation, crossover and selection.

Mutation is carried out by the mutation scheme. For each vector x_i at any time or generation t, we first randomly choose three distinct vectors x_p, x_q and x_r at t, and then generate a so-called donor vector by the mutation scheme

$$v_i^{t+1} = x_p^t + F(x_q^t - x_r^t), \tag{2.26}$$

where $F \in [0,2]$ is a parameter, often referred to as the differential weight. This requires that the minimum number of population size is $n \geq 4$. In principle, $F \in [0,2]$, but in practice, a scheme with $F \in [0,1]$ is more efficient and stable.

The crossover is controlled by a crossover probability $C_r \in [0,1]$ and actual crossover can be carried out in two ways: binomial and exponential. Selection is essentially the same as that used in genetic algorithms. It is to select the most fittest, and for minimization problem, the minimum objective value. Therefore, we have

$$x_i^{t+1} = \begin{cases} u_i^{t+1} & \text{if } f(u_i^{t+1}) \leq f(x_i^t), \\ x_i^t & \text{otherwise.} \end{cases} \tag{2.27}$$

Most studies have focused on the choice of F, C_r and n as well as the modification of (2.26). In fact, when generating mutation vectors, we can use many different ways of formulating (2.26), and this leads to various schemes with the naming convention: DE/x/y/z where x is the mutation scheme (rand or best), y is the number of difference vectors, and z is the crossover scheme (binomial or exponential). The basic DE/Rand/1/Bin scheme is given in (2.26). Following a similar strategy, we can design various schemes. In fact, 10 different schemes have been formulated, and for details, readers can refer to [23].

2.4.3 Particle Swarm Optimization

Particle swarm optimization (PSO) was developed by Kennedy and Eberhart in 1995 [15], based on the swarm behaviour such as fish and bird schooling in nature. Since then, PSO has generated much wider interests, and forms an exciting, ever-expanding research subject, called swarm intelligence. PSO has been applied to almost every area in optimization, computational intelligence, and design/scheduling applications. There are at least two dozens of PSO variants, and hybrid algorithms by combining PSO with other existing algorithms are also increasingly popular.

This algorithm searches the space of an objective function by adjusting the trajectories of individual agents, called particles, as the piecewise paths formed by positional vectors in a quasi-stochastic manner. The movement of a swarming particle consists of two major components: a stochastic component and a deterministic component. Each particle is attracted toward the position of the current global best g^* and its own best location x_i^* in history, while at the same time it has a tendency to move randomly.

Let x_i and v_i be the position vector and velocity for particle i, respectively. The new velocity vector is determined by the following formula

$$v_i^{t+1} = v_i^t + \alpha\varepsilon_1 \odot [g^* - x_i^t] + \beta\varepsilon_2 \odot [x_i^* - x_i^t]. \quad (2.28)$$

where ε_1 and ε_2 are two random vectors, and each entry taking the values between 0 and 1. The Hadamard product of two matrices $u \odot v$ is defined as the entrywise product, that is $[u \odot v]_{ij} = u_{ij}v_{ij}$. The parameters α and β are the learning parameters or acceleration constants, which can typically be taken as, say, $\alpha \approx \beta \approx 2$.

The initial locations of all particles should distribute relatively uniformly so that they can sample over most regions, which is especially important for multimodal problems. The initial velocity of a particle can be taken as zero, that is, $v_i^{t=0} = 0$. The new position can then be updated by

$$x_i^{t+1} = x_i^t + v_i^{t+1}. \quad (2.29)$$

Although v_i can be any values, it is usually bounded in some range $[0, v_{\max}]$.

There are many variants which extend the standard PSO algorithm [15, 30, 31], and the most noticeable improvement is probably to use inertia function $\theta(t)$ so that v_i^t is replaced by $\theta(t)v_i^t$

$$v_i^{t+1} = \theta v_i^t + \alpha\varepsilon_1 \odot [g^* - x_i^t] + \beta\varepsilon_2 \odot [x_i^* - x_i^t], \quad (2.30)$$

where θ takes the values between 0 and 1. In the simplest case, the inertia function can be taken as a constant, typically $\theta \approx 0.5 \sim 0.9$. This is equivalent to introducing a virtual mass to stabilize the motion of the particles, and thus the algorithm is expected to converge more quickly.

2.4.4 *Harmony Search*

Harmony Search (HS) is a relatively new heuristic optimization algorithm and it was first developed by Z. W. Geem et al. in 2001 [9]. Harmony search can be explained in more detail with the aid of the discussion of the improvisation process by a musician. When a musician is improvising, he or she has three possible choices: (1) play any famous piece of music (a series of pitches in harmony) exactly from his or her memory; (2) play something similar to a known piece (thus adjusting the pitch slightly); or (3) compose new or random notes. If we formalize these three options for optimization, we have three corresponding components: usage of harmony memory, pitch adjusting, and randomization.

The usage of harmony memory is important as it is similar to choose the best fit individuals in the genetic algorithms. This will ensure the best harmonies will be carried over to the new harmony memory. In order to use this memory more effectively, we can assign a parameter $r_{\text{accept}} \in [0,1]$, called harmony memory accepting or considering rate. If this rate is too low, only few best harmonies are selected and it may converge too slowly. If this rate is extremely high (near 1), almost all the harmonies are used in the harmony memory, then other harmonies are not explored well, leading to potentially wrong solutions. Therefore, typically, $r_{\text{accept}} = 0.7 \sim 0.95$.

To adjust the pitch slightly in the second component, we have to use a method such that it can adjust the frequency efficiently. In theory, the pitch can be adjusted linearly or nonlinearly, but in practice, linear adjustment is used. If x_{old} is the current solution (or pitch), then the new solution (pitch) x_{new} is generated by

$$x_{\text{new}} = x_{\text{old}} + b_p\,(2\,\varepsilon - 1), \tag{2.31}$$

where ε is a random number drawn from a uniform distribution $[0,1]$. Here b_p is the bandwidth, which controls the local range of pitch adjustment. In fact, we can see that the pitch adjustment (2.31) is a random walk.

Pitch adjustment is similar to the mutation operator in genetic algorithms. We can assign a pitch-adjusting rate (r_{pa}) to control the degree of the adjustment. If r_{pa} is too low, then there is rarely any change. If it is too high, then the algorithm may not converge at all. Thus, we usually use $r_{pa} = 0.1 \sim 0.5$ in most simulations.

The third component is the randomization, which is to increase the diversity of the solutions. Although adjusting pitch has a similar role, but it is limited to certain local pitch adjustment and thus corresponds to a local search. The use of randomization can drive the system further to explore various regions with high solution diversity so as to find the global optimality. HS has been applied to solve many optimization problems including function optimization, water distribution network, groundwater modelling, energy-saving dispatch, structural design, vehicle routing, and others.

2.4.5 Firefly Algorithm

Firefly Algorithm (FA) was developed by Xin-She Yang in 2007 [29, 32], which was based on the flashing patterns and behaviour of fireflies. In essence, FA uses the following three idealized rules:

- Fireflies are unisex so that one firefly will be attracted to other fireflies regardless of their sex.
- The attractiveness is proportional to the brightness and they both decrease as their distance increases. Thus for any two flashing fireflies, the less brighter one will move towards the brighter one. If there is no brighter one than a particular firefly, it will move randomly.
- The brightness of a firefly is determined by the landscape of the objective function.

As a firefly's attractiveness is proportional to the light intensity seen by adjacent fireflies, we can now define the variation of attractiveness β with the distance r by

$$\beta = \beta_0 e^{-\gamma r^2}, \tag{2.32}$$

where β_0 is the attractiveness at $r = 0$.

The movement of a firefly i is attracted to another more attractive (brighter) firefly j is determined by

$$x_i^{t+1} = x_i^t + \beta_0 e^{-\gamma r_{ij}^2}(x_j^t - x_i^t) + \alpha\, \varepsilon_i^t, \tag{2.33}$$

where the second term is due to the attraction. The third term is randomization with α being the randomization parameter, and ε_i^t is a vector of random numbers drawn from a Gaussian distribution or uniform distribution at time t. If $\beta_0 = 0$, it becomes a simple random walk. Furthermore, the randomization ε_i^t can easily be extended to other distributions such as Lévy flights.

The Lévy flight essentially provides a random walk whose random step length is drawn from a Lévy distribution

$$\text{Lévy} \sim u = t^{-\lambda}, \quad (1 < \lambda \leq 3), \tag{2.34}$$

which has an infinite variance with an infinite mean. Here the steps essentially form a random walk process with a power-law step-length distribution with a heavy tail. Some of the new solutions should be generated by Lévy walk around the best solution obtained so far, this will speed up the local search.

A demo version of firefly algorithm implementation, without Lévy flights, can be found at Mathworks file exchange web site.[1] Firefly algorithm has attracted much attention [1, 24]. A discrete version of FA can efficiently solve NP-hard scheduling problems [24], while a detailed analysis has demonstrated the efficiency of FA over a wide range of test problems, including multobjective load dispatch problems [1].

2.4.6 *Cuckoo Search*

Cuckoo search (CS) is one of the latest nature-inspired metaheuristic algorithms, developed in 2009 by Xin-She Yang and Suash Deb [34]. CS is based on the brood parasitism of some cuckoo species. In addition, this algorithm is enhanced by the so-called Lévy flights [21], rather than by simple isotropic random walks. Recent studies show that CS is potentially far more efficient than PSO and genetic algorithms [35].

Cuckoo are fascinating birds, not only because of the beautiful sounds they can make, but also because of their aggressive reproduction strategy. Some species such as the *ani* and *Guira* cuckoos lay their eggs in communal nests, though they may remove others' eggs to increase the hatching probability of their own eggs. Quite a number of species engage the obligate brood parasitism by laying their eggs in the nests of other host birds (often other species).

There are three basic types of brood parasitism: intraspecific brood parasitism, cooperative breeding, and nest takeover. Some host birds can engage direct conflict with the intruding cuckoos. If a host bird discovers the eggs are not their owns, they

[1] http://www.mathworks.com/matlabcentral/fileexchange/29693-firefly-algorithm

will either get rid of these alien eggs or simply abandon its nest and build a new nest elsewhere. Some cuckoo species such as the New World brood-parasitic *Tapera* have evolved in such a way that female parasitic cuckoos are often very specialized in the mimicry in colour and pattern of the eggs of a few chosen host species. This reduces the probability of their eggs being abandoned and thus increases their reproductivity.

In addition, the timing of egg-laying of some species is also amazing. Parasitic cuckoos often choose a nest where the host bird just laid its own eggs. In general, the cuckoo eggs hatch slightly earlier than their host eggs. Once the first cuckoo chick is hatched, the first instinct action it will take is to evict the host eggs by blindly propelling the eggs out of the nest, which increases the cuckoo chick's share of food provided by its host bird. Studies also show that a cuckoo chick can also mimic the call of host chicks to gain access to more feeding opportunity.

For simplicity in describing the Cuckoo Search, we now use the following three idealized rules:

- Each cuckoo lays one egg at a time, and dumps it in a randomly chosen nest;
- The best nests with high-quality eggs will be carried over to the next generations;
- The number of available host nests is fixed, and the egg laid by a cuckoo is discovered by the host bird with a probability $p_a \in [0, 1]$. In this case, the host bird can either get rid of the egg, or simply abandon the nest and build a completely new nest.

As a further approximation, this last assumption can be approximated by a fraction p_a of the n host nests are replaced by new nests (with new random solutions).

For a maximization problem, the quality or fitness of a solution can simply be proportional to the value of the objective function. Other forms of fitness can be defined in a similar way to the fitness function in genetic algorithms.

For the implementation point of view, we can use the following simple representations that each egg in a nest represents a solution, and each cuckoo can lay only one egg (thus representing one solution), the aim is to use the new and potentially better solutions (cuckoos) to replace a not-so-good solution in the nests. Obviously, this algorithm can be extended to the more complicated case where each nest has multiple eggs representing a set of solutions. For this present work, we will use the simplest approach where each nest has only a single egg. In this case, there is no distinction between egg, nest or cuckoo, as each nest corresponds to one egg which also represents one cuckoo.

Based on these three rules, the basic steps of the Cuckoo Search (CS) can be summarized as the pseudo code shown in Fig. 2.1.

When generating new solutions $x^{(t+1)}$ for, say, a cuckoo i, a Lévy flight is performed

$$x_i^{(t+1)} = x_i^{(t)} + \alpha \oplus \text{Lévy}(\lambda), \tag{2.35}$$

where $\alpha > 0$ is the step size which should be related to the scales of the problem of interests. In most cases, we can use $\alpha = O(L/10)$ where L is the characteristic scale of the problem of interest, while in some case $\alpha = O(L/100)$ can be more effective and avoid flying to far. The above equation is essentially the stochastic

Objective function $f(x)$, $x = (x_1, ..., x_d)^T$
Generate initial population of n host nests x_i
while ($t <$ MaxGeneration) or (stop criterion)
Get a cuckoo randomly/generate a solution by Lévy flights
and then evaluate its quality/fitness F_i
Choose a nest among n (say, j) randomly
if $(F_i > F_j)$,
Replace j by the new solution
end
A fraction (p_a) of worse nests are abandoned
and new ones/solutions are built/generated
Keep best solutions (or nests with quality solutions)
Rank the solutions and find the current best
end while

Fig. 2.1 Pseudo code of the Cuckoo Search (CS)

equation for a random walk. In general, a random walk is a Markov chain whose next status/location only depends on the current location (the first term in the above equation) and the transition probability (the second term). The product $\oplus$ means entrywise multiplications. This entrywise product is similar to those used in PSO, but here the random walk via Lévy flight is more efficient in exploring the search space, as its step length is much longer in the long run. However, a substantial fraction of the new solutions should be generated by far field randomization and whose locations should be far enough from the current best solution, this will make sure that the system will not be trapped in a local optimum [35].

The pseudo code given here is sequential, however, vectors should be used from an implementation point of view, as vectors are more efficient than loops. A Matlab implementation is given by the author, and can be downloaded.[2]

2.5 A Unified Approach to Metaheuristics

2.5.1 Characteristics of Metaheuristics

There are many other metaheuristic algorithms which are equally popular and powerful, and these include Tabu search [11], ant colony optimization[7], artificial immune system [8], bee algorithms, bat algorithm [33] and others [18, 31].

The efficiency of metaheuristic algorithms can be attributed to the fact that they imitate the best features in nature, especially the selection of the fittest in biological systems which have evolved by natural selection over millions of years.

Two important characteristics of metaheuristics are: intensification and diversification [2]. Intensification intends to search locally and more intensively, while

[2] www.mathworks.com/matlabcentral/fileexchange/29809-cuckoo-search-cs-algorithm

diversification makes sure the algorithm explores the search space globally (hopefully also efficiently).

Furthermore, intensification is also called exploitation, as it typically searches around the current best solutions and selects the best candidates or solutions. Similarly, diversification is also called exploration, as it tends to explore the search space more efficiently, often by large-scale randomization.

A fine balance between these two components is very important to the overall efficiency and performance of an algorithm. Too little exploration and too much exploitation could cause the system to be trapped in local optima, which makes it very difficult or even impossible to find the global optimum. On the other hand, if there is too much exploration but too little exploitation, it may be difficult for the system to converge and thus slows down the overall search performance. A proper balance itself is an optimization problem, and one of the main tasks of designing new algorithms is to find a certain balance concerning this optimality and/or tradeoff.

Furthermore, just exploitation and exploration are not enough. During the search, we have to use a proper mechanism or criterion to select the best solutions. The most common criterion is to use the *Survival of the Fittest*, that is to keep updating the the current best found so far. In addition, certain elitism is often used, and this is to ensure the best or fittest solutions are not lost, and should be passed onto the next generations.

2.6 Generalized Evolutionary Walk Algorithm (GEWA)

From the above discussion of all the major components and their characteristics, we realized that a good combination of local search and global search with a proper selection mechanism should produce a good metaheuristic algorithm, whatever the name it may be called.

In principle, the global search should be carried out more frequently at the initial stage of the search or iterations. Once a number of good quality solutions are found, exploration should be sparse on the global scale, but frequent enough so as to escape any local trap if necessary. On the other hand, the local search should be carried out as efficient as possible, so a good local search method should be used. The proper balance of these two is paramount.

Using these basic components, we can now design a generic, metaheuristic algorithm for optimization, we can call it the Generalized Evolutional Walk Algorithm (GEWA), which was first formulated by Yang [30] in 2010. Evolutionary walk is a random walk, but with a biased selection towards optimality. This is a generalized framework for global optimization.

There are three major components in this algorithm: 1) global exploration by randomization, 2) intensive local search by random walk, and 3) the selection of the best with some elitism. The pseudo code of GEWA is shown in Fig. 2.2. The random walk should be carried out around the current global best g_* so as to exploit the system information such as the current best more effectively. We have

$$x_{t+1} = g_* + w, \tag{2.36}$$

and

$$w = \varepsilon d, \tag{2.37}$$

where ε is drawn from a Gaussian distribution or normal distribution $\mathrm{N}(0, \sigma^2)$, and d is the step length vector which should be related to the actual scales of independent variables. For simplicity, we can take $\sigma = 1$.

Initial a population of n walkers x_i $(i = 1, 2, ..., n)$;
Evaluate fitness F_i of n walkers & find the current best g_*;
while (t <MaxGeneration) or (stop criterion);
  Discard the worst solution and replace it by (2.38) or (2.39);
  if (rand $< \alpha$),
    Local search: random walk around the best

$$x_{t+1} = g_* + \varepsilon d \tag{2.38}$$

  else
    Global search: randomization (Uniform, Lévy flights etc)

$$x_{t+1} = L + (U - L)\varepsilon \quad \text{(uniform)} \tag{2.39}$$

  end
  Evaluate new solutions and find the current best g_*^t;
  $t = t + 1$;
end while
Postprocess results and visualization;

Fig. 2.2 Generalized Evolutionary Walk Algorithm (GEWA)

The randomization step can be achieved by

$$x_{t+1} = L + (U - L)\varepsilon_u, \tag{2.40}$$

where ε_u is drawn from a uniform distribution Unif[0,1]. U and L are the upper and lower bound vectors, respectively.

Typically, $\alpha \approx 0.25 \sim 0.7$. We will use $\alpha = 0.5$ in our implementation. Interested readers can try to do some parametric studies.

Again two important issues are: 1) the balance of intensification and diversification controlled by a single parameter α, and 2) the choice of the step size of the random walk. Parameter α is typically in the range of 0.25 to 0.7. The choice of the right step size is also important. Simulations suggest that the ratio of the step size to its length scale can typically be around 0.001 to 0.01 for most applications.

Another important issue is the selection of the best and/or elitism, as we intend to discard the worst solution and replace it by generating new solution. This may implicitly weed out the least-fit solutions, while the solution with the highest fitness remains in the population. The selection of the best and elitism can be guaranteed implicitly in the evolutionary walkers.

Furthermore, the number (n) of random walkers is also important. Too few walkers are not efficient, while too many may lead to slow convergence. In general, the choice of n should follow the similar guidelines as those for all population-based algorithms. Typically, we can use $n = 15$ to 50 for most applications.

2.6.1 To Be Inspired or Not to Be Inspired

We have seen that nature-inspired algorithms are always based on a particular (often most successful) mechanism of the natural world. Nature has evolved over billions of years, she has found almost perfect solutions to every problem she has met. Almost all the not-so-good solutions have been discarded via natural selection. The optimal solutions seem (often after a huge number of generations) to appear at the evolutionarilly stable equilibrium, even though we may not understand how the perfect solutions are reached. When we try to solve engineering problems, why not try to be inspired by nature's success? The simple answer to the question 'To be inspired or not to be inspired?' is 'why not?'. If we do not have good solutions at hand, it is always a good idea to learn from nature.

Nature provides almost unlimited ways for problem-solving. If we can observe carefully, we are surely inspired to develop more powerful and efficient new generation algorithms. Intelligence is a product of biological evolution in nature. Ultimately some intelligent algorithms (or systems) may appear in the future, so that they can evolve and optimally adapt to solve NP-hard optimization problems efficiently and intelligently.

References

1. Apostolopoulos, T., Vlachos, A.: Application of the Firefly Algorithm for Solving the Economic Emissions Load Dispatch Problem. International Journal of Combinatorics 2011 Article ID 523806 (2011), http://www.hindawi.com/journals/ijct/2011/523806.html
2. Blum, C., Roli, A.: Metaheuristics in combinatorial optimization: Overview and conceptural comparision. ACM Comput. Surv. 35, 268–308 (2003)
3. Cox, M.G., Forbes, A.B., Harris, P.M.: Discrete Modelling, SSfM Best Practice Guide No. 4, National Physical Laboratory, UK (2002)
4. Boyd, S.P., Vandenberghe, L.: Convex Optimization. Cambridge University Press, Cambridge (2004)
5. Celis, M., Dennis, J.E., Tapia, R.A.: A trust region strategy for nonlinear equality constrained optimization. In: Boggs, P., Byrd, R., Schnabel, R. (eds.) Numerical Optimization 1994, pp. 71–82. SIAM, Philadelphia (1994)
6. Conn, A.R., Gould, N.I.M., Toint, P.L.: Trust-region methods. SIAM&MPS (2000)
7. Dorigo, M., Stütle, T.: Ant Colony Optimization. MIT Press, Cambridge (2004)
8. Farmer, J.D., Packard, N., Perelson, A.: The immune system, adapation and machine learning. Physica D 2, 187–204 (1986)

9. Geem, Z.W., Kim, J.H., Loganathan, G.V.: A new heuristic optimization: Harmony search. Simulation 76, 60–68 (2001)
10. Gill, P.E., Murray, W., Wright, M.H.: Practical optimization. Academic Press Inc, London (1981)
11. Glover, F., Laguna, M.: Tabu Search. Kluwer Academic Publishers, Boston (1997)
12. Goldberg, D.E.: Genetic Algorithms in Search, Optimization and Machine Learning. Addison Wesley, Reading (1989)
13. Hestenes, M.R., Stiefel, E.: Methods of conjugate gradients for solving linear systems. Journal of Research of the National Bureaus of Standards 49(6), 409–436 (1952)
14. Holland, J.: Adaptation in Natural and Artificial Systems. University of Michigan Press, Ann Anbor (1975)
15. Kennedy, J., Eberhart, R.C.: Particle swarm optimization. In: Proc. of IEEE International Conference on Neural Networks, Piscataway, NJ, pp. 1942–1948 (1995)
16. Karmarkar, N.: A new polynomial-time algorithm for linear programming. Combinatorica 4(4), 373–395 (1984)
17. Kirkpatrick, S., Gelatt, C.D., Vecchi, M.P.: Optimization by simulated annealing. Science 220(4598), 671–680 (1983)
18. Koziel, S., Yang, X.S.: Computational Optimization and Applications in Engineering and Industry. Springer, Germany (2011)
19. Nelder, J.A., Mead, R.: A simplex method for function optimization. Computer Journal 7, 308–313 (1965)
20. Matthews, C., Wright, L., Yang, X.S.: Sensitivity Analysis, Optimization, and Sampling Methodds Applied to Continous Models. National Physical Laboratory Report, UK (2009)
21. Pavlyukevich, I.: Lévy flights, non-local search and simulated annealing. J. Computational Physics 226, 1830–1844 (2007)
22. Powell, M.J.D.: A new algorithm for unconstrained optimization. In: Rosen, J.B., Mangasarian, O.L., Ritter, K. (eds.) Nonlinear Programming, pp. 31–65 (1970)
23. Price, K., Storn, R., Lampinen, J.: Differential Evolution: A Practical Approach to Global Optimization. Springer, Heidelberg (2005)
24. Sayadi, M.K., Ramezanian, R., Ghaffari-Nasab, N.: A discrete firefly meta-heuristic with local search for makespan minimization in permutation flow shop scheduling problems. Int. J. of Industrial Engineering Computations 1, 1–10 (2010)
25. Storn, R.: On the usage of differential evolution for function optimization. In: Biennial Conference of the North American Fuzzy Information Processing Society (NAFIPS), pp. 519–523 (1996)
26. Storn, R., Price, K.: Differential evolution - a simple and efficient heuristic for global optimization over continuous spaces. Journal of Global Optimization 11, 341–359 (1997)
27. Talbi, E.G.: Metaheuristics: From Design to Implementation. John Wiley & Sons, Chichester (2009)
28. Yang, X.S.: Introduction to Computational Mathematics. World Scientific Publishing, Singapore (2008)
29. Yang, X.S.: Nature-Inspired Metaheuristic Algorithms, 1st edn. Lunver Press, UK (2008)
30. Yang, X.S.: Nature-Inspired Metaheuristic Algoirthms, 2nd edn. Luniver Press, UK (2010)
31. Yang, X.S.: Engineering Optimization: An Introduction with Metaheuristic Applications. John Wiley & Sons, Chichester (2010)
32. Yang, X.-S.: Firefly algorithms for multimodal optimization. In: Watanabe, O., Zeugmann, T. (eds.) SAGA 2009. LNCS, vol. 5792, pp. 169–178. Springer, Heidelberg (2009)

33. Yang, X.-S.: A new metaheuristic bat-inspired algorithm. In: González, J.R., Pelta, D.A., Cruz, C., Terrazas, G., Krasnogor, N. (eds.) NICSO 2010. SCI, vol. 284, pp. 65–74. Springer, Heidelberg (2010)
34. Yang, X.S., Deb, S.: Cuckoo search via Lévy flights. In: Proc. of World Congress on Nature & Biologically Inspired Computing (NaBic 2009), pp. 210–214. IEEE Publications, USA (2009)
35. Yang, X.S., Deb, S.: Engineering optimization by cuckoo search. Int. J. Math. Modelling Num. Optimisation 1(4), 330–343 (2010)

Chapter 3
Surrogate-Based Methods

Slawomir Koziel, David Echeverría Ciaurri, and Leifur Leifsson

Abstract. Objective functions that appear in engineering practice may come from measurements of physical systems and, more often, from computer simulations. In many cases, optimization of such objectives in a straightforward way, i.e., by applying optimization routines directly to these functions, is impractical. One reason is that simulation-based objective functions are often analytically intractable (discontinuous, non-differentiable, and inherently noisy). Also, sensitivity information is usually unavailable, or too expensive to compute. Another, and in many cases even more important, reason is the high computational cost of measurement/simulations. Simulation times of several hours, days or even weeks per objective function evaluation are not uncommon in contemporary engineering, despite the increase of available computing power. Feasible handling of these unmanageable functions can be accomplished using surrogate models: the optimization of the original objective is replaced by iterative re-optimization and updating of the analytically tractable and computationally cheap surrogate. This chapter briefly describes the basics of surrogate-based optimization, various ways of creating surrogate models, as well as several examples of surrogate-based optimization techniques.

Keywords: Surrogate-based optimization, multi-fidelity optimization, surrogate models, simulation-driven design, trust-region methods, function approximation, design of experiments.

Slawomir Koziel · Leifur Leifsson
Engineering Optimization & Modeling Center, School of Science and Engineering, Reykjavik University, Menntavegur 1, 101 Reykjavik, Iceland
email: {koziel,leifurth}@ru.is

David Echeverría Ciaurri
Department of Energy Resources Engineering,
Stanford University, Stanford, CA 94305-2220, USA
email: echeverr@stanford.edu

S. Koziel & X.-S. Yang (Eds.): Comput. Optimization, Methods and Algorithms, SCI 356, pp. 33–59.
springerlink.com

3.1 Introduction

Contemporary engineering is more and more dependent on computer-aided design (CAD). In most engineering fields, numerical simulations are used extensively, not only for design verification but also directly in the design process. As a matter of fact, because of increasing system complexity, ready-to-use theoretical (e.g., analytical) models are not available in many cases. Thus, simulation-driven design and design optimization becomes the only option to meet the specifications prescribed, improve the system reliability, or reduce the fabrication cost.

The simulation-driven design can be formulated as a nonlinear minimization problem of the following form

$$\boldsymbol{x}^* = \arg\min_{\boldsymbol{x}} f(\boldsymbol{x}), \tag{3.1}$$

where $f(\boldsymbol{x})$ denotes the objective function to be minimized evaluated at the point $\boldsymbol{x} \in R^n$ ($\boldsymbol{x}$ is the design variable vector). In many engineering problems f is of the form $f(\boldsymbol{x}) = U(\boldsymbol{R}_f(\boldsymbol{x}))$, where $\boldsymbol{R}_f \in R^m$ denotes the response vector of the system of interest (in particular, one may have $m > n$ or even $m >> n$ [1]), whereas U is a given scalar merit function. In particular, U can be defined through a norm that measures the distance between $\boldsymbol{R}_f(\boldsymbol{x})$ and a target vector $\boldsymbol{y}$. An optimal design vector is denoted by $\boldsymbol{x}^*$. In many cases, $\boldsymbol{R}_f$ is obtained through computationally expensive computer simulations. We will refer to it as a high-fidelity or fine model. To simplify notation, f itself will also be referred to as the high-fidelity (fine) model.

Unfortunately, a direct attempt to solve (3.1) by embedding the simulator directly in the optimization loop may be impractical. The underlying simulations can be very time-consuming (in some instances, the simulation time can be as long as several hours, days or even weeks per single design), and the presence of massive computing resources is not always translated in computational speedup. This latter fact is due to a growing demand for simulation accuracy, both by including multiphysics and second-order effects, and by using finer discretization of the structure under consideration. As conventional optimization algorithms (e.g., gradient-based schemes with numerical derivatives) require tens, hundreds or even thousands of objective function calls per run (that depends on the number of design variables), the computational cost of the whole optimization process may not be acceptable.

Another problem is that objective functions coming from computer simulations are often analytically intractable (i.e., discontinuous, non-differentiable, and inherently noisy). Moreover, sensitivity information is frequently unavailable, or too expensive to compute. While in some cases it is possible to obtain derivative information inexpensively through adjoint sensitivities [2], numerical noise is an important issue that can complicate simulation-driven design. We should also mention that adjoint-based sensitivities require detailed knowledge of and access to the simulator source code, and this is something that cannot be assumed to be generally available.

Surrogate-based optimization (SBO) [1,3,4] has been suggested as an effective approach for the design with time-consuming computer models. The basic concept of SBO is that the direct optimization of the computationally expensive model is replaced by an iterative process that involves the creation, optimization and updating of a fast and analytically tractable surrogate model. The surrogate should be a reasonably accurate representation of the high-fidelity model, at least locally. The design obtained through optimizing the surrogate model is verified by evaluating the high-fidelity model. The high-fidelity model data obtained in this verification process is then used to update the surrogate. SBO proceeds in this predictor-corrector fashion iteratively until some termination criterion is met. Because most of the operations are performed on the surrogate model, SBO reduces the computational cost of the optimization process when compared to optimizing the high-fidelity model directly, without resorting to any surrogate.

In this chapter, we review the basics of surrogate-based optimization. We briefly present various ways of generating surrogate models, and we emphasize on the distinction between models based on function approximations of sampled high-fidelity model data and models constructed from physically-based low-fidelity models. A few selected surrogate-based optimization algorithms including space mapping [1,5,6], approximation model management [7], manifold mapping [8], and the surrogate-management framework [9], are also discussed. We conclude the chapter with some final remarks.

3.2 Surrogate-Based Optimization

As mentioned in the introduction, there are several reasons why the straightforward optimization of the high-fidelity model may not work or can be impractical. These reasons include high computational cost of each model evaluation, numerical noise and discontinuities in the cost function. Surrogate-based optimization [3,5] aims at alleviating such problems by using an auxiliary model, the surrogate, that is preferably fast, amenable to optimization, and yet reasonably accurate. One popular approach for constructing surrogate models is through approximations of high-fidelity model data obtained by sampling the design space using appropriate design of experiments methodologies [3]. Some of these strategies for allocating samples [10], generating approximations [3,4,10], as well as validating the surrogates are discussed in Section 3.3.

The surrogate model optimization yields an approximation of the minimizer associated to the high-fidelity model. This approximation has to be verified by evaluating the high-fidelity model at the predicted high-fidelity model minimizer. Depending on the result of this verification, the optimization process may be terminated. Otherwise, the surrogate model is updated using the new available high-fidelity model data, and then re-optimized to obtain a new, and hopefully better, approximation of the high-fidelity model minimizer.

The surrogate-based optimization process can be summarized as follows (Fig. 3.1):

1. Generate the initial surrogate model.
2. Obtain an approximate solution to (3.1) by optimizing the surrogate.

3. Evaluate the high-fidelity model at the approximate solution computed in Step 2.
4. Update the surrogate model using the new high-fidelity model data.
5. Stop if the termination condition is satisfied; otherwise go to Step 2.

The SBO framework can be formulated as an iterative procedure [3,5]:

$$\boldsymbol{x}^{(i+1)} = \arg\min_{\boldsymbol{x}} s^{(i)}(\boldsymbol{x}) \,. \tag{3.2}$$

This scheme generates a sequence of points (designs) $\boldsymbol{x}^{(i)}$ that (hopefully) converge to a solution (or a good approximation) of the original design problem (3.1). Each $\boldsymbol{x}^{(i+1)}$ is the optimal design of the surrogate model $s^{(i)}$, which is assumed to be a computationally cheap and sufficiently reliable representation of the fine model f, particularly in the neighborhood of the current design $\boldsymbol{x}^{(i)}$. Under these assumptions, the algorithm (3.2) aims at a sequence of designs to quickly approach $\boldsymbol{x}^*$. Typically, and for verification purposes, the high-fidelity model is evaluated only once per iteration (at every new design $\boldsymbol{x}^{(i+1)}$). The data obtained from the validation is used to update the surrogate model. Because the surrogate model is computationally cheap, the optimization cost associated with (3.2) can—in many cases—be viewed as negligible, so that the total optimization cost is determined by the evaluation of the high-fidelity model. Normally, the number of iterations often needed within a surrogate-based optimization algorithm is substantially smaller than for any method that optimizes the high-fidelity model directly (e.g., gradient-based schemes with numerical derivatives) [5].

If the surrogate model satisfies zero- and first-order consistency conditions with the high-fidelity model (i.e., $s^{(i)}(\boldsymbol{x}^{(i)}) = f(\boldsymbol{x}^{(i)})$ and $\nabla s^{(i)}(\boldsymbol{x}^{(i)}) = \nabla f(\boldsymbol{x}^{(i)})$ [7]; it should be noticed that the verification of the latter requires high-fidelity model sensitivity data), and the surrogate-based algorithm is enhanced by, for example, a trust region method [11] (see Section 3.4.1), then the sequence of intermediate solutions is provably convergent to a local optimizer of the fine model [12] (some standard assumptions concerning the smoothness of the functions involved are also necessary) [13]. Convergence can also be guaranteed if the SBO algorithm is embedded within the framework given in [5,14] (space mapping), [13] (manifold mapping) or [9] (surrogate management framework). A more detailed description of several surrogate-based optimization techniques is given in Section 3.4.

Space mapping [1,5,6] is an example of a surrogate-based methodology that does not normally rely on using sensitivity data or trust region convergence safeguards; however, it requires the surrogate model to be constructed from a physically-based coarse model [1]. This usually gives remarkably good performance in the sense of the algorithm being able to locate a satisfactory design quickly. Unfortunately space mapping suffers from convergence problems [14] and it is sensitive to the quality of the coarse model and the specific analytical formulation of the surrogate [15,16].

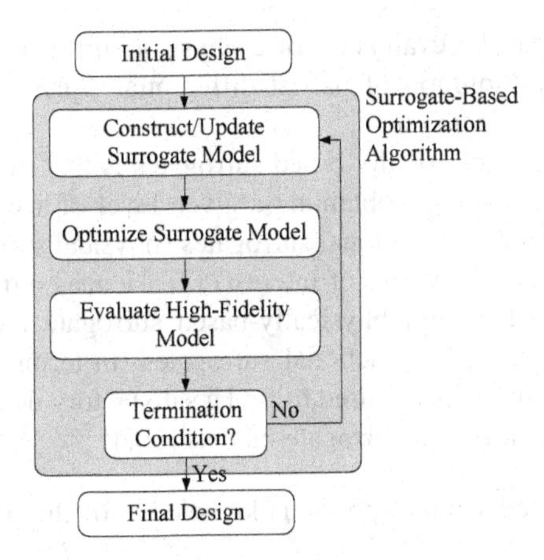

Fig. 3.1 Flowchart of the surrogate-based optimization process. An approximate high-fidelity model minimizer is obtained iteratively by optimizing the surrogate model. The high-fidelity model is evaluated at each new design for verification purposes. If the termination condition is not satisfied, the surrogate model is updated and the search continues. In most cases the high-fidelity model is evaluated only once per iteration. The number of iterations needed in SBO is often substantially smaller than for conventional (direct) optimization techniques.

3.3 Surrogate Models

The surrogate model is a key component of any SBO algorithm. It has to be computationally cheap, preferably smooth, and, at the same time, reasonably accurate, so that it can be used to predict the location of high-fidelity model minimizers. We can clearly distinguish between physical and functional surrogate models.

Physical (or physically-based) surrogates are constructed from an underlying low-fidelity (coarse) model. The low-fidelity model is a representation of the system of interest with relaxed accuracy [1]. Coarse models are computationally cheaper than high-fidelity models and, in many cases, have better analytical properties. The low-fidelity model can be obtained, for example, from the same simulator as the one used for the high-fidelity model but using a coarse discretization [17]. Alternatively, the low-fidelity model can be based on simplified physics (e.g., by exploiting simplified equations [1], or by neglecting certain second-order effects) [18], or on a significantly different physical description (e.g., lumped parameter versus partial differential equation based models [1]). In some cases, low-fidelity models can be formulated using analytical or semi-empirical formulas [19]. The coarse model can be corrected if additional data from the high-fidelity model is available (for example, during the course of the optimization).

In general, physical surrogate models are:

- based on particular knowledge about the physical system of interest,
- dedicated (reuse across different designs is uncommon),
- more expensive to evaluate and more accurate (in a global sense) than functional surrogates.

It should be noticed that the evaluation of a physical surrogate may involve, for example, the numerical solution of partial differential equations or even actual measurements of the physical system.

The main advantage of physically-based surrogates is that the amount of high-fidelity model data necessary for obtaining a given level of accuracy is generally substantially smaller than for functional surrogates (physical surrogates inherently embed knowledge about the system of interest) [1]. Hence, surrogate-based optimization algorithms that exploit physically-based surrogate models are usually more efficient than those using functional surrogates (in terms of the number of high-fidelity model evaluations required to find a satisfactory design) [5].

Functional (or approximation) surrogate models [20,4]:

- can be constructed without previous knowledge of the physical system of interest,
- are generic, and therefore applicable to a wide class of problems,
- are based on (usually simple) algebraic models,
- are often very cheap to evaluate but require considerable amount of data to ensure reasonable general accuracy.

An initial functional surrogate can be generated using high-fidelity model data obtained through sampling of the design space. Figure 3.2 shows the model construction flowchart for a functional surrogate. Design of experiments involves the use of strategies for allocating samples within the design space. The particular choice depends on the number of samples one can afford (in some occasions only a few points may be allowed), but also on the specific modeling technique that will be used to create the surrogate. Though in some cases the surrogate can be found using explicit formulas (e.g., polynomial approximation) [3], in most situations it is computed by means of a separate minimization problem (e.g., when using kriging [21] or neural networks [22]). The accuracy of the model should be tested in order to estimate its prediction/generalization capability. The main difficulty in obtaining a good functional surrogate lies in keeping a balance between accuracy at the known and at the unknown data (training and testing set, respectively). The surrogate could be subsequently updated using new high-fidelity model data that is accumulated during the run of the surrogate-based optimization algorithm.

In this section we first describe the fundamental steps for generating functional surrogates. Various sampling techniques are presented in Section 3.3.1. The surrogate creation and the model validations steps are tackled in Section 3.3.2 and Section 3.3.3, respectively. If the quality of the surrogate is not sufficient, more data points can be added, and/or the model parameters can be updated to improve accuracy. Several correction methods, both for functional and physical surrogates, are described in Section 3.3.4.

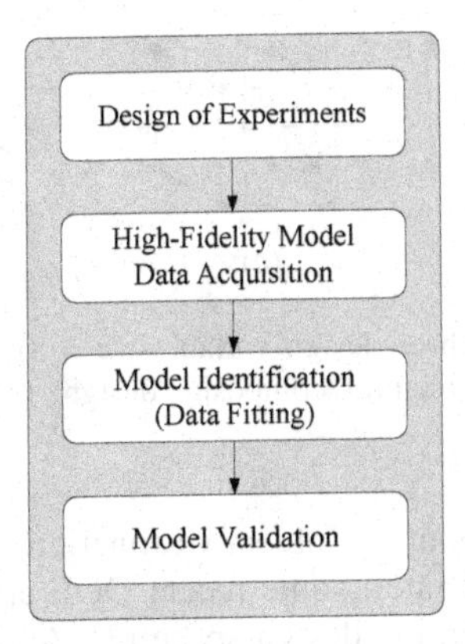

Fig. 3.2 Surrogate model construction flowchart. If the quality of the model is not satisfactory, the procedure can be iterated (more data points will be required).

3.3.1 Design of Experiments

Design of experiments (DOE) [23,24,25] is a strategy for allocating samples (points) in the design space that aims at maximizing the amount of information acquired. The high-fidelity model is evaluated at these points to create the training data set that is subsequently used to construct the functional surrogate model. When sampling, there is a clear trade-off between the number of points used and the amount of information that can be extracted from these points. The samples are typically spread apart as much as possible in order to capture global trends in the design space.

Factorial designs [23] are classical DOE techniques that, when applied to discrete design variables, explore a large region of the search space. The sampling of all possible combinations is called full factorial design. Fractional factorial designs can be used when model evaluation is expensive and the number of design variables is large (in full factorial design the number of samples increases exponentially with the number of design variables). Continuous variables, once discretized, can be easily analyzed through factorial design. Full factorial two-level and three-level design (also known as 2^k and 3^k design) allows us to estimate main effects and interactions between design variables, and quadratic effects and interactions, respectively. Figures 3.3(a) and 3.3(b) show examples of full two-level and fractional two-level design, respectively, for three design variables (i.e., $n = 3$). Alternative factorial designs can be found in practice: central composite design (see Figure 3.3(c)), star design (frequently used in combination with space mapping [26]; see Figure 3.3(d)), or Box-Behnken design (see Figure 3.3(e)).

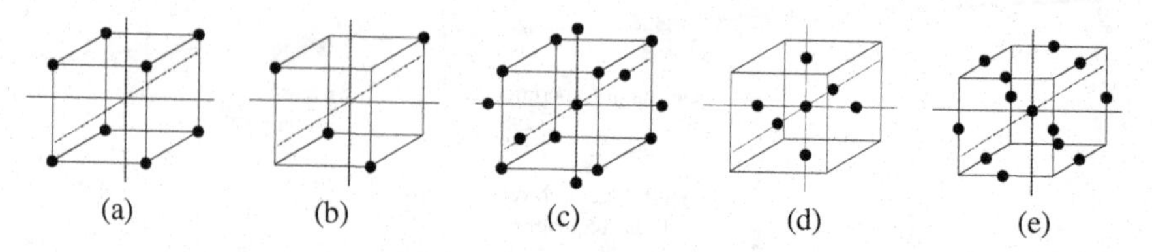

Fig. 3.3 Factorial designs for three design variables (n = 3): (a) full factorial design, (b) fractional factorial design, (c) central composite design, (d) star design, and (e) Box-Behnken design.

If no prior knowledge about the objective function is available (typical while constructing the initial surrogate), some recent DOE approaches tend to allocate the samples uniformly within the design space [3]. A variety of space filling designs are available. The simplest ones do not ensure sufficient uniformity (e.g., pseudo-random sampling [23]) or are not practical (e.g., stratified random sampling, where the number of samples needed is on the order of 2^n). One of the most popular DOE for (relatively) uniform sample distributions is Latin hypercube sampling (LHS) [27]. In order to allocate p samples with LHS, the range for each parameter is divided into p bins, which for n design variables, yields a total number of p^n bins in the design space. The samples are randomly selected in the design space so that (i) each sample is randomly placed inside a bin, and (ii) for all one-dimensional projections of the p samples and bins, there is exactly one sample in each bin. Figure 3.4 shows a LHS realization of 15 samples for two design variables (n = 2). It should be noted that the standard LHS may lead to non-uniform distributions (for example, samples allocated along the design space diagonal satisfy conditions (i) and (ii)). Numerous improvements of standard LHS, e.g., [28]-[31], provide more uniform sampling distributions.

Other DOE methodologies commonly used include orthogonal array sampling [3], quasi-Monte Carlo sampling [23], or Hammersley sampling [23]. Sample distribution can be improved through the incorporation of optimization techniques that minimize a specific non-uniformity measure, e.g., $\sum_{i=1}^{p}\sum_{j=i+1}^{p} d_{ij}^{-2}$ [29], where d_{ij} is the Euclidean distance between samples i and j.

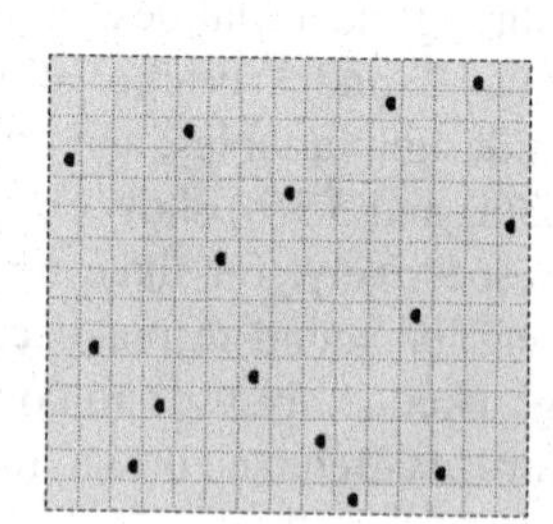

Fig. 3.4 Latin hypercube sampling realization of 15 samples in a two-dimensional design space.

3.3.2 *Surrogate Modeling Techniques*

Having selected the design of experiments technique and sampled the data, the next step is to choose an approximation model and a fitting methodology. In this section, we describe in some detail the most popular surrogate modeling techniques, and we briefly mention alternatives.

3.3.2.1 Polynomial Regression

Polynomial regression [3] assumes the following relation between the function of interest f and K polynomial basis functions v_j using p samples $f(\boldsymbol{x}^{(i)})$, $i = 1, \ldots, p$:

$$f(\boldsymbol{x}^{(i)}) = \sum_{j=1}^{K} \beta_j v_j(\boldsymbol{x}^{(i)}) \cdot \tag{3.3}$$

These equations can be represented in matrix form

$$\boldsymbol{f} = \boldsymbol{X}\boldsymbol{\beta}, \tag{3.4}$$

where $\boldsymbol{f} = [f(\boldsymbol{x}^{(1)})\ f(\boldsymbol{x}^{(2)})\ \ldots\ f(\boldsymbol{x}^{(p)})]^T$, $\boldsymbol{X}$ is a $p \times K$ matrix containing the basis functions evaluated at the sample points, and $\boldsymbol{\beta} = [\beta_1\ \beta_2\ \ldots\ \beta_K]^T$. The number of sample points p should be consistent with the number of basis functions considered K (typically $p \geq K$). If the sample points and basis function are taken arbitrarily, some columns of $\boldsymbol{X}$ can be linearly dependent. If $p \geq K$ and rank($\boldsymbol{X}$) = K, a solution of (3.4) in the least-squares sense can be computed through $\boldsymbol{X}^+$, the pseudoinverse (or generalized inverse) of $\boldsymbol{X}$ [32]:

$$\boldsymbol{\beta} = \boldsymbol{X}^+ = (\boldsymbol{X}^T \boldsymbol{X})^{-1} \boldsymbol{X}^T \boldsymbol{f}. \tag{3.5}$$

The simplest examples of regression models are the first- and second-order order polynomial models

$$s(\boldsymbol{x}) = s([x_1\ x_2 \ldots x_n]^T) = \beta_0 + \sum_{j=1}^{n} \beta_j x_j\ , \tag{3.6}$$

$$s(\boldsymbol{x}) = s([x_1\ x_2 \ldots x_n]^T) = \beta_0 + \sum_{j=1}^{n} \beta_j x_j + \sum_{i=1}^{n} \sum_{j \leq i}^{n} \beta_{ij} x_i x_j \cdot \tag{3.7}$$

Polynomial interpolation/regression appears naturally and is crucial in developing robust and efficient derivative-free optimization algorithms. For more details, please refer to [33].

3.3.2.2 Radial Basis Functions

Radial basis function interpolation/approximation [4,34] exploits linear combinations of K radially symmetric functions ϕ

$$s(\boldsymbol{x}) = \sum_{j=1}^{K} \lambda_j \phi(\| \boldsymbol{x} - \boldsymbol{c}^{(j)} \|), \tag{3.8}$$

where $\boldsymbol{\lambda} = [\lambda_1\ \lambda_2\ \ldots\ \lambda_K]^T$ is the vector of model parameters, and $\boldsymbol{c}^{(j)}$, $j = 1, \ldots, K$, are the (known) basis function centers.

As in polynomial regression the model parameters $\boldsymbol{\lambda}$ can be computed by

$$\boldsymbol{\lambda} = \boldsymbol{\Phi}^{+} = (\boldsymbol{\Phi}^T \boldsymbol{\Phi})^{-1} \boldsymbol{\Phi}^T \boldsymbol{f}, \tag{3.9}$$

where again $\boldsymbol{f} = [f(\boldsymbol{x}^{(1)})\ f(\boldsymbol{x}^{(2)})\ \ldots\ f(\boldsymbol{x}^{(p)})]^T$, and the $p\times K$ matrix $\boldsymbol{\Phi}$ is defined as

$$\boldsymbol{\Phi} = \begin{bmatrix} \phi(\|\boldsymbol{x}^{(1)} - \boldsymbol{c}^{(1)}\|) & \phi(\|\boldsymbol{x}^{(1)} - \boldsymbol{c}^{(2)}\|) & \cdots & \phi(\|\boldsymbol{x}^{(1)} - \boldsymbol{c}^{(K)}\|) \\ \phi(\|\boldsymbol{x}^{(2)} - \boldsymbol{c}^{(1)}\|) & \phi(\|\boldsymbol{x}^{(2)} - \boldsymbol{c}^{(2)}\|) & \cdots & \phi(\|\boldsymbol{x}^{(2)} - \boldsymbol{c}^{(K)}\|) \\ \vdots & \vdots & \ddots & \vdots \\ \phi(\|\boldsymbol{x}^{(p)} - \boldsymbol{c}^{(1)}\|) & \phi(\|\boldsymbol{x}^{(p)} - \boldsymbol{c}^{(2)}\|) & \cdots & \phi(\|\boldsymbol{x}^{(p)} - \boldsymbol{c}^{(K)}\|) \end{bmatrix}. \tag{3.10}$$

If we select $p = K$ (i.e., the number of basis functions is equal to the number of samples), and if the centers of the basis functions coincide with the data points (and these are all different), $\boldsymbol{\Phi}$ is a regular square matrix (and thus, $\boldsymbol{\lambda} = \boldsymbol{\Phi}^{-1}\boldsymbol{f}$).

Typical choices for the basis functions are $\phi(r) = r$, $\phi(r) = r^3$, or $\phi(r) = r^2 \ln r$ (thin plate spline). More flexibility can be obtained by using parametric basis functions such as $\phi(r) = \exp(-r^2/2\sigma^2)$ (Gaussian), $\phi(r) = (r^2 + \sigma^2)^{1/2}$ (multi-quadric), or $\phi(r) = (r^2 + \sigma^2)^{-1/2}$ (inverse multi-quadric).

3.3.2.3 Kriging

Kriging is a popular technique to interpolate deterministic noise-free data [35,10,21,36]. Kriging is a Gaussian process [37] based modeling method, which is compact and cheap to evaluate. Kriging has been proven to be useful in a wide variety of fields (see, e.g., [4,38] for applications in optimization).

In its basic formulation, kriging [35,10] assumes that the function of interest f is of the form

$$f(\boldsymbol{x}) = \boldsymbol{g}(\boldsymbol{x})^T \boldsymbol{\beta} + Z(\boldsymbol{x}), \tag{3.11}$$

where $\boldsymbol{g}(\boldsymbol{x}) = [g_1(\boldsymbol{x})\ g_2(\boldsymbol{x})\ \ldots\ g_K(\boldsymbol{x})]^T$ are known (e.g., constant) functions, $\boldsymbol{\beta} = [\beta_1\ \beta_2\ \ldots\ \beta_K]^T$ are the unknown model parameters, and $Z(\boldsymbol{x})$ is a realization of a normally distributed Gaussian random process with zero mean and variance σ^2.

The regression part $\boldsymbol{g}(\boldsymbol{x})^T\boldsymbol{\beta}$ approximates globally the function f, and $Z(\boldsymbol{x})$ takes into account localized variations. The covariance matrix of $Z(\boldsymbol{x})$ is given as

$$Cov[Z(\boldsymbol{x}^{(i)})Z(\boldsymbol{x}^{(j)})] = \sigma^2 \boldsymbol{R}([R(\boldsymbol{x}^{(i)}, \boldsymbol{x}^{(j)})]), \tag{3.12}$$

where $\boldsymbol{R}$ is a $p\times p$ correlation matrix with $R_{ij} = R(\boldsymbol{x}^{(i)}, \boldsymbol{x}^{(j)})$. Here, $R(\boldsymbol{x}^{(i)}, \boldsymbol{x}^{(j)})$ is the correlation function between sampled data points $\boldsymbol{x}^{(i)}$ and $\boldsymbol{x}^{(j)}$. The most popular choice is the Gaussian correlation function

$$R(\boldsymbol{x}, \boldsymbol{y}) = \exp\left[-\sum_{k=1}^{n} \theta_k |x_k - y_k|^2\right], \tag{3.13}$$

where θ_k are unknown correlation parameters, and x_k and y_k are the kth component of the vectors $\boldsymbol{x}$ and $\boldsymbol{y}$, respectively.

The kriging predictor [10,35] is defined as

$$s(\boldsymbol{x}) = \boldsymbol{g}(\boldsymbol{x})^T \boldsymbol{\beta} + \boldsymbol{r}^T(\boldsymbol{x})\boldsymbol{R}^{-1}(\boldsymbol{f} - \boldsymbol{G}\boldsymbol{\beta}), \tag{3.14}$$

where $\boldsymbol{r}(\boldsymbol{x}) = [R(\boldsymbol{x}, \boldsymbol{x}^{(1)}) \ \ldots \ R(\boldsymbol{x}, \boldsymbol{x}^{(p)})]^T$, $\boldsymbol{f} = [f(\boldsymbol{x}^{(1)}) \ f(\boldsymbol{x}^{(2)}) \ \ldots \ f(\boldsymbol{x}^{(p)})]^T$, and $\boldsymbol{G}$ is a $p \times K$ matrix with $G_{ij} = g_j(\boldsymbol{x}^{(i)})$.

The vector of model parameters $\boldsymbol{\beta}$ can be computed as

$$\boldsymbol{\beta} = (\boldsymbol{G}^T \boldsymbol{R}^{-1} \boldsymbol{G})^{-1} \boldsymbol{G}^T \boldsymbol{R}^{-1} \boldsymbol{f}. \tag{3.15}$$

An estimate of the variance σ^2 is given by

$$\hat{\sigma}^2 = \frac{1}{p}(\boldsymbol{f} - \boldsymbol{G}\boldsymbol{\beta})^T \boldsymbol{R}^{-1}(\boldsymbol{f} - \boldsymbol{G}\boldsymbol{\beta}). \tag{3.16}$$

Model fitting is accomplished by maximum likelihood for θ_k [35]. In particular, the n-dimensional unconstrained nonlinear maximization problem with cost function

$$-(p \ln(\hat{\sigma}^2) + \ln |\boldsymbol{R}|)/2, \tag{3.17}$$

where the variance σ^2 and $|\boldsymbol{R}|$ are both functions of θ_k, is solved for positive values of θ_k as optimization variables.

It should be noted that, once the kriging-based surrogate has been obtained, the random process $Z(\boldsymbol{x})$ gives valuable information regarding the approximation error that can be used for improving the surrogate [4,35].

3.3.2.4 Neural Networks

The basic structure in a neural network [39,40] is the neuron (or single-unit perceptron). A neuron performs an affine transformation followed by a nonlinear operation (see Fig. 3.5(a)). If the inputs to a neuron are denoted as $x_1, \ldots, x_n$, the neuron output y is computed as

$$y = \frac{1}{1 + \exp(-\eta / T)}, \tag{3.18}$$

where $\eta = w_1 x_1 + \ldots + w_n x_n + \gamma$, with $w_1, \ldots, w_n$ being regression coefficients. Here, γ is the bias value of a neuron, and T is a user-defined (slope) parameter. Neurons can be combined in multiple ways [39]. The most common neural network architecture is the multi-layer feed-forward network (see Fig. 3.5(b)).

The construction of a functional surrogate based on a neural network requires two main steps: (i) architecture selection, and (ii) network training. The network training can be stated as a nonlinear least-squares regression problem for a number of training points. Since the optimization cost function is nonlinear in all the optimization variables (neurons coefficients), the solution cannot be written using a closed-form expression, as it was the case before in (3.5) or in (3.9). A very popular technique for solving this regression problem is the error back-propagation algorithm [10,39]. If the network architecture is sufficiently complex, a neural network can approximate a general set of functions [10]. However, in complicated cases (e.g., nonsmooth functions with a large number of variables) the underlying regression problem may be significantly involved.

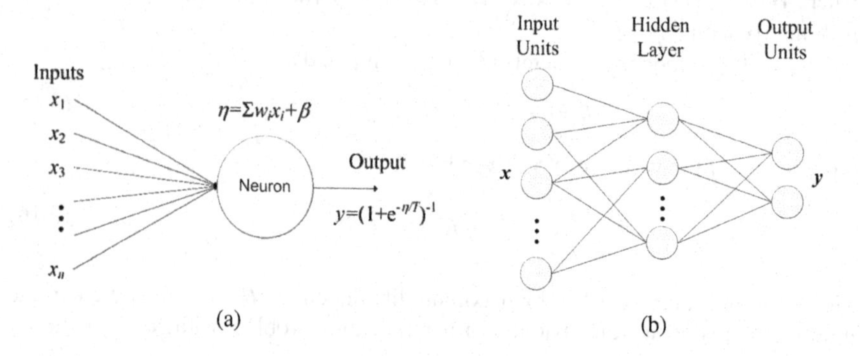

Fig. 3.5. Neural networks: (a) neuron basic structure; (b) two-layer feed-forward neural network architecture.

3.3.2.5 Other Techniques

The techniques described in this section refer to some other approaches that are gaining popularity recently. One of the most prominent approaches, which has been observed as a very general approximation tool, is support vector regression (SVR) [41,42]. SVR resorts to quadratic programming for a robust solving of the underlying optimization in the approximation procedure [43]. SVR is a variant of the support vector machines (SVMs) methodology developed by Vapnik [44], which was originally applied to classification problems. SVR/SVM exploits the structural risk minimization (SRM) principle, which has been shown (see, e.g., [41]) to be superior to the traditional empirical risk minimization (ERM) principle employed by several modeling technologies (e.g., neural networks). ERM is based on minimizing an error function for the set of training points. When the model structure is complex (e.g., higher order polynomials), ERM-based surrogates often result in overfitting. SRM incorporates the model complexity in the regression, and therefore yields surrogates that may be more accurate outside of the training set.

Moving least squares (MLS) [45] is a technique particularly popular in aerospace engineering. MLS is formulated as weighted least squares (WLS) [46]. In MLS, the error contribution from each training point $\boldsymbol{x}^{(i)}$ is multiplied by a weight ω_i that depends on the distance between $\boldsymbol{x}$ and $\boldsymbol{x}^{(i)}$. A common choice for the weights is

$$\omega_i(\|\boldsymbol{x}-\boldsymbol{x}^{(i)}\|)=\exp(-\|\boldsymbol{x}-\boldsymbol{x}^{(i)}\|^2)\ . \tag{3.19}$$

MLS is essentially an adapting surrogate, and this additional flexibility can be translated in more appealing designs (especially in computer graphics applications). However, MLS is computationally more expensive than WLS, since computing the approximation for each point $\boldsymbol{x}$ requires solving a new optimization problem.

Gaussian process regression (GPR) [47] is a surrogate modeling technique that, as kriging, addresses the approximation problem from a stochastic point view. From this perspective, and since Gaussian processes are mathematically tractable, it is relatively easy to compute error estimations for GPR-based surrogates in the form of uncertainty distributions. Under appropriate conditions, Gaussian processes can be shown to be equivalent to large neural networks [47]. Nevertheless, Gaussian process modeling typically requires much less regression parameters than neural networks.

3.3.3 Model Validation

Some of the methodologies described above determine a surrogate model together with some estimation of the attendant approximation error (e.g., kriging or Gaussian process regression). Alternatively, there are procedures that can be used in a stand-alone manner to validate the prediction capability of a given model beyond the set of training points. A simple way for validating a model is the split-sample method [3]. In this algorithm, the set of available data samples is divided into two subsets. The first subset is called the training subset and contains the points considered for the construction of the surrogate. The second subset is the testing subset and serves purely as a model validation objective. In general, the error estimated by a split-sample method depends strongly on how the set of data samples is partitioned. We also note that in this approach the samples available do not appear to be put to good use, since the surrogate is based on only a subset of them.

Cross-validation [3,48] is an extremely popular methodology for verifying the prediction capabilities of a model generated from a set of samples. In cross-validation the data set is divided into L subsets, and each of these subsets is sequentially used as testing set for a surrogate constructed on the other L–1 subsets. If the number of subsets L is equal to the sample size p, the approach is called leave-one-out cross-validation [3]. The prediction error can be estimated with all the L error measures obtained in this process (for example, as an average value). Cross-validation provides an error estimation that is less biased than with the split-sample method [3]. The disadvantage of this method is that the surrogate has to be constructed more than once. However, having multiple approximations may improve the robustness of the whole surrogate generation and validation approach, since all the data available is used with both training and testing purposes.

3.3.4 Surrogate Correction

In the first stages on any surrogate-based optimization procedure, it is desirable to use a surrogate that is valid globally in the search space [4] in order to avoid being trapped in local solutions with unacceptable cost function values. Once the search starts becoming local, the global accuracy of the initial surrogate may not be beneficial for making progress in the optimization[1]. For this reason, surrogate correction is crucial within any SBO methodology.

---

[1] As mentioned in Section 3.2, when solving the original optimization problem in (3.1) using a surrogate-based optimization framework, zero- and first-order local consistency conditions are essential for obtaining convergence to a first-order stationary point.

In this section we will describe two strategies for improving surrogates locally. The corrections described in Section 3.3.4.1 are based on mapping objective function values. In some occasions, the cost function can be expressed as a function of a model response. Section 3.3.4.2 presents the space-mapping concept that gives rise to a whole surrogate-based optimization paradigm (see Section 3.4.2).

3.3.4.1 Objective Function Correction

Most of the objective function corrections used in practice can be identified in one of these three groups: compositional, additive or multiplicative corrections. We will briefly illustrate each of these categories for correcting the surrogate $s^{(i)}(\boldsymbol{x})$, and discuss if zero- and first-order consistency conditions with $f(\boldsymbol{x})$ [7] can be satisfied.

The following compositional correction [20]

$$s^{(i+1)}(\boldsymbol{x}) = g(s^{(i)}(\boldsymbol{x})) \tag{3.20}$$

represents a simple scaling of the objective function. Since the mapping g is a real-valued function of a real variable, a compositional correction will not in general yield first-order consistency conditions. By selecting a mapping g that satisfies

$$g'(s^{(i)}(\boldsymbol{x}^{(i)})) = \frac{\nabla f(\boldsymbol{x}^{(i)}) \nabla s^{(i)}(\boldsymbol{x}^{(i)})^T}{\nabla s^{(i)}(\boldsymbol{x}^{(i)}) \nabla s^{(i)}(\boldsymbol{x}^{(i)})^T}, \tag{3.21}$$

the discrepancy between $\nabla f(\boldsymbol{x}^{(i)})$ and $\nabla s^{(i+1)}(\boldsymbol{x}^{(i)})$ (expressed in Euclidean norm) is minimized. It should be noticed that the correction in (3.21), as many transformations that ensure first-order consistency, requires a high-fidelity gradient, which may be expensive to compute. However, numerical estimates of $\nabla f(\boldsymbol{x}^{(i)})$ may yield in practice acceptable results.

The compositional correction can be also introduced in the parameter space [1]

$$s^{(i+1)}(\boldsymbol{x}) = s^{(i)}(\boldsymbol{p}(\boldsymbol{x})). \tag{3.22}$$

If $f(\boldsymbol{x}^{(i)})$ is not in the range of $s^{(i)}(\boldsymbol{x})$, then the condition $s^{(i)}(\boldsymbol{p}(\boldsymbol{x}^{(i)})) = f(\boldsymbol{x}^{(i)})$ is not achievable. We can overcome that issue by combining both compositional corrections. In that case, the following selection for g and $\boldsymbol{p}$

$$g(t) = t - s^{(i)}(\boldsymbol{x}^{(i)}) + f(\boldsymbol{x}^{(i)}), \tag{3.23}$$

$$\boldsymbol{p}(\boldsymbol{x}) = \boldsymbol{x}^{(i)} + \boldsymbol{J}_p(\boldsymbol{x} - \boldsymbol{x}^{(i)}), \tag{3.24}$$

where $\boldsymbol{J}_p$ is a $n \times n$ matrix for which $\boldsymbol{J}_p^T \nabla s^{(i)} = \nabla f(\boldsymbol{x}^{(i)})$, guarantees consistency.

Additive and multiplicative corrections allow obtaining first-order consistency conditions. For the additive case we can generally express the correction as

$$s^{(i+1)}(\boldsymbol{x}) = \lambda(\boldsymbol{x}) + s^{(i)}(\boldsymbol{x}). \tag{3.25}$$

The associated consistency conditions require that $\lambda(\boldsymbol{x})$ satisfies

$$\lambda(\boldsymbol{x}^{(i)}) = f(\boldsymbol{x}^{(i)}) - s^{(i)}(\boldsymbol{x}^{(i)}), \tag{3.26}$$

and

$$\nabla\lambda(\boldsymbol{x}^{(i)}) = \nabla f(\boldsymbol{x}^{(i)}) - \nabla s^{(i)}(\boldsymbol{x}^{(i)}). \tag{3.27}$$

Those requirements can be obtained by the following linear additive correction:

$$s^{(i+1)}(\boldsymbol{x}) = f(\boldsymbol{x}^{(i)}) - s^{(i)}(\boldsymbol{x}^{(i)}) + (\nabla f(\boldsymbol{x}^{(i)}) - \nabla s^{(i)}(\boldsymbol{x}^{(i)}))(\boldsymbol{x} - \boldsymbol{x}^{(i)}) + s^{(i)}(\boldsymbol{x}). \tag{3.28}$$

Multiplicative corrections (also known as the β-correlation method [20]) can be represented generically by

$$s^{(i+1)}(\boldsymbol{x}) = \alpha(\boldsymbol{x}) s^{(i)}(\boldsymbol{x}). \tag{3.29}$$

Assuming that $s^{(i)}(\boldsymbol{x}^{(i)}) \neq 0$, zero- and first-order consistency can be achieved if

$$\alpha(\boldsymbol{x}^{(i)}) = \frac{f(\boldsymbol{x}^{(i)})}{s^{(i)}(\boldsymbol{x}^{(i)})}, \tag{3.30}$$

and

$$\nabla\alpha(\boldsymbol{x}^{(i)}) = [\nabla f(\boldsymbol{x}^{(i)}) - f(\boldsymbol{x}^{(i)}) / s^{(i)}(\boldsymbol{x}^{(i)}) \nabla s^{(i)}(\boldsymbol{x}^{(i)})] / s^{(i)}(\boldsymbol{x}^{(i)}). \tag{3.31}$$

The requirement $s^{(i)}(\boldsymbol{x}^{(i)}) \neq 0$ is not strong in practice since very often the range of $f(\boldsymbol{x})$ (and thus, of the surrogate $s^{(i)}(\boldsymbol{x})$) is known beforehand, and hence, a bias can be introduced both for $f(\boldsymbol{x})$ and $s^{(i)}(\boldsymbol{x})$ to avoid cost function values equal to zero. In these circumstances the following multiplicative correction

$$s^{(i+1)}(\boldsymbol{x}) = \left[\frac{f(\boldsymbol{x}^{(i)})}{s^{(i)}(\boldsymbol{x}^{(i)})} + \frac{\nabla f(\boldsymbol{x}^{(i)}) s^{(i)}(\boldsymbol{x}^{(i)}) - f(\boldsymbol{x}^{(i)}) \nabla s^{(i)}(\boldsymbol{x}^{(i)})}{(s^{(i)}(\boldsymbol{x}^{(i)}))^2} (\boldsymbol{x} - \boldsymbol{x}^{(i)}) \right] s^{(i)}(\boldsymbol{x}), \tag{3.32}$$

is consistent with conditions (3.30) and (3.31).

3.3.4.2 Space Mapping Concept

Space mapping (SM) [1,5,6] is a well-known methodology for correcting a given (either functional or physical) surrogate. SM algorithms aim at objective functions $f(\boldsymbol{x})$ that can be written as a functional U of a so-called system response $\boldsymbol{R}_f(\boldsymbol{x}) \in R^m$

$$f(\boldsymbol{x}) = U(\boldsymbol{R}_f(\boldsymbol{x})). \tag{3.33}$$

The fine model response $\boldsymbol{R}_f(\boldsymbol{x})$ is assumed to be accurate but computationally expensive. The coarse model response $\boldsymbol{R}_c(\boldsymbol{x}) \in R^m$ is much cheaper to evaluate than the fine model response at the expense of being an approximation of it. SM establishes a correction between model responses rather than between objective functions. The corrected model response will be denoted as $\boldsymbol{R}_s(\boldsymbol{x}; \boldsymbol{p}_{SM}) \in R^m$, and $\boldsymbol{p}_{SM}$ represents a set of parameters that describes the type of correction performed.

We can find in the literature four different groups of coarse model response corrections [1,5]:

1. Input space mapping [1]. The response correction is based on an affine transformation on the low-fidelity model parameter space. Example: $\boldsymbol{R}_s(\boldsymbol{x}; \boldsymbol{p}_{SM}) = \boldsymbol{R}_s(\boldsymbol{x}; \boldsymbol{B},\boldsymbol{c}) = \boldsymbol{R}_c(\boldsymbol{B}\,\boldsymbol{x} + \boldsymbol{c})$.
2. Output space mapping [5]. The response correction is based on an affine transformation on the low-fidelity model response. Example: $\boldsymbol{R}_s(\boldsymbol{x}; \boldsymbol{p}_{SM}) = \boldsymbol{R}_s(\boldsymbol{x}; \boldsymbol{A},\boldsymbol{d}) = \boldsymbol{A}\,\boldsymbol{R}_c(\boldsymbol{x})+\boldsymbol{d}$. Manifold-mapping (see Section 3.4.3) is a particular case of output space mapping.
3. Implicit space mapping [49]. In some cases, there are additional parameters $\boldsymbol{x}_p \in R^{n_p}$ in the coarse model response $\boldsymbol{R}_c(\boldsymbol{x}; \boldsymbol{x}_p)$ that can be tuned for better aligning of the fine and coarse model responses. Example: $\boldsymbol{R}_s(\boldsymbol{x}; \boldsymbol{p}_{SM}) = \boldsymbol{R}_s(\boldsymbol{x}; \boldsymbol{x}_p) = \boldsymbol{R}_c(\boldsymbol{x}; \boldsymbol{x}_p)$. These additional parameters are known in SM lexicon as pre-assigned parameters, and are in general different from the optimization variables $\boldsymbol{x}$.
4. Custom corrections that exploit the structure of the given design problem [1]. In many occasions the model responses are obtained through the sweeping of some parameter t:

$$\boldsymbol{R}_f(\boldsymbol{x}) = \boldsymbol{R}_f(\boldsymbol{x};t) = [R_f(\boldsymbol{x};t_1)\;\; R_f(\boldsymbol{x};t_2)\;\; \ldots\;\; R_f(\boldsymbol{x};t_m)]^T, \tag{3.34}$$

$$\boldsymbol{R}_c(\boldsymbol{x}) = \boldsymbol{R}_c(\boldsymbol{x};t) = [R_c(\boldsymbol{x};t_1)\;\; R_c(\boldsymbol{x};t_2)\;\; \ldots\;\; R_c(\boldsymbol{x};t_m)]^T. \tag{3.35}$$

Examples of this situation appear when the parameter t represents time or frequency. The response correction considered in this case[2] could be based on an affine transformation on the sweeping parameter space:

$$\boldsymbol{R}_s(\boldsymbol{x}; \boldsymbol{p}_{SM}) = \boldsymbol{R}_s(\boldsymbol{x}; r_0, r_1) = \boldsymbol{R}_c(\boldsymbol{x}; r_0 + r_1 t). \tag{3.36}$$

In Fig. 3.6 we illustrate by means of block diagrams the four SM-based correction strategies introduced above, together with a combination of three of them.

The surrogate response is usually optimized with respect to the SM parameters $\boldsymbol{p}_{SM}$ in order to reduce the model discrepancy for all or part of the data available $\boldsymbol{R}_f(\boldsymbol{x}^{(1)})$, $\boldsymbol{R}_f(\boldsymbol{x}^{(2)})$, … , $\boldsymbol{R}_f(\boldsymbol{x}^{(p)})$:

$$\boldsymbol{p}_{SM} = \arg\min_{\boldsymbol{p}_{SM}} \sum_{k=1}^{p} \omega^{(k)} \| \boldsymbol{R}_f(\boldsymbol{x}^{(k)}) - \boldsymbol{R}_s(\boldsymbol{x}^{(k)}; \boldsymbol{p}_{SM}) \|, \tag{3.37}$$

where $0 \le \omega^{(k)} \le 1$ are weights for each of the samples. The corrected surrogate $\boldsymbol{R}_s(\boldsymbol{x}; \boldsymbol{p}_{SM})$ can be used as an approximation to the fine response $\boldsymbol{R}_f(\boldsymbol{x})$ in the vicinity of the sampled data. The minimization in (3.37) is known in SM literature as parameter extraction [1]. The solving of this optimization process is not exempt from difficulties, since in many cases the problem is ill-conditioned. We can find in [1] a number of techniques for addressing parameter extraction in a robust manner.

[2] This type of space mapping is known as frequency space mapping [4], and it was originally proposed in microwave engineering applications (in these applications t usually refers to frequency).

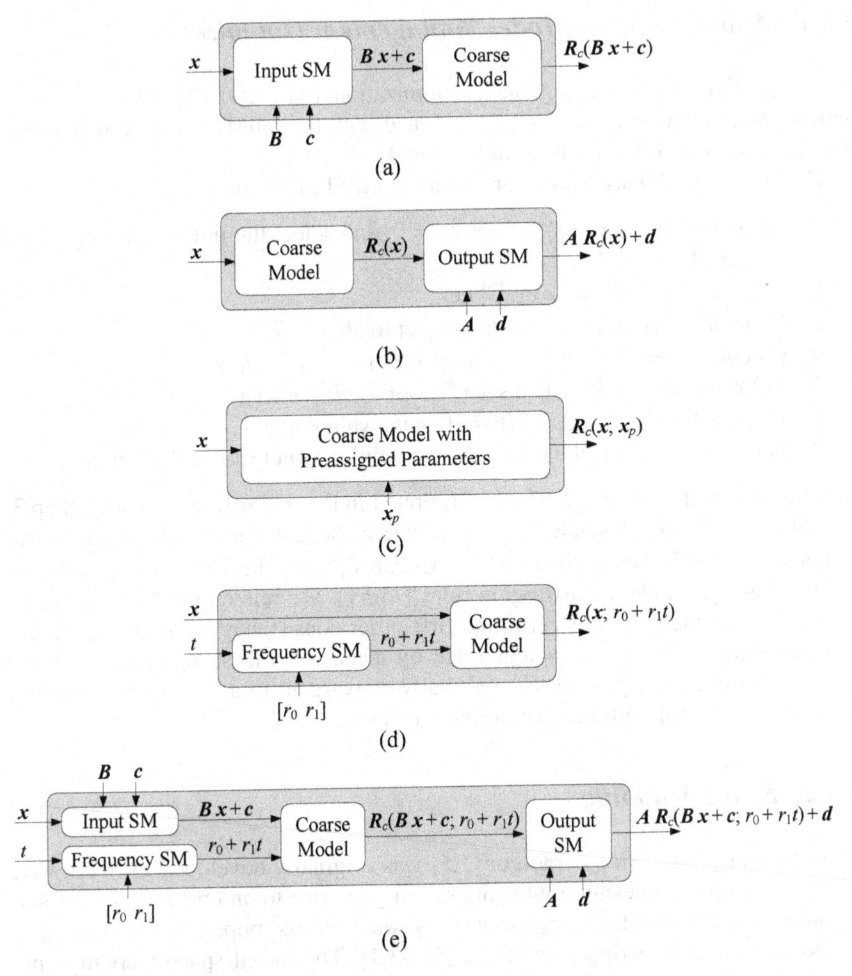

Fig. 3.6 Basic space-mapping surrogate correction types: (a) input SM, (b) output SM, (c) implicit SM, (d) frequency SM, and (e) composite using input, output and frequency SM.

3.4 Surrogate-Based Optimization Techniques

In this section, we will introduce several optimization strategies that exploit surrogate models. More specifically, we will describe approximation model management optimization [7], space mapping [5], manifold mapping [8], and the surrogate management framework [9]. The first three approaches follow the surrogate-based optimization framework presented in Section 3.2. We will conclude the section with a brief discussion on addressing the tradeoff between exploration and/or exploitation in the optimization process.

3.4.1 Approximation Model Management Optimization

Approximation model management optimization (AMMO) [7] relies on trust-region gradient-based optimization combined with the multiplicative linear surrogate correction (3.32) introduced in Section 3.3.4.1.

The basic AMMO algorithm can be summarized as follows:

1. Set initial guess $\boldsymbol{x}^{(0)}$, $s^{(0)}(\boldsymbol{x})$, and $i = 0$, and select the initial trust-region radius $\delta > 0$.
2. If $i > 0$, then $s^{(i)}(\boldsymbol{x}) = \alpha(\boldsymbol{x})\, s^{(i-1)}(\boldsymbol{x})$.
3. Solve $\boldsymbol{h}^* = \mathrm{argmin}\; s^{(i)}(\boldsymbol{x}^{(i)} + \boldsymbol{h})$ subject to $||\boldsymbol{h}||_\infty \leq \delta$.
4. Calculate $\rho = (f(\boldsymbol{x}^{(i)}) - f(\boldsymbol{x}^{(i)} + \boldsymbol{h}^*))/(s(\boldsymbol{x}^{(i)}) - s^{(i)}(\boldsymbol{x}^{(i)} + \boldsymbol{h}^*))$.
5. If $f(\boldsymbol{x}^{(i)}) > f(\boldsymbol{x}^{(i)} + \boldsymbol{h}^*)$, then set $\boldsymbol{x}^{(i+1)} = \boldsymbol{x}^{(i)} + \boldsymbol{h}^*$; otherwise $\boldsymbol{x}^{(i+1)} = \boldsymbol{x}^{(i)}$.
6. Update the search radius δ based on the value of ρ.
7. Set $i = i + 1$, and if the termination condition is not satisfied, go to Step 2.

Additional constraints can also be incorporated in the optimization through Step 3. AMMO can also be extended to cases where the constraints are expensive to evaluate and can be approximated by surrogates [50]. The search radius δ is updated using the standard trust-region rules [11,51]. We reiterate that the surrogate correction considered yields zero- and first-order consistency with $f(\boldsymbol{x})$. Since this surrogate-based approach is safeguarded by means of a trust-region method, the whole scheme can be proven to be globally convergent to a first-order stationary point of the original optimization problem (3.1).

3.4.2 Space Mapping

The space mapping (SM) paradigm [1,5] was originally developed in microwave engineering optimal design applications, and gave rise to an entire family of surrogate-based optimization approaches. Nowadays, its popularity is spreading across several engineering disciplines [52,53,1]. The initial space-mapping optimization methodologies were based on input SM [1], i.e., a linear correction of the coarse model design space. This kind of correction is well suited for many engineering problems, particularly in electrical engineering, where the model discrepancy is mostly due to second-order effects (e.g., presence of parasitic components). In these applications the model response ranges are often similar in shape, but slightly distorted and/or shifted with respect to a sweeping parameter (e.g., signal frequency).

Space mapping can be incorporated in the SBO framework by just identifying the sequence of surrogates with

$$s^{(0)}(\boldsymbol{x}) = U(R_c(\boldsymbol{x})), \tag{3.38}$$

and

$$s^{(i)}(\boldsymbol{x}) = U(R_s(\boldsymbol{x}; \boldsymbol{p}_{SM}^{(i)})), \tag{3.39}$$

for $i > 0$. The parameters $p_{SM}^{(i)}$ are obtained by parameter extraction as in (3.37). The accuracy of the corrected surrogate will clearly depend on the quality of the coarse model response [16]. In microwave design applications it has been many times observed that the number of points p needed for obtaining a satisfactory SM-based corrected surrogate is on the order of the number of optimization variables n [1]. Though output SM can be used to obtain both zero- and first-order consistency conditions with $f(\boldsymbol{x})$, many other SM-based optimization algorithms that have been applied in practice do not satisfy those conditions, and in some occasions convergence problems have been identified [14]. Additionally, the choice of an adequate SM correction approach is not always obvious [14]. However, in multiple occasions and in several different disciplines [52,53,1], space mapping has been reported as a very efficient means for obtaining satisfactory optimal designs.

Convergence properties of space-mapping optimization algorithms can be improved when these are safeguarded by a trust region [54]. Similarly to AMMO, the SM surrogate model optimization is restricted to a neighborhood of $\boldsymbol{x}^{(i)}$ (this time by using the Euclidean norm) as follows

$$\boldsymbol{x}^{(i+1)} = \arg\min_{\boldsymbol{x}} s^{(i)}(\boldsymbol{x}) \quad \text{subject to} \quad \| \boldsymbol{x} - \boldsymbol{x}^{(i)} \|_2 \leq \delta^{(i)}, \tag{3.40}$$

where $\delta^{(i)}$ denotes the trust-region radius at iteration i. The trust region is updated at every iteration by means of precise criteria [11]. A number of enhancements for space mapping have been suggested recently in the literature (e.g., zero-order and aproximate/exact first order consistency conditions with $f(\boldsymbol{x})$ [54], or adaptively constrained parameter extraction [55]).

The quality of a surrogate within space mapping can be assessed by means of the techniques described in [14,16]. These methods are based on evaluating the high-fidelity model at several points (and thus, they require some extra computational effort). With that information, some conditions required for convergence are approximated numerically, and as a result, low-fidelity models can be compared based on these approximate conditions. The quality assessment algorithms presented in [14,16] can also be embedded into SM optimization algorithms in order to throw some light on the delicate issue of selecting the most adequate SM surrogate correction.

It should be emphasized that space mapping is not a general-purpose optimization approach. The existence of the computationally cheap and sufficiently accurate low-fidelity model is an important prerequisite for this technique. If such a coarse model does exist, satisfactory designs are often obtained by space mapping after a relatively small number of evaluations of the high-fidelity model. This number is usually on the order of the number of optimization variables n [14], and very frequently represents a dramatic reduction in the computational cost required for solving the same optimization problem with other methods that do not rely on surrogates. In the absence of the above-mentioned low-fidelity model, space-mapping optimization algorithms may not perform efficiently.

3.4.3 Manifold Mapping

Manifold mapping (MM) [8,56] is a particular case of output space mapping, that is supported by convergence theory [13,56], and does not require the parameter extraction step shown in (3.37). Manifold mapping can be integrated in the SBO framework by just considering $s^{(i)}(\boldsymbol{x}) = U(\boldsymbol{R}_s^{(i)}(\boldsymbol{x}))$ with the response correction for $i \geq 0$ defined as

$$\boldsymbol{R}_s^{(i)}(\boldsymbol{x}) = \boldsymbol{R}_f(\boldsymbol{x}^{(i)}) + \boldsymbol{S}^{(i)}\left(\boldsymbol{R}_c(\boldsymbol{x}) - \boldsymbol{R}_c(\boldsymbol{x}^{(i)})\right), \tag{3.41}$$

where $\boldsymbol{S}^{(i)}$, for $i \geq 1$, is the following $m \times m$ matrix

$$\boldsymbol{S}^{(i)} = \Delta \boldsymbol{F} \Delta \boldsymbol{C}^{\dagger}, \tag{3.42}$$

with

$$\Delta \boldsymbol{F} = [\boldsymbol{R}_f(\boldsymbol{x}^{(i)}) - \boldsymbol{R}_f(\boldsymbol{x}^{(i-1)}) \ \ \ldots \ \ \boldsymbol{R}_f(\boldsymbol{x}^{(i)}) - \boldsymbol{R}_f(\boldsymbol{x}^{(\max\{i-n,0\})})], \tag{3.43}$$

$$\Delta \boldsymbol{C} = [\boldsymbol{R}_c(\boldsymbol{x}^{(i)}) - \boldsymbol{R}_c(\boldsymbol{x}^{(i-1)}) \ \ \ldots \ \ \boldsymbol{R}_c(\boldsymbol{x}^{(i)}) - \boldsymbol{R}_c(\boldsymbol{x}^{(\max\{i-n,0\})})]. \tag{3.44}$$

The matrix $\boldsymbol{S}^{(0)}$ is typically taken as the identity matrix $\boldsymbol{I}_m$. Here, † denotes the pseudoinverse operator defined for $\Delta \boldsymbol{C}$ as

$$\Delta \boldsymbol{C}^{\dagger} = \boldsymbol{V}_{\Delta C} \boldsymbol{\Sigma}_{\Delta C}^{\dagger} \boldsymbol{U}_{\Delta C}^{T}, \tag{3.45}$$

where $\boldsymbol{U}_{\Delta C}$, $\Sigma_{\Delta C}$, and $\boldsymbol{V}_{\Delta C}$ are the factors in the singular value decomposition of ΔC. The matrix $\boldsymbol{\Sigma}_{\Delta C}{}^{\dagger}$ is the result of inverting the nonzero entries in $\boldsymbol{\Sigma}_{\Delta C}$, leaving the zeroes invariant [8]. Some mild general assumptions on the model responses are made in theory [56] so that every pseudoinverse introduced is well defined.

The response correction $\boldsymbol{R}_s^{(i)}(\boldsymbol{x})$ is an approximation of

$$\boldsymbol{R}_s^{*}(\boldsymbol{x}) = \boldsymbol{R}_f(\boldsymbol{x}^{*}) + \boldsymbol{S}^{*}(\boldsymbol{R}_c(\boldsymbol{x}) - \boldsymbol{R}_c(\boldsymbol{x}^{*})), \tag{3.46}$$

with $\boldsymbol{S}^{*}$ being the $m \times m$ matrix defined as

$$\boldsymbol{S}^{*} = J_f(\boldsymbol{x}^{*}) J_c^{\dagger}(\boldsymbol{x}^{*}), \tag{3.47}$$

where $J_f(\boldsymbol{x}^*)$ and $J_c(\boldsymbol{x}^*)$ stand for the fine and coarse model response Jacobian, respectively, evaluated at $\boldsymbol{x}^*$. Obviously, neither $\boldsymbol{x}^*$ nor $\boldsymbol{S}^*$ is known beforehand. Therefore, one needs to use an iterative approximation, such as the one in (3.41)-(3.45), in the actual manifold-mapping algorithm.

The manifold-mapping model alignment is illustrated in Fig. 3.7 for the least-squares optimization problem

$$U(\boldsymbol{R}_f(\boldsymbol{x})) = \| \boldsymbol{R}_f(\boldsymbol{x}) - \boldsymbol{y} \|_2^2, \tag{3.48}$$

with $\boldsymbol{y} \in R^m$ being the design specifications given. In that figure the point $\boldsymbol{x}_c^*$ denotes the minimizer corresponding to the coarse model cost function $U(\boldsymbol{R}_c(\boldsymbol{x}))$. We note that, in absence of constraints, the optimality associated to (3.48) is translated into the orthogonality between the tangent plane for $\boldsymbol{R}_f(\boldsymbol{x})$ at $\boldsymbol{x}^*$ and the vector $\boldsymbol{R}_f(\boldsymbol{x}^*) - \boldsymbol{y}$.

If the low-fidelity model has a negligible computational cost when compared to the high-fidelity one, the MM surrogate can be explored globally. The MM algorithm is in this case endowed with some robustness with respect to being trapped in unsatisfactory local minima.

For least-squares optimization problems as in (3.48), manifold mapping is supported by mathematically sound convergence theory [13]. We can identify four factors relevant for the convergence of the scheme above to the fine model optimizer $\boldsymbol{x}^*$:

1. The model responses being smooth.
2. The surrogate optimization in (3.2) being well-posed.
3. The discrepancy of the optimal model response $\boldsymbol{R}_f(\boldsymbol{x}^*)$ with respect to the design specifications being sufficiently small.
4. The low-fidelity model response being a sufficiently good approximation of the high-fidelity model response.

In most practical situations the requirements associated to the first three factors are satisfied, and since the low-fidelity models often considered are based on expert knowledge accumulated over the years, the similarity between the model responses can be frequently good enough for having convergence.

Manifold-mapping algorithms can be expected to converge for a merit function U sufficiently smooth. Since the correction in (3.41) does not involve U, if the model responses are smooth enough, and even when U is not differentiable, manifold mapping may still yield satisfactory solutions. The experimental evidence given in [57] for designs based on minimax objective functions indicates that the MM approach can be used successfully in more general situations than those for which theoretical results have been obtained.

The basic manifold-mapping algorithm can be modified in a number of ways. Convergence appears to improve if derivative information is introduced in the algorithm [13]. The incorporation of a Levenberg-Marquardt strategy in manifold mapping [58] can be seen as a convergence safeguard analogous to a trust-region method [11]. Manifold mapping can also be extended to designs where the constraints are determined by time-consuming functions, and for which surrogates are available as well [59].

3.4.4 Surrogate Management Framework

The surrogate management framework (SMF) [9] is mainly based on pattern search. Pattern search [60] is a general set of derivative-free optimizers that can be proven to be globally convergent to first-order stationary points. A pattern search optimization algorithm is based on exploring the search space by means of a structured set of points (pattern or stencil) that is modified along iterations. The pattern search scheme considered in [9] has two main steps per iteration: search and poll. Each iteration starts with a pattern of size Δ centered at $\boldsymbol{x}^{(i)}$. The search step is optional and is always performed before the poll step. In the search stage a (small) number of points are selected from the search space (typically by means of a surrogate), and the cost function $f(\boldsymbol{x})$ is evaluated at these points. If the cost function

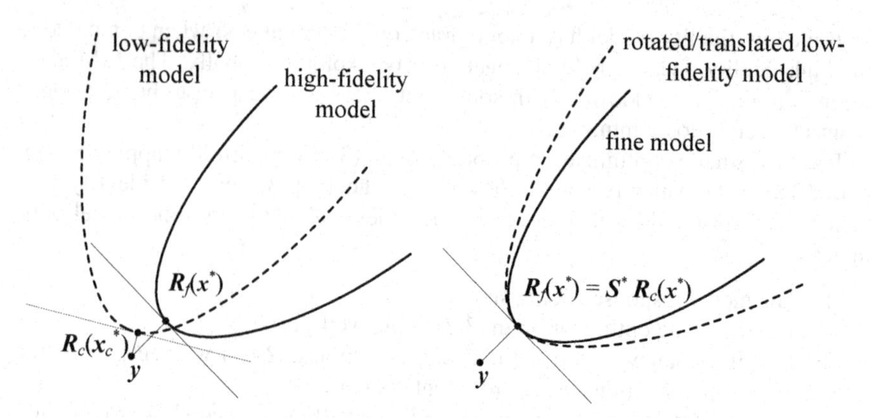

Fig. 3.7 Illustration of the manifold-mapping model alignment for a least-squares optimization problem. The point x_c^* denotes the minimizer corresponding to the coarse model response, and the point y is the vector of design specifications. Thin solid and dashed straight lines denote the tangent planes for the fine and coarse model response at their optimal designs, respectively. By the linear correction S^*, the point $R_c(x^*)$ is mapped to $R_f(x^*)$, and the tangent plane for $R_c(x)$ at $R_c(x^*)$ to the tangent plane for $R_f(x)$ at $R_f(x^*)$ [13].

for some of them improves on $f(x^{(i)})$ the search step is declared successful, the current pattern is centered at this new point, and a new search step is started. Otherwise a poll step is taken. Polling requires computing $f(x)$ for points in the pattern. If one of these points is found to improve on $f(x^{(i)})$, the poll step is declared successful, the pattern is translated to this new point, and a new search step is performed. Otherwise the whole pattern search iteration is considered unsuccessful and the termination condition is checked. This stopping criterion is typically based on the pattern size Δ [9,61]. If, after the unsuccessful pattern search iteration another iteration is needed, the pattern size Δ is decreased, and a new search step is taken with the pattern centered again at $x^{(i)}$. Surrogates are incorporated in the SMF through the search step. For example, kriging (with Latin hypercube sampling) is considered in the SMF application studied in [61].

In order to guarantee convergence to a stationary point, the set of vectors formed by each pattern point and the pattern center should be a generating (or positive spanning) set [60,61]. A generating set for R^n consists of a set of vectors whose non-negative linear combinations span R^n. Generating sets are crucial in proving convergence (for smooth objective functions) due to the following property: if a generating set is centered at $x^{(i)}$ and $\nabla f(x^{(i)}) \neq 0$, then at least one of the vectors in the generating set defines a descent direction [60]. Therefore, if $f(x)$ is smooth and $\nabla f(x^{(i)}) \neq 0$, we can expect that for a pattern size Δ small enough, some of the points in the associated stencil will improve on $f(x^{(i)})$.

Though pattern search optimization algorithms typically require many more function evaluations than gradient-based techniques, the computations in both the search and poll steps can be performed in a distributed fashion. On top of that, the use of surrogates, as is the case for the SMF, generally accelerates noticeably the entire optimization process.

3.4.5 Exploitation versus Exploration

The surrogate-based optimization framework starts from an initial surrogate model which is updated using the high-fidelity model data that is accumulated in the optimization process. In particular, the high-fidelity model has to be evaluated for verification at any new design $\boldsymbol{x}^{(i)}$ provided by the surrogate model. The new points at which we evaluate the high-fidelity model are sometimes referred to as infill points [4]. We reiterate that this data can be used to enhance the surrogate. The selection of the infill points is also known as adaptive sampling [4].

Infill points in approximation model management optimization, space mapping and manifold mapping are in practice selected through local optimization of the surrogate (global optimization for problems with a medium/large number of variables and even relatively inexpensive surrogates can be a time-consuming procedure). The new infill points in the surrogate management framework are taken based only on high-fidelity cost function improvement. As we have seen in this section, the four surrogate-based optimization approaches discussed are supported by local optimality theoretical results. In other words, these methodologies intrinsically aim at the exploitation of certain region of the design space (the neighborhood of a first-order stationary point). If the surrogate is valid globally, the first iterations of these four optimization approaches can be used to avoid being trapped in unsatisfactory local solutions (i.e., global exploration steps).

The exploration of the design space implies in most cases a global search. If the underlying objective function is non-convex, exploration usually boils down to performing a global sampling of the search space, for example, by selecting those points that maximize some estimation of the error associated to the surrogate considered [4]. It should be stressed that global exploration is often impractical, especially for computationally expensive cost functions with a medium/large number of optimization variables (more than a few tens). Additionally, pure exploration may not be a good approach for updating the surrogate in an optimization context, since a great amount of computing resources can be spent in modeling parts of the search space that are not interesting from an optimal design point of view.

Therefore, it appears that in optimization there should be a balance between exploitation and exploration. As suggested in [4], this tradeoff could be formulated in the context of surrogate-based optimization, for example, by means of a bi-objective optimization problem (with global measure of the error associated to the surrogate as second objective function), by maximizing the probability of improvement upon the best observed objective function value, or through the maximization of the expected cost function improvement. As mentioned above, these hybrid approaches will find difficulties in performing an effective global search in designs with a medium/large number of optimization variables.

3.5 Final Remarks

In this chapter, an overview of surrogate modeling, with an emphasis on optimization, has been presented. Surrogate-based optimization plays an important role in contemporary engineering design, and the importance of this role will most likely increase in the near future. One of the reasons for this increase is the fact that computer simulations have become a major design tool in most engineering areas. In order for these simulations to be sufficiently accurate, more and more phenomena have to be captured. This level of sophistication renders simulations computationally expensive, particularly when they deal with the time-varying three-dimensional structures considered in many engineering fields. Hence, evaluation times of several days, or even weeks, are nowadays not uncommon. The direct use of CPU-intensive numerical models in some off-the-shelf automated optimization procedures (e.g., gradient-based techniques with approximate derivatives) is very often prohibitive. Surrogate-based optimization can be a very useful approach in this context, since, apart from reducing significantly the number of high-fidelity expensive simulations in the whole design process, it also helps in addressing important high-fidelity cost function issues (e.g., presence of discontinuities and/or multiple local optima).

References

1. Bandler, J.W., Cheng, Q.S., Dakroury, S.A., Mohamed, A.S., Bakr, M.H., Madsen, K., Søndergaard, J.: Space mapping: the state of the art. IEEE Trans. Microwave Theory Tech. 52, 337–361 (2004)
2. Pironneau, O.: On optimum design in fluid mechanics. J. Fluid Mech. 64, 97–110 (1974)
3. Queipo, N.V., Haftka, R.T., Shyy, W., Goel, T., Vaidynathan, R., Tucker, P.K.: Surrogate-based analysis and optimization. Progress in Aerospace Sciences 41, 1–28 (2005)
4. Forrester, A.I.J., Keane, A.J.: Recent advances in surrogate-based optimization. Prog. Aerospace Sciences 45, 50–79 (2009)
5. Koziel, S., Bandler, J.W., Madsen, K.: A space mapping framework for engineering optimization: theory and implementation. IEEE Trans. Microwave Theory Tech. 54, 3721–3730 (2006)
6. Koziel, S., Cheng, Q.S., Bandler, J.W.: Space mapping. IEEE Microwave Magazine 9, 105–122 (2008)
7. Alexandrov, N.M., Lewis, R.M.: An overview of first-order model management for engineering optimization. Optimization and Engineering 2, 413–430 (2001)
8. Echeverria, D., Hemker, P.W.: Space mapping and defect correction. CMAM Int. Mathematical Journal Computational Methods in Applied Mathematics 5, 107–136 (2005)
9. Booker, A.J., Dennis, J.E., Frank, P.D., Serafini, D.B., Torczon, V., Trosset, M.W.: A rigorous framework for optimization of expensive functions by surrogates. Structural Optimization 17, 1–13 (1999)
10. Simpson, T.W., Peplinski, J., Koch, P.N., Allen, J.K.: Metamodels for computer-based engineering design: survey and recommendations. Engineering with Computers 17, 129–150 (2001)
11. Conn, A.R., Gould, N.I.M., Toint, P.L.: Trust Region Methods. MPS-SIAM Series on Optimization (2000)

12. Alexandrov, N.M., Dennis, J.E., Lewis, R.M., Torczon, V.: A trust region framework for managing use of approximation models in optimization. Struct. Multidisciplinary Optim. 15, 16–23 (1998)
13. Echeverría, D., Hemker, P.W.: Manifold mapping: a two-level optimization technique. Computing and Visualization in Science 11, 193–206 (2008)
14. Koziel, S., Bandler, J.W., Madsen, K.: Quality assessment of coarse models and surrogates for space mapping optimization. Optimization Eng. 9, 375–391 (2008)
15. Koziel, S., Bandler, J.W.: Coarse and surrogate model assessment for engineering design optimization with space mapping. In: IEEE MTT-S Int. Microwave Symp. Dig, Honolulu, HI, pp. 107–110 (2007)
16. Koziel, S., Bandler, J.W.: Space-mapping optimization with adaptive surrogate model. IEEE Trans. Microwave Theory Tech. 55, 541–547 (2007)
17. Alexandrov, N.M., Nielsen, E.J., Lewis, R.M., Anderson, W.K.: First-order model management with variable-fidelity physics applied to multi-element airfoil optimization. In: AIAA/USAF/NASA/ISSMO Symposium on Multidisciplinary Design and Optimization, Long Beach, CA, AIAA Paper 2000-4886 (2000)
18. Wu, K.-L., Zhao, Y.-J., Wang, J., Cheng, M.K.K.: An effective dynamic coarse model for optimization design of LTCC RF circuits with aggressive space mapping. IEEE Trans. Microwave Theory Tech. 52, 393–402 (2004)
19. Robinson, T.D., Eldred, M.S., Willcox, K.E., Haimes, R.: Surrogate-based optimization using multifidelity models with variable parameterization and corrected space mapping. AIAA Journal 46, 2814–2822 (2008)
20. Søndergaard, J.: Optimization using surrogate models – by the space mapping technique. Ph.D. Thesis, Informatics and Mathematical Modelling, Technical University of Denmark, Lyngby (2003)
21. Kleijnen, J.P.C.: Kriging metamodeling in simulation: a review. European Journal of Operational Research 192, 707–716 (2009)
22. Rayas-Sanchez, J.E.: EM-based optimization of microwave circuits using artificial neural networks: the state-of-the-art. IEEE Trans. Microwave Theory Tech. 52, 420–435 (2004)
23. Giunta, A.A., Wojtkiewicz, S.F., Eldred, M.S.: Overview of modern design of experiments methods for computational simulations. American Institute of Aeronautics and Astronautics, paper AIAA 2003–0649 (2003)
24. Santner, T.J., Williams, B., Notz, W.: The Design and Analysis of Computer Experiments. Springer, Heidelberg (2003)
25. Koehler, J.R., Owen, A.B.: Computer experiments. In: Ghosh, S., Rao, C.R. (eds.) Handbook of Statistics, vol. 13, pp. 261–308. Elsevier Science B.V., Amsterdam (1996)
26. Cheng, Q.S., Koziel, S., Bandler, J.W.: Simplified space mapping approach to enhancement of microwave device models. Int. J. RF and Microwave Computer-Aided Eng. 16, 518–535 (2006)
27. McKay, M., Conover, W., Beckman, R.: A comparison of three methods for selecting values of input variables in the analysis of output from a computer code. Technometrics 21, 239–245 (1979)
28. Beachkofski, B., Grandhi, R.: Improved distributed hypercube sampling. American Institute of Aeronautics and Astronautics, Paper AIAA 2002–1274 (2002)
29. Leary, S., Bhaskar, A., Keane, A.: Optimal orthogonal-array-based latin hypercubes. Journal of Applied Statistics 30, 585–598 (2003)

30. Ye, K.Q.: Orthogonal column latin hypercubes and their application in computer experiments. Journal of the American Statistical Association 93, 1430–1439 (1998)
31. Palmer, K., Tsui, K.-L.: A minimum bias latin hypercube design. IIE Transactions 33, 793–808 (2001)
32. Golub, G.H., Van Loan, C.F.: Matrix Computations, 3rd edn. The Johns Hopkins University Press, Baltimore (1996)
33. Conn, A.R., Scheinberg, K., Vicente, L.N.: Introduction to Derivative-Free Optimization. MPS-SIAM Series on Optimization, MPS-SIAM (2009)
34. Wild, S.M., Regis, R.G., Shoemaker, C.A.: ORBIT: Optimization by radial basis function interpolation in trust-regions. SIAM J. Sci. Comput. 30, 3197–3219 (2008)
35. Journel, A.G., Huijbregts, C.J.: Mining Geostatistics. Academic Press, London (1981)
36. O'Hagan, A.: Curve fitting and optimal design for predictions. Journal of the Royal Statistical Society B 40, 1–42 (1978)
37. Rasmussen, C.E., Williams, C.K.I.: Gaussian Processes for Machine Learning. MIT Press, Cambridge (2006)
38. Jones, D., Schonlau, M., Welch, W.: Efficient global optimization of expensive black-box functions. Journal of Global Optimization 13, 455–492 (1998)
39. Haykin, S.: Neural Networks: A Comprehensive Foundation, 2nd edn. Prentice-Hall, Englewood Cliffs (1998)
40. Minsky, M.I., Papert, S.A.: Perceptrons: An Introduction to Computational Geometry. MIT Press, Cambridge (1969)
41. Gunn, S.R.: Support vector machines for classification and regression. Technical Report. School of Electronics and Computer Science, University of Southampton (1998)
42. Angiulli, G., Cacciola, M., Versaci, M.: Microwave devices and antennas modeling by support vector regression machines. IEEE Trans. Magn. 43, 1589–1592 (2007)
43. Smola, A.J., Schölkopf, B.: A tutorial on support vector regression. Statistics and Computing 14, 199–222 (2004)
44. Vapnik, V.N.: The Nature of Statistical Learning Theory. Springer, New York (1995)
45. Levin, D.: The approximation power of moving least-squares. Mathematics of Computation 67, 1517–1531 (1998)
46. Aitken, A.C.: On least squares and linear combinations of observations. Proceedings of the Royal Society of Edinburgh 55, 42–48 (1935)
47. Rasmussen, C.E., Williams, C.K.I.: Gaussian Processes for Machine Learning. MIT Press, Massachussets (2006)
48. Geisser, S.: Predictive Inference. Chapman and Hall, Boca Raton (1993)
49. Koziel, S., Cheng, Q.S., Bandler, J.W.: Implicit space mapping with adaptive selection of preassigned parameters. IET Microwaves, Antennas & Propagation 4, 361–373 (2010)
50. Alexandrov, N.M., Lewis, R.M., Gumbert, C.R., Green, L.L., Newman, P.A.: Approximation and model management in aerodynamic optimization with variable-fidelity models. AIAA Journal of Aircraft 38, 1093–1101 (2001)
51. Moré, J.J.: Recent developments in algorithms and software for trust region methods. In: Bachem, A., Grötschel, M., Korte, B. (eds.) Mathematical Programming. The State of Art, pp. 258–287. Springer, Heidelberg (1983)
52. Leary, S.J., Bhaskar, A., Keane, A.J.: A constraint mapping approach to the structural optimization of an expensive model using surrogates. Optimization and Engineering 2, 385–398 (2001)

53. Redhe, M., Nilsson, L.: Optimization of the new Saab 9-3 exposed to impact load using a space mapping technique. Structural and Multidisciplinary Optimization 27, 411–420 (2004)
54. Koziel, S., Bandler, J.W., Cheng, Q.S.: Robust trust-region space-mapping algorithms for microwave design optimization. IEEE Trans. Microwave Theory and Tech. 58, 2166–2174 (2010)
55. Koziel, S., Bandler, J.W., Cheng, Q.S.: Adaptively constrained parameter extraction for robust space mapping optimization of microwave circuits. IEEE MTT-S Int. Microwave Symp. Dig., 205–208 (2010)
56. Echeverría, D.: Multi-Level optimization: space mapping and manifold mapping. Ph.D. Thesis, Faculty of Science, University of Amsterdam (2007)
57. Koziel, S., Echeverría Ciaurri, D.: Reliable simulation-driven design optimization of microwave structures using manifold mapping. Progress in Electromagnetics Research B 26, 361–382 (2010)
58. Hemker, P.W., Echeverría, D.: A trust-region strategy for manifold mapping optimization. JCP Journal of Computational Physics 224, 464–475 (2007)
59. Echeverría, D.: Two new variants of the manifold-mapping technique. COMPEL The International Journal for Computation and Mathematics in Electrical Engineering 26, 334–344 (2007)
60. Kolda, T.G., Lewis, R.M., Torczon, V.: Optimization by direct search: new perspectives on some classical and modern methods. SIAM Review 45, 385–482 (2003)
61. Marsden, A.L., Wang, M., Dennis, J.E., Moin, P.: Optimal aeroacoustic shape design using the surrogate management framework. Optimization and Engineering 5, 235–262 (2004)

Chapter 4
Derivative-Free Optimization

Oliver Kramer, David Echeverría Ciaurri, and Slawomir Koziel

Abstract. In many engineering applications it is common to find optimization problems where the cost function and/or constraints require complex simulations. Though it is often, but not always, theoretically possible in these cases to extract derivative information efficiently, the associated implementation procedures are typically non-trivial and time-consuming (e.g., adjoint-based methodologies). Derivative-free (non-invasive, black-box) optimization has lately received considerable attention within the optimization community, including the establishment of solid mathematical foundations for many of the methods considered in practice. In this chapter we will describe some of the most conspicuous derivative-free optimization techniques. Our depiction will concentrate first on local optimization such as pattern search techniques, and other methods based on interpolation/approximation. Then, we will survey a number of global search methodologies, and finally give guidelines on constraint handling approaches.

4.1 Introduction

Efficient optimization very often hinges on the use of derivative information of the cost function and/or constraints with respect to the design variables. In the last

Oliver Kramer
UC Berkeley, Berkeley, CA 94704, USA
e-mail: okramer@icsi.berkeley.edu

David Echeverría Ciaurri
Department of Energy Resources Engineering,
Stanford University, Stanford, CA 94305-2220, USA
e-mail: echeverr@stanford.edu

Slawomir Koziel
Engineering Optimization & Modeling Center,
School of Science and Engineering, Reykjavik University, Menntavegur 1,
101 Reykjavik, Iceland
e-mail: koziel@ru.is

S. Koziel & X.-S. Yang (Eds.): Comput. Optimization, Methods and Algorithms, SCI 356, pp. 61–83.
springerlink.com

decades, the computational models used in design have increased in sophistication to such an extent that it is common to find situations where (reliable) derivative information is not available. Although in simulation-based design there are methodologies that allow one to extract derivatives with a modest amount of additional computation, these approaches are in general invasive with respect to the simulator (e.g., adjoint-based techniques [1]), and thus, require precise knowledge of the simulation code and access to it. Moreover, obtaining derivatives in this intrusive way often implies significant coding (not only at the code development stage, but also subsequently, when maintaining or upgrading the software), and consequently, many simulators simply yield, as output, the data needed for the cost function and/or constraint values. Furthermore, optimal design has currently a clear multidisciplinary nature, so it is reasonable to expect that some components of the overall simulation do not include derivatives. This situation is even more likely when commercial software is used, since then the source code is typically simply inaccessible.

In this chapter we review a number of techniques that can be applied to generally constrained continuous optimization problems for which the cost function and constraint computation can be considered as a black box system. We wish to clearly distinguish between methods that aim at providing just a solution (local optimization; see Section 4.3), and approaches that try to avoid being trapped in local optima (global optimization; see Section 4.4). Local optimization is much easier to handle than global optimization, since, in general, there is no algorithmically suitable characterization of global optima. As a consequence, there are more theoretical results of practical relevance for local than for global optimizers (e.g., convergence conditions and rate). For more details on theoretical aspects of derivative-free optimization we strongly recommend both the review [2] and the book [3]. The techniques are described for continuous variables, but it is possible to apply, with care, extensions of many of them to mixed-integer scenarios. However, since mixed-integer nonlinear programming is still an emergent area (especially in simulated-based optimization), we prefer not to include recommendations in this case.

In some situations, numerical derivatives can be computed fairly efficiently (e.g., via a computer cluster), and still yield results that can be acceptable in practice. However, if the function/constraint evaluations are even moderately noisy, numerical derivatives are usually not useful. Though methods that rely on approximate derivatives are not derivative-free techniques per se, for example, in the absence of noise, they can address optimization in a black box approach. We should note that in addition to their inherent additional computational costs, numerical derivatives very often imply the tuning of the derivative approximation together with the simulation tolerances, and this is not always easy to do. Implicit filtering [4, 5] may somehow alleviate some of these issues. This approach is essentially a gradient-based procedure where the derivative approximation is improved as the optimization progresses. Implicit filtering has been recommended for problems with multiple local optima (e.g., noisy cost functions). For more details on gradient-based methodologies the reader is encouraged to regard nonlinear optimization references (for example, [6, 7]).

Many derivative-free methods are easy to implement, and this feature makes them attractive when approximate solutions are required in a short time frame. An obvious statement that is often neglected is that the computational cost of an iteration of an algorithm is not always a good estimate of the time needed within a project (measured from its inception) to obtain results that are satisfactory. However, one important drawback of derivative-free techniques (when compared, for example, with adjoint-based approaches) is the limitation on the number of optimization variables that can be handled. For example, in [3] and [2] the limit given is a few hundred variables. However, this limit in the problem size can be overcome, at least to some extent, if one is not restricted to a single sequential environment. For some of the algorithms though, adequately exploiting parallelism may be difficult or even impossible. When distributed computing resources are scarce or not available, and for simulation-based designs with significantly more than a hundred optimization variables, some form of parameter reduction is mandatory. In these cases, surrogates or reduced order models [8] for the cost function and constraints are desirable approaches. Fortunately, suitable parameter and model order reduction techniques can often be found in many engineering applications, although they may give rise to inaccurate models. We should add that even in theory, as long as a problem with nonsmooth/noisy cost functions/constraints can be reasonably approximated by a smooth function (see [9], Section 10.6), some derivative-free optimization algorithms perform well with nonsmooth/noisy cost functions, as has been observed in practice [2, 3].

In the last decade, there has been a renaissance of gradient-free optimization methodologies, and they have been successfully applied in a number of areas. Examples of this are ubiquitous; to name a few, derivative-free techniques have been used within molecular geometry [10], aircraft design [11, 12], hydrodynamics [13, 14], medicine [15, 16] and earth sciences [17, 18, 19, 20]. These references include generally constrained cases with derivative-free objective functions and constraints, continuous and integer optimization variables, and local and global approaches. In spite of all this apparent abundance of results, we should not disregard the general recommendation (see [3, 2]) of strongly preferring gradient-based methods if accurate derivative information can be computed reasonably efficiently and globally.

This chapter is structured as follows. In Section 4.2 we introduce the general problem formulation and notation. A number of derivative-free methodologies for unconstrained continuous optimization are presented in the next two sections. Section 4.3 refers to local optimization, and Section 4.4 is devoted to global optimization. Guidelines for extending all these algorithms to generally constrained optimization are given in Section 4.5. We bring the chapter to an end with some conclusions and recommendations.

4.2 Derivative-Free Optimization

A general single-objective optimization problem can be formally stated as:

$$\min_{\mathbf{x}\in\Omega\subset\mathbb{R}^n} f(\mathbf{x}) \quad \text{subject to} \quad \mathbf{g}(\mathbf{x}) \leq 0, \tag{4.1}$$

where $f(\mathbf{x})$ is the objective function, $\mathbf{x} \in \mathbb{R}^n$ is the vector of control variables, and $\mathbf{g} : \mathbb{R}^n \rightarrow \mathbb{R}^m$ represents the nonlinear constraints in the problem. Bound and linear constraints are included in the set $\Omega \subset \mathbb{R}^n$. For many approaches it is natural to treat any constraints for which derivatives are available separately. In particular, bounds and linear constraints, and any other structure than can be exploited, should be. So for example, nonlinear least-squares problems should exploit that inherent structure whenever possible (see e.g. [21]). We are interested in applications for which the objective function and constraint variables are computed using the output from a simulator, rendering function evaluations expensive and derivatives unavailable.

We will begin by discussing some general issues with respect to optimization with derivatives since they have important relevancy to the derivative-free case. Essentially all approaches to the former are somewhere between steepest descent and Newton's method, or equivalently use something that is between a linear and a quadratic model. This is reinforced by the realization that almost all practical computation is linear at its core, and (unconstrained) minima are characterized by the gradient being zero, and quadratic models give rise to linear gradients. In fact, theoretically at least, steepest descent is robust but slow (and in fact sometimes so slow that in practice it is not robust) whereas Newton's method is fast but may have a very small radius of convergence. That is, one needs to start close to the solution. It is also computationally more demanding. Thus in a sense, most practical unconstrained algorithms are intelligent compromises between these two extremes. Although, somewhat oversimplified, one can say that the constrained case is dealt with by being feasible, determining which constraints are tight, linearizing these constraints and then solving the reduced problem determined by these linearizations. Therefore, some reliable first-order model is essential, and for faster convergence, something more like a second-order model is desirable. In the unconstrained case with derivatives these are typically provided by a truncated Taylor series model (in the first-order case) and some approximation to a truncated second-order Taylor series model. A critical property of such models is that as the step sizes become small the models become more accurate. In the case where derivatives, or good approximations to them, are not available, clearly, one cannot use truncated Taylor series models. It thus transpires that, if for example, one uses interpolation or regression models, that depend only on function values, one can no longer guarantee that as the step sizes become small the models become more accurate. Thus one has to have some explicit way to make this guarantee, at least approximately. It turns out that this is usually done by considering the geometry of the points at which the function is evaluated, at least, before attempting to decrease the effective maximum step size. In pattern search methods, this is done by explicitly using a pattern with good geometry, for example, a regular mesh that one only scales while maintaining the a priori good geometry.

In the derivative case the usual stopping criteria relates to the first-order optimality conditions. In the derivative-free case, one does not explicitly have these, since they require (approximations to) the derivatives. At this stage we just remark that any criteria used should relate to the derivative case conditions, so, for example one needs something like a reasonable first-order model, at least asymptotically.

4.3 Local Optimization

The kernel of many optimizers are local methods. This is not surprising, since, as we already mentioned, there is no suitable algorithmic characterization of global optima unless one considers special situations such as where all local optima are global, as for example in convex minimization problems. In this section we concentrate on local search methods based on pattern search and also on interpolation and approximation models. Some constraint handling procedures are described in Section 4.5.

4.3.1 Pattern Search Methods

Pattern search methods are optimization procedures that evaluate the cost function in a stencil-based fashion determined by a set of directions with intrinsic properties meant to be desirable from a geometric/algebraic point of view. This stencil is sequentially modified as iterations proceed. The recent popularity of these schemes is due in part to the development of a mathematically sound convergence theory [2, 3]. Moreover, they are attractive because they can relatively easily leverage the widespread availability of parallel computing resources. However, most published computational results are not parallel exploiting.

4.3.1.1 Generalized Pattern Search

Generalized pattern search (GPS; [22, 23]) refers to a whole family of optimization methods. GPS relies on polling (local exploration of the cost function on the pattern) but may be enhanced by additional searches, see [23]. At any particular iteration a stencil (pattern) is centered at the current solution. The stencil comprises a set of directions such that at least one direction is a descent direction. This is also called a generating set (see e.g. [2]). If any of the points in the stencil represent an improvement in the cost function, the stencil is moved to one of them. Otherwise, the stencil size is decreased. The optimization progresses until some stopping criterion is satisfied (typically, a minimum stencil size). Generalized pattern search can be further generalized by polling in an asymptotically dense set of directions (this set varies with the iterations). The resulting algorithm is the mesh adaptive direct search (MADS; [24]). In particular, some generalization of a simple fixed pattern is essential for constrained problems. The GPS method parallelizes naturally since, at a particular iteration, the objective function evaluations at the polling points can be accomplished in a distributed fashion. The method typically requires on the order of n function evaluations per iteration (where n is the number of optimization variables).

4.3.1.2 Hooke-Jeeves Direct Search

The Hooke-Jeeves direct search (HJDS; [25]) is another pattern search method and was the first to use the term 'direct search' method and take advantage of the idea

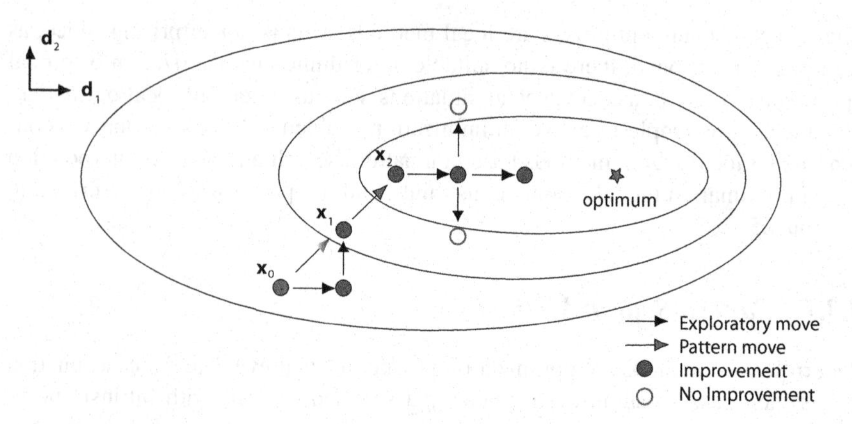

Fig. 4.1 Illustration of exploratory and pattern moves in Hooke-Jeeves direct search (modified from [19]). The star represents the optimum.

of a pattern. HJDS is based on two types of moves: exploratory and pattern. These moves are illustrated in Figure 4.1 for some optimization iterations in $\mathbb{R}^2$.

The iteration starts with a base point $\mathbf{x}_0$ and a given step size. During the exploratory move, the objective function is evaluated at successive changes of the base point in the search (for example coordinate) directions. All the directions are polled sequentially and in an opportunistic way. This means that if $\mathbf{d}_1 \in \mathbb{R}^n$ is the first search direction, the first function evaluation is at $\mathbf{x}_0 + \mathbf{d}_1$. If this represents an improvement in the cost function, the next point polled will be, assuming $n > 1$, $\mathbf{x}_0 + \mathbf{d}_1 + \mathbf{d}_2$, where $\mathbf{d}_2$ is the second search direction. Otherwise the point $\mathbf{x}_0 - \mathbf{d}_1$ is polled. Upon success at this last point, the search proceeds with $\mathbf{x}_0 - \mathbf{d}_1 + \mathbf{d}_2$, and alternatively with $\mathbf{x}_0 + \mathbf{d}_2$. The exploration continues until all search directions have been considered. If after the exploratory step no improvement in the cost function is found, the step size is reduced. Otherwise, a new point $\mathbf{x}_1$ is obtained, but instead of centering another exploratory move at $\mathbf{x}_1$, the algorithm performs the pattern move, which is a more aggressive step that moves further in the underlying successful direction. After the pattern move, the next polling center $\mathbf{x}_2$ is set at $\mathbf{x}_0 + 2(\mathbf{x}_1 - \mathbf{x}_0)$. If the exploratory move at $\mathbf{x}_2$ fails to improve upon $\mathbf{x}_1$, a new polling is performed around $\mathbf{x}_1$. If this again yields no cost function decrease, the step size is reduced, keeping the polling center at $\mathbf{x}_1$.

Notice the clear serial nature of the algorithm. This makes HJDS a reasonable pattern search option when distributed computing resources are not available. Because of the pattern move, HJDS may also be beneficial in situations where an optimum is far from the initial guess. One could argue that initially pattern search techniques should use a relatively large stencil size on the hope that this feature enables them to avoid some local minima and, perhaps, some robustness against noisy cost functions.

4.3.2 *Derivative-Free Optimization with Interpolation and Approximation Models*

The other major approach to derivative-free optimization is based on building models that are meant to approximate the functions and then make use of derivative methods on the models. The advantage is that one is trying to take account of the shape of the function rather than naively just using the function evaluations alone. As our introductory remarks in Section 4.2 suggest we can expect our models to be at least first-order models or better still, second-order.

A major drawback of this approach is that, since the models are not based upon an a priori pattern, as with just polling, the geometry of the sample points used requires special attention. Additionally, one pays for the extra sophistication of these methods in that they are not obviously parallelizable. Some of the better known algorithms in this class include DFO [3], NEWUOA [26] and BOOSTERS [27]. The basic ideas will be given here but it is recommended that the diligent reader consult Chapters 3-6 of [3].

First of all, what does good geometry mean? Essentially, for example, if one wants to consider interpolation by a polynomial of degree d, where $d = 1$, that is linear interpolation, one needs $n+1$ points and good geometry means they do not lie on or close to a linear surface. Similarly, if one wants to consider interpolation by a polynomial of degree d, where $d = 2$, that is quadratic interpolation, one needs $(n+1)(n+2)/2$ points and good geometry means they do not lie on or close to a quadratic or linear surface. The extension to higher degree is clear. One can also see why the problem goes away if one works with a suitable pattern, as in a pattern search method.

Now, all three methods mentioned above are trust-region based. For an introduction to trust-region techniques the readers are referred to [7], or [9] for a monographic volume. In the case with derivatives the essential ingredients are the following. Starting at a given point $\mathbf{x}_0$ one has a region about that point, coined the trust region and denoted by Δ_0. The trust region is typically a sphere in the Euclidean or in the infinity norm. One then requires a model $m(\mathbf{x})$ for the true objective function that is relatively easy to minimize within the trust region (e.g., a truncated first-order Taylor series or an approximation to a truncated second-order Taylor series, about the current point). A search direction from the current point is determined based upon the model and one (approximately) minimizes the model within the trust region.

The trust region can be updated in the following manner. Suppose $\mathbf{y}_1$ is the approximate minimizer of the model within the trust region Δ_0. We then compare the predicted reduction to truth in the sense that we consider

$$\rho = \frac{f(\mathbf{x}_0) - f(\mathbf{y}_1)}{m(\mathbf{x}_0) - m(\mathbf{y}_1)}.$$

Then typically one assigns some updating strategy to the trust-region radius Δ_0 like

$$\Delta_1 = \begin{cases} 2 \cdot \Delta_0, & \text{if } \rho > 0.9, \\ \Delta_0, & \text{if } 0.1 \leq \rho \leq 0.9, \\ 0.5 \cdot \Delta_0 & \text{if } \rho < 0.1, \end{cases}$$

where Δ_1 denotes the updated radius. In the first two cases $\mathbf{x}_1 = \mathbf{y}_1$ and in the third case $\mathbf{x}_1 = \mathbf{x}_0$.

Thus, although oversimplified, if we are using Taylor series approximations for our models, within the trust management scheme one can ensure convergence to a solution satisfying first-order optimality conditions [9]. Perhaps the most important difference once derivatives are not available is that we cannot take Taylor series models and so, in general, optimality can no longer be guaranteed. In fact, we have to be sure that when we reduce the trust-region radius it is because of the problem and not just a consequence of having a bad model as a result of poor geometry of the sampling points. So it is here that one has to consider the geometry. Fortunately, it can be shown that one can constructively ensure good geometry, and with that, support the whole derivative-free approach with convergence to solutions that satisfy first-order optimality conditions. For details see [3], Chapter 6.

4.4 Global Optimization

In the previous section we have concentrated on local search methods. Unfortunately, most real-world problems are multimodal, and global optima are generally extremely difficult to obtain. Local search methods find local optima that are not guaranteed to be global. Here we will give a short survey of global optimization methods. However, the reader should take note of the following. In practice, often good local optima suffice. If one is considering even a modest number of variables, say fifty, it is generally very difficult, if not impossible, to ensure convergence to a provable global solution, in a reasonable length of time, even if derivatives are available, not to mention in the derivative-free case. Almost all algorithms designed to determine local optima are significantly more efficient than global methods.

Many successful methods in global optimization are based on stochastic components, as they allow to escape from local optima and overcome premature stagnation. Famous classes of families of stochastic global optimization methods are evolutionary algorithms, estimation of distribution algorithms, particle swarm optimization, and differential evolution. Further heuristics known in literature are simulated annealing [28, 29], tabu search [30, 31], ant colony optimization [32, 33], and artificial immune systems [34, 35]. In this section, we concentrate on the first four classes of methods that have been successful in a number of practical applications.

4.4.1 Evolutionary Algorithms

A history of more than forty years of active research on evolutionary computation indicates that stochastic optimization algorithms are an important class of

```
1   Start
2     Initialize solutions x_i of population 𝒫
3     Evaluate objective function for the solutions x_i in 𝒫
4     Repeat
5       For i = 0 To λ
6         Select ρ parents from 𝒫
7         Create new x_i by recombination
8         Mutate x_i
9         Evaluate objective function for x_i
10        Add x_i to 𝒫'
11      Next
12      Select μ parents from 𝒫' and form new 𝒫
13    Until termination condition
14  End
```

Fig. 4.2 Pseudocode of a generic evolutionary algorithm.

derivative-free search methodologies. The separate development of evolutionary algorithms (EAs) in the United States and Europe led to different kinds of algorithmic variants. Genetic algorithms were developed by John Holland in the United States at the beginning of the seventies. Holland's intention was to exploit adaptive behavior. In his book *Adaptation in Natural and Artificial Systems* [36] he describes the development of genetic algorithms (GAs). His original algorithm is today known as simple GA. Evolutionary programming by Fogel, Owens and Walsh [37] was originally designed for optimization of the evolvement of deterministic finite automata, but has today been extended to numerical optimization. Evolution strategies (ES) were developed by Rechenberg and Schwefel in the middle of the sixties in Germany [38, 39, 40]. In the following, we introduce the idea of evolutionary optimization, that is closely related to evolution strategies.

4.4.1.1 Algorithmic Framework

The basis of evolutionary search is a population $\mathscr{P} := \{\mathbf{x}_1, \ldots, \mathbf{x}_\lambda\}$ of candidate solutions, also called individuals. Figure 4.2 shows the pseudocode of a general evolutionary algorithm. The optimization process takes three steps. In the first step the recombination operator (see Section 4.4.1.2) selects ρ parents and combines them to obtain new solutions. In the second step the mutation operator (see Section 4.4.1.3) adds random noise to the preliminary candidate solution. The objective function $f(\mathbf{x})$ is interpreted in terms of the quality of the individuals, and in EA lexicon is called fitness. The fitness of the new offspring solution is evaluated. All individuals of a generation form the new population $\mathscr{P}'$. In the third step, when λ solutions have been produced, μ individuals, with $\mu < \lambda$, are selected (see Section 4.4.1.4), and form the new parental population of the following generation. The process starts again until a termination condition is reached. Typical termination conditions are the accomplishment of a certain solution quality, or an upper bound on the number of generations. We now concentrate on the stochastic operators that are often used in evolutionary computation.

4.4.1.2 Recombination

In biological systems recombination, also known as crossover, mixes the genetic material of two parents. Most EAs also make use of a recombination operator and combine the information of two or more individuals into a new offspring solution. Hence, the offspring carries parts of the genetic material of its parents. The use of recombination is discussed controversially within the building block hypothesis by Goldberg [41, 42], and the genetic repair effect by Beyer [43].

Typical recombination operators for continuous representations are dominant and intermediary recombination. Dominant recombination randomly combines the genes of all parents. If we consider parents of the form $\mathbf{x} = (x_1, \ldots x_n)$, dominant recombination with ρ parents $\mathbf{x}^1, \ldots, \mathbf{x}^\rho$ creates the offspring vector $\mathbf{x}' = (x'_1, \ldots, x'_n)$ by random choice of the i-th component x'_i:

$$x'_i := x^k_i, \quad k \in \text{random}\,\{1, \ldots, \rho\}. \tag{4.2}$$

Intermediate recombination is appropriate for integer and real-valued solution spaces. Given ρ parents $\mathbf{x}^1, \ldots, \mathbf{x}^\rho$ each component of the offspring vector $\mathbf{x}'$ is the arithmetic mean of the components of all ρ parents. Thus, the characteristics of descendant solutions lie between their parents:

$$x'_i := \frac{1}{\rho} \sum_{k=1}^{\rho} x^k_i. \tag{4.3}$$

Integer representations may require rounding procedures to produce intermediate integer solutions.

4.4.1.3 Mutation

Mutation is the second main source for evolutionary changes. According to Beyer and Schwefel [38], a mutation operator is supposed to fulfill three conditions. First, from each point in the solution space each other point must be reachable. Second, in unconstrained solution spaces a bias is disadvantageous, because the direction to the optimum is not known. And third, the mutation strength should be adjustable in order to adapt to solution space conditions. In the following, we concentrate on the well-known Gaussian mutation operator. We assume that solutions are vectors of real values. Random numbers based on the Gaussian distribution $\mathcal{N}(0, 1)$ satisfy these conditions in continuous domains. The Gaussian distribution can be used to describe many natural and artificial processes. By isotropic Gaussian mutation each component of $\mathbf{x}$ is perturbed independently with a random number from a Gaussian distribution with zero mean and standard deviation σ.

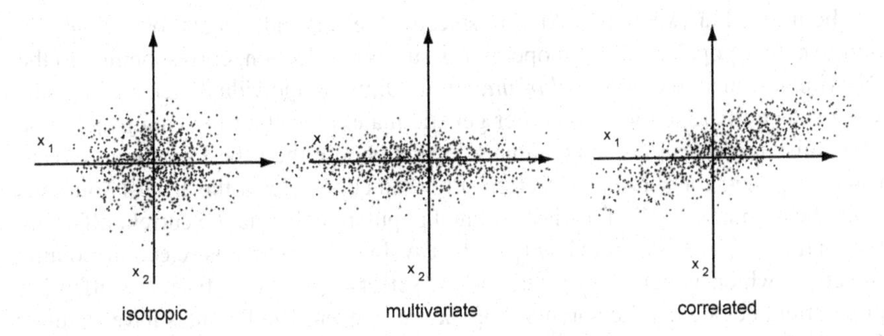

Fig. 4.3 Gaussian mutation: isotropic Gaussian mutation (left) uses one step size σ for each dimension, multivariate Gaussian mutation (middle) allows independent step sizes for each dimension, and correlated mutation (right) introduces an additional rotation of the coordinate system

The standard deviation σ plays the role of the mutation strength, and is also known as step size. The step size σ can be kept constant, but convergence can be improved by adapting σ according to the local solution space characteristics. In case of high success rates, i.e., a high number of offspring solutions being better than their parents, large step sizes are advantageous in order to promote the exploration of the search space. This is often the case at the beginning of the search. Small step sizes are appropriate for low success rates. This is frequently adequate in later phases of the search, when the optimization history can be exploited while the optimum is approximated. An example for an adaptive control of step sizes is the $1/5$-th success rule by Rechenberg [39] that increases the step size if the success rate is over $1/5$-th, and decreases it, if the success rate is lower.

The isotropic Gaussian mutation can be extended to the multivariate Gaussian mutation by introducing a step size vector σ with independent step sizes σ_i. Figure 4.3 illustrates the differences between isotropic Gaussian mutation (left) and the multivariate Gaussian mutation (middle). The multivariate variant considers a mutation ellipsoid that adapts flexibly to local solution space characteristics.

Even more flexibility can be obtained through the correlated mutation proposed by Schwefel [44] that aligns the coordinate system to the solution space characteristics. The mutation ellipsoid is rotated by means of an orthogonal matrix, and this rotation can be modified along iterations. The rotated ellipsoid is also shown in Figure 4.3 (right). The covariance matrix adaptation evolution strategies (CMA-ES) and derivates [45, 46] are self-adapting control strategies based on an automatic alignment of the coordinate system.

4.4.1.4 Selection

The counterpart of the variation operators mutation and recombination is selection. Selection gives the evolutionary search a direction. Based on the fitness, a subset of the population is selected, while the rest is rejected. In EAs the selection operator can be utilized at two points. Mating selection selects individuals for recombination. Another popular selection operator is survivor selection, corresponding to the Darwinian principle of *survival of the fittest*. Only the individuals selected by survivor selection are allowed to confer genetic material to the following generation. The elitist strategies plus and comma selection choose the μ best solutions and are usually applied for survivor selection. Plus selection selects the μ best solutions from the union $\mathscr{P} \cup \mathscr{P}'$ of the last parental population $\mathscr{P}$ and the current offspring population $\mathscr{P}'$, and is denoted by $(\mu + \lambda)$-EA. In contrast to plus selection, comma selection, which is denoted by (μ, λ)-EA, selects exclusively from the offspring population, neglecting the parental population – even if individuals have superior fitness. Though disregarding these apparently promising solutions may seem to be disadvantageous, this strategy that prefers the new population to the old population can be useful to avoid being trapped in unfavorable local optima.

The deterministic selection scheme described in the previous paragraph is a characteristic feature of ES. Most evolutionary algorithms use selection schemes containing random components. An example is fitness proportionate selection (also called roulette-wheel selection) popular in the early days of genetic algorithms [41]. Another example is tournament selection, a widely used selection scheme for EAs. Here, the candidate with the highest fitness out of a randomly chosen subset of the population is selected to the new population. The stochastic-based selection schemes permit survival of not-so-fit individuals and thus helps with preventing premature convergence and preserving the genetic material that may come in handy at later stages of the optimization process.

4.4.2 Estimation of Distribution Algorithms

Related to evolutionary algorithms are estimation of distribution algorithms (EDAs). They also operate with a set of candidate solutions. Similar to ES, a random set of points is initially generated, and the objective function is computed for all these points. The core of EDAs are successive steps where distributions of the best solutions within a population are estimated, and a new population is sampled according to the previous distribution estimation.

The principle has been extended in a number of different manners. Most EDAs make use of parametric distributions, i.e., the parameters of distribution functions are determined in the estimation step. The assumption of a Gaussian distribution is frequent in EDAs. EDAs may suffer from premature convergence. The weighted variance estimator introduced in [47] has been observed to alleviate that convergence issue. Adaptive variance scaling [48], i.e., the variance can be increased if good solutions are found, otherwise it is decreased, has also been suggested to avoid early stagnation. The sampling process can be enhanced by anticipated mean shift (AMS; [49]). In this approach, about two thirds of the population are sampled regularly, and the rest is shifted in the direction of a previously estimated gradient. If this estimate is accurate, all the shifted individuals, together with part of the non-shifted individuals, may survive, and the variance estimate in the direction of the gradient could be larger than without AMS.

4.4.3 Particle Swarm Optimization

Similar to evolutionary algorithms, particle swarm optimization (PSO) is a population approach with stochastic components. Introduced by Kennedy and Eberhart [50], it is inspired by the movement of natural swarms and flocks. The algorithm utilizes particles with a position $\mathbf{x}$ that corresponds to the optimization variables, and a speed $\mathbf{v}$ which is similar to the mutation strength in evolutionary computation. The principle of particle swarm optimization is based on the idea that the particles move in the solution space, influencing each other with stochastic changes, while previous successful solutions act as attractors.

In each iteration the position of particle $\mathbf{x}$ is updated by adding the current velocity $\mathbf{v}$

$$\mathbf{x}' := \mathbf{x} + \mathbf{v}. \tag{4.4}$$

The velocity is updated as follows

$$\mathbf{v}' := \mathbf{v} + c_1 r_1 (\mathbf{x}_p^* - \mathbf{x}) + c_2 r_2 (\mathbf{x}_s^* - \mathbf{x}), \tag{4.5}$$

where $\mathbf{x}_p^*$ and $\mathbf{x}_s^*$ denote the best previous positions of the particle and of the swarm, respectively. The weights c_1 and c_2 are acceleration coefficients that determine the bias of the particle towards its own or the swarm history. The recommendation given by Kennedy and Eberhart is to set both parameters to one. The stochastic components r_1 and r_2 are uniformly drawn from the interval $[0,1]$, and can be used to promote the global exploration of the search space.

4.4.4 Differential Evolution

Another population-based optimization approach is differential evolution (DE), originally introduced by Storn and Price [51]. As the algorithms in the previous three subsections, DE exploits a set of candidate solutions (agents in DE lexicon). New agents are allocated in the search space by combining the positions of other existing agents. More specifically, an intermediate agent is generated from two agents randomly chosen from the current population. This temporary agent is then mixed with a predetermined target agent. The new agent is accepted for the next generation if and only if it yields reduction in objective function.

The basic DE algorithm uses a random initialization. A new agent $\mathbf{y} = [y_1, \ldots, y_n]$ is created from the existing one $\mathbf{x} = [x_1, \ldots, x_n]$ as indicated below.

1. Three agents $\mathbf{a} = [a_1, \ldots, a_n]$, $\mathbf{b} = [b_1, \ldots, b_n]$ and $\mathbf{c} = [c_1, \ldots, c_n]$ are randomly extracted from the population (all distinct from each other and from $\mathbf{x}$).
2. A position index $p \in \{1, \ldots, N\}$ is determined randomly.

3. The position of the new agent $\mathbf{y}$ is computed by means of the following iteration over $i \in \{1,\ldots,n\}$:

   i) select a random number $r_i \in (0,1)$ with uniform probability distribution;
   ii) if $i = p$ or $r_i < CR$ let $y_i = a_i + F(b_i - c_i)$, otherwise let $y_i = x_i$; here, $F \in [0,2]$ is the differential weight and $CR \in [0,1]$ is the crossover probability, both defined by the user;
   iii) if $f(\mathbf{y}) < f(\mathbf{x})$ then replace $\mathbf{x}$ by $\mathbf{y}$; otherwise reject $\mathbf{y}$ and keep $\mathbf{x}$.

Although DE resembles some other stochastic optimization techniques, unlike traditional EAs, DE perturbs the solutions in the current generation vectors with scaled differences of two randomly selected agents. As a consequence, no separate probability distribution has to be used, and thus the scheme presents some degree of self-organization. Additionally, DE is simple to implement, uses very few control parameters, and has been observed to perform satisfactorily in a number of multi-modal optimization problems [52].

4.5 Guidelines for Generally Constrained Optimization

We now describe nonlinear constraint handling techniques that can be combined with the optimization methods presented in Sections 4.3 and 4.4.

4.5.1 Penalty Functions

The penalty function method (cf. [7]) for general optimization constraints involves modifying the objective function with a penalty term that depends on the constraint violation $h : \mathbb{R}^n \rightarrow \mathbb{R}$. The original optimization problem in (4.1) is thus modified as follows:

$$\min_{\mathbf{x} \in \Omega \subset \mathbb{R}^n} f(\mathbf{x}) + \rho\, h(\mathbf{x}), \tag{4.6}$$

where $\rho > 0$ is a penalty parameter. The modified optimization problem may still have constraints that are straightforward to handle.

If the penalty parameter is iteratively increased (tending to infinity), the solution of (4.6) converges to that of the original problem in (4.1). However, in certain cases, a finite (and fixed) value of the penalty parameter ρ also yields the correct solution (this is the so-called *exact* penalty; see [7]). For exact penalties, the modified cost function is not smooth around the solution [7], and thus the corresponding optimization problem can be significantly more involved than that in (4.6). However, one can argue that in the derivative-free case exact penalty functions may in some cases be attractive. Common definitions of $h(\mathbf{x})$, where I and J denote the indices that refer to inequality and equality constraints, respectively, are

$$h(\mathbf{x}) = \frac{1}{2}\left(\sum_{i \in I} \max(0, g_i(\mathbf{x}))^2 + \sum_{j \in J} g_i^2(\mathbf{x})\right)$$

the quadratic penalty and

$$h(\mathbf{x}) = \sum_{i \in I} \max(0, g_i(\mathbf{x})) + \sum_{j \in J} |g_i(\mathbf{x})|$$

an exact penalty. It should be noticed that by these penalties, the search considers both feasible and infeasible points. Those optimization methodologies where the optimum can be approached from outside the feasible region are known as exterior methods.

The log-barrier penalty (for inequality constraints)

$$h(\mathbf{x}) = -\sum_{i \in I} \log(-g_i(\mathbf{x}))$$

has to be used with a decreasing penalty parameter (tending to zero). This type of penalty methods (also known as barrier methods) confines the optimization to the feasible region of the search space. Interior methods aim at reaching the optimum from inside the feasible region.

In [53], non-quadratic penalties have been suggested for pattern search techniques. However, the optimizations presented in that work are somewhat simpler than those found in many practical situations, so the recommendations given might not be generally applicable. In future research, it will be useful to explore further the performance of different penalty functions in the context of simulation-based optimization.

4.5.2 Augmented Lagrangian Method

As mentioned above, in exterior penalty function methods, as $\rho \to \infty$ the local minimum is approached from outside the feasible region. Not surprisingly, there is a way to shift the feasible region so one is able to determine the local solution for a finite penalty parameter. See, for example, [54, 55] for original references, and also [7], Chapter 17.

Augmented Lagrangian methods [56, 57] aim at minimizing, in the equality constraint case, the following extended cost function

$$\min_{\mathbf{x} \in \Omega \subset \mathbb{R}^n} \ f(\mathbf{x}) + \tfrac{1}{2}\rho \, \|\mathbf{g}(\mathbf{x})\|_2^2 + \boldsymbol{\lambda}^T \mathbf{g}(\mathbf{x}), \tag{4.7}$$

where $\rho > 0$ is a penalty parameter, and $\boldsymbol{\lambda} \in \mathbb{R}^m$ are Lagrange multipliers. This cost function can indeed be interpreted as a quadratic penalty with the constraints shifted by some constant term [56]. As in penalty methods, the penalty parameter and the Lagrange multipliers are iteratively updated. It turns out that if one is sufficiently stationary for Equation (4.7), which is exactly when we have good approximations for the Lagrange multipliers, then $\boldsymbol{\lambda}$ can be updated via

$$\boldsymbol{\lambda}^+ = \boldsymbol{\lambda} + \rho \, \mathbf{g}(\mathbf{x}), \tag{4.8}$$

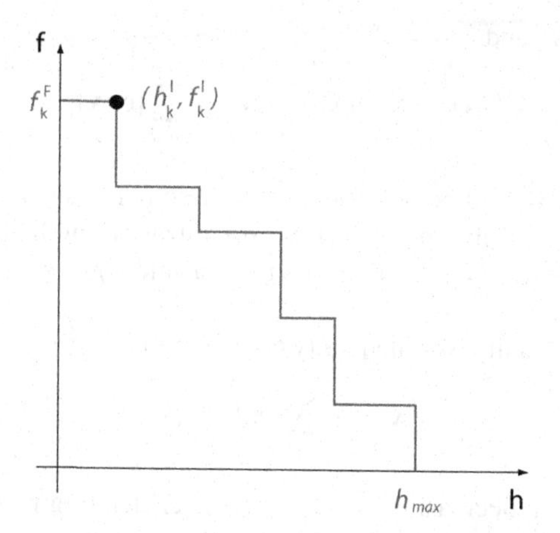

Fig. 4.4 An idealized (pattern search) filter at iteration k (modified from [19])

where $\boldsymbol{\lambda}^+$ denotes the updated Lagrange multipliers. Otherwise one should increase the penalty parameter ρ (say by multiplying it by 10). The Lagrange multipliers are typically initialized to zero. What is significant is that one can prove (see e.g. [56]) that after a finite number of iterations the penalty parameter is never updated, and that the whole scheme eventually converges to a solution of the original optimization problem in (4.1). Inequality constraints can also be incorporated in the augmented Lagrangian framework by introducing slack variables and simple bounds [56]. The augmented Lagrangian approach can be combined with most optimization algorithms. For example, refer to [58] for a nonlinear programming methodology based on generalized pattern search.

4.5.3 Filter Method

A relatively recent approach that avoids using a penalty parameter and has been rather successful is the class of so-called filter methods [59, 7]. Using filters, the original problem (4.1) is typically viewed as a bi-objective optimization problem. Besides minimizing the cost function $f(\mathbf{x})$, one also seeks to reduce the constraint violation $h(\mathbf{x})$. The concept of dominance, crucial in multi-objective optimization, is defined as follows: the point $\mathbf{x}_1 \in \mathbb{R}^n$ dominates $\mathbf{x}_2 \in \mathbb{R}^n$ if and only if either $f(\mathbf{x}_1) \leq f(\mathbf{x}_2)$ and $h(\mathbf{x}_1) < h(\mathbf{x}_2)$, or $f(\mathbf{x}_1) < f(\mathbf{x}_2)$ and $h(\mathbf{x}_1) \leq h(\mathbf{x}_2)$. A filter is a set of pairs $(h(\mathbf{x}), f(\mathbf{x}))$, such that no pair dominates another pair. In practice, a maximum allowable constraint violation $h_{\max}$ is specified. This is accomplished by introducing the pair $(h_{\max}, -\infty)$ in the filter. An idealized filter (at iteration k) is shown in Figure 4.4.

A filter can be understood as essentially an add-on for an optimization procedure. The intermediate solutions proposed by the optimization algorithm at a given iteration are accepted if they are not dominated by any point in the filter. The filter is updated at each iteration based on all the points evaluated by the optimizer. We reiterate that, as for exterior methods, the optimization search is enriched by considering infeasible points, although the ultimate solution is intended to be feasible (or very nearly so). Filters are often observed to lead to faster convergence than methods that rely only on feasible iterates.

Pattern search optimization techniques have been previously combined with filters [60]. In Hooke-Jeeves direct search, the filter establishes the acceptance criterion for each (unique) new solution. For schemes where, in each iteration, multiple solutions can be accepted by the filter (such as in GPS), the new polling center must be selected from the set of validated points. When the filter is not updated in a particular iteration (and thus the best feasible point is not improved), the pattern size is decreased. As in [60], when we combine GPS with a filter, the polling center at a given iteration will be the feasible point with lowest cost function or, if no feasible points remain, it will be the infeasible point with lowest constraint violation. These two points, $(0, f_k^{\mathrm{F}})$ and $(h_k^{\mathrm{I}}, f_k^{\mathrm{I}})$, respectively, are shown in Figure 4.4 (it is assumed that both points have just been accepted by the filter, and thus it makes sense to use one of them as the new polling center). Refer to [60] and [61] for more details on pattern search filter methods.

4.5.4 Other Approaches

We will now briefly overview a number of constraint handling methodologies that have been proposed for evolutionary algorithms. Repair algorithms [62, 63] project infeasible solutions back to the feasible space. This projection is in most cases accomplished in an approximate manner, and can be as complex as solving the optimization problem itself. Repair algorithms can be seen as local procedures that aim at reducing constraint violation. In the so-called Baldwinian case, the fitness of the repaired solution replaces the fitness of the original (infeasible) solution. In the Lamarckian case, feasible solutions prevail over infeasible solutions.

Constraint-handling techniques borrowed from multi-objective optimization are based on the idea of dealing with each constraint as an additional objective [64, 65, 66, 67, 68, 69]. Under this assumption, multi-objective optimization methods such as NSGA-II [70] or SPEA [71] can be applied. The output of a multi-objective approach for constrained optimization is an approximation of a Pareto set that involves the objective function and the constraints. The user may then select one or more solutions from the Pareto set. A simpler but related and computationally less expensive procedure is the behavioral memory method presented in [72]. This evolutionary method concentrates on minimizing the constraint violation of each constraint sequentially, and the objective function is addressed separately afterwards. However, treating objective function and constraints independently may yield in many cases infeasible solutions.

Further constraint handling methods have been proposed in EA literature that do not rely either on repair algorithms or multi-objective approaches. In [73] a technique based on a multi-membered evolution strategy with a feasibility comparison mechanism is introduced. The dynamic multi-swarm particle optimizer studied in [74] makes use of a set of sub-swarms that focus on different constraints, and is coupled with a local search algorithm (sequential quadratic programming).

4.6 Concluding Remarks

In this chapter, we have concentrated on methods for solving optimization problems without derivates. The existence of local optima makes a hard optimization problem even harder. Many methods have been proposed to solve non-convex optimization problems. The approaches range from pattern search for local optimization problems to stochastic bio-inspired search heuristics for multi-modal problems. Deterministic local methods are guaranteed to find local optima, and restart variants can be applied to avoid unsatisfactory solutions. Stochastic methods are not guaranteed to find the global optimum, but in some practical cases they can be beneficial.

The hybridization between local and global optimizers has led to a paradigm sometimes called memetic algorithms or hybrid metaheuristics [75, 76]. A number of hybridizations have been proposed, but they are often tailored to specific problem types and search domains due to their specific operators and methods. In the memetic method introduced in [77] for continuous search spaces, a gradient-based scheme is combined with a deterministic perturbation component. The local optimization procedure for real-valued variables described in [78] is based on variable neighborhood search. It would be very useful if in future research some effort is dedicated to better understand from a theoretical point of view the hybridization of local and global optimization algorithms.

Most problems that can be found in practice present constraints. We have outlined a number of constraint handling techniques that can be incorporated in a derivative-free optimization framework. Though penalty functions are appealing due to their simplicity, some of the other approaches mentioned here may be more efficient and still of a relatively easy implementation.

Multi-objective optimization is an important challenge for derivative-free methodologies. Some of the evolutionary techniques mentioned above have performed successfully in some not especially involved multi-objective test cases. Other areas where derivative-free optimization could potentially be very helpful include dynamic optimization, mixed-integer nonlinear programming, and optimization under uncertainty (stochastic programming).

Acknowledgements. We are grateful to the industry sponsors of the Stanford Smart Fields Consortium for partial funding of this work, and also to J. Smith for his valuable suggestions.

References

[1] Pironneau, O.: On optimum design in fluid mechanics. Journal of Fluid Mechanics 64, 97–110 (1974)

[2] Kolda, T.G., Lewis, R.M., Torczon, V.: Optimization by direct search: new perspectives on some classical and modern methods. SIAM Review 45(3), 385–482 (2003)

[3] Conn, A.R., Scheinberg, K., Vicente, L.N.: Introduction to Derivative-Free Optimization. MPS-SIAM Series on Optimization. MPS-SIAM (2009)

[4] Gilmore, P., Kelley, C.T.: An implicit filtering algorithm for optimization of functions with many local minima. SIAM Journal on Optimization 5, 269–285 (1995)

[5] Kelley, C.T.: Iterative Methods for Optimization. In: Frontiers in Applied Mathematics, SIAM, Philadelphia (1999)

[6] Dennis Jr., J.E., Schnabel, R.B.: Numerical Methods for Unconstrained Optimization and Nonlinear Equations. SIAM's Classics in Applied Mathematics Series. SIAM, Philadelphia (1996)

[7] Nocedal, J., Wright, S.J.: Numerical Optimization, 2nd edn. Springer, Heidelberg (2006)

[8] Schilders, W.H.A., van der Vorst, H.A., Rommes, J.: Model Order Reduction: Theory, Research Aspects and Applications. Mathematics in Industry Series. Springer, Heidelberg (2008)

[9] Conn, A.R., Gould, N.I.M.: Toint, Ph.L.: Trust-Region Methods. MPS-SIAM Series on Optimization. MPS-SIAM (2000)

[10] Meza, J.C., Martinez, M.L.: On the use of direct search methods for the molecular conformation problem. Journal of Computational Chemistry 15, 627–632 (1994)

[11] Booker, A.J., Dennis Jr., J.E., Frank, P.D., Moore, D.W., Serafini, D.B.: Optimization using surrogate objectives on a helicopter test example. In: Borggaard, J.T., Burns, J., Cliff, E., Schreck, S. (eds.) Computational Methods for Optimal Design and Control, pp. 49–58. Birkháuser, Basel (1998)

[12] Marsden, A.L., Wang, M., Dennis Jr., J.E., Moin, P.: Trailing-edge noise reduction using derivative-free optimization and large-eddy simulation. Journal of Fluid Mechanics 572, 13–36 (2003)

[13] Duvigneau, R., Visonneau, M.: Hydrodynamic design using a derivative-free method. Structural and Multidisciplinary Optimization 28, 195–205 (2004)

[14] Fowler, K.R., Reese, J.P., Kees, C.E., Dennis Jr., J.E., Kelley, C.T., Miller, C.T., Audet, C., Booker, A.J., Couture, G., Darwin, R.W., Farthing, M.W., Finkel, D.E., Gablonsky, J.M., Gray, G., Kolda, T.G.: Comparison of derivative-free optimization methods for groundwater supply and hydraulic capture community problems. Advances in Water Resources 31(5), 743–757 (2008)

[15] Oeuvray, R., Bierlaire, M.: A new derivative-free algorithm for the medical image registration problem. International Journal of Modelling and Simulation 27, 115–124 (2007)

[16] Marsden, A.L., Feinstein, J.A., Taylor, C.A.: A computational framework for derivative-free optimization of cardiovascular geometries. Computational Methods in Applied Mechanics and Engineering 197, 1890–1905 (2008)

[17] Artus, V., Durlofsky, L.J., Onwunalu, J.E., Aziz, K.: Optimization of nonconventional wells under uncertainty using statistical proxies. Computational Geosciences 10, 389–404 (2006)

[18] Dadashpour, M., Echeverría Ciaurri, D., Mukerji, T., Kleppe, J., Landrø, M.: A derivative-free approach for the estimation of porosity and permeability using time-lapse seismic and production data. Journal of Geophysics and Engineering 7, 351–368 (2010)

[19] Echeverría Ciaurri, D., Isebor, O.J., Durlofsky, L.J.: Application of derivativefree methodologies for generally constrained oil production optimization problems. International Journal of Mathematical Modelling and Numerical Optimisation 2(2), 134–161 (2011)

[20] Onwunalu, J.E., Durlofsky, L.J.: Application of a particle swarm optimization algorithm for determining optimum well location and type. Computational Geosciences 14, 183–198 (2010)

[21] Zhang, H., Conn, A.R., Scheinberg, K.: A derivative-free algorithm for leastsquares minimization. SIAM Journal on Optimization 20(6), 3555–3576 (2010)

[22] Torczon, V.: On the convergence of pattern search algorithms. SIAM Journal on Optimization 7(1), 1–25 (1997)

[23] Audet, C., Dennis Jr., J.E.: Analysis of generalized pattern searches. SIAM Journal on Optimization 13(3), 889–903 (2002)

[24] Audet, C., Dennis Jr., J.E.: Mesh adaptive direct search algorithms for constrained optimization. SIAM Journal on Optimization 17(1), 188–217 (2006)

[25] Hooke, R., Jeeves, T.A.: Direct search solution of numerical and statistical problems. Journal of the ACM 8(2), 212–229 (1961)

[26] Powell, M.J.D.: The NEWUOA software for unconstrained optimization without derivatives. Technical report DAMTP 2004/NA5, Dept. of Applied Mathematics and Theoretical Physics, University of Cambridge (2004)

[27] Oeuvray, R., Bierlaire, M.: BOOSTERS: a derivative-free algorithm based on radial basis functions. International Journal of Modelling and Simulatio 29(1), 26–36 (2009)

[28] Metropolis, N., Rosenbluth, A., Teller, A., Teller, E.: Equation of state calculations by fast computing machines. Chemical Physics 21(6), 1087–1092 (1953)

[29] Kirkpatrick, S., Gelatt Jr., C.D., Vecchi, M.: Optimization by simulated annealing. Science 220(4498), 671–680 (1983)

[30] Glover, F.: Tabu search – part I. ORSA Journal on Computing 1(3), 190–206 (1990)

[31] Glover, F.: Tabu search – part II. ORSA Journal on Computing 2(1), 4–32 (1990)

[32] Dorigo, M.: Optimization, Learning and Natural Algorithms. PhD thesis, Dept. of Electronics, Politecnico di Milano (1992)

[33] Dorigo, M., Stützle, T.: Ant Colony Optimization. Prentice-Hall, Englewood Cliffs (2004)

[34] Farmer, J., Packard, N., Perelson, A.: The immune system, adaptation and machine learning. Physica 2, 187–204 (1986)

[35] Castro, L.N.D., Timmis, J.: Artificial Immune Systems: A New Computational Intelligence. Springer, Heidelberg (2002)

[36] Holland, J.H.: Adaptation in Natural and Artificial Systems. University of Michigan Press (1975)

[37] Fogel, D.B.: Artificial Intelligence through Simulated Evolution. Wiley, Chichester (1966)

[38] Beyer, H.-G., Schwefel, H.-P.: Evolution strategies - a comprehensive introduction. Natural Computing 1, 3–52 (2002)

[39] Rechenberg, I.: Evolutionsstrategie: Optimierung Technischer Systeme nach Prinzipien der Biologischen Evolution. Frommann-Holzboog (1973)

[40] Schwefel, H.-P.: Numerische Optimierung von Computer-Modellen mittel der Evolutionsstrategie. Birkhäuser, Basel (1977)
[41] Goldberg, D.E.: Genetic Algorithms in Search, Optimization and Machine Learning. Addison-Wesley, Reading (1989)
[42] Holland, J.H.: Hidden Order: How Adaptation Builds Complexity. Addison- Wesley, London (1995)
[43] Beyer, H.-G.: An alternative explanation for the manner in which genetic algorithms operate. BioSystems 41(1), 1–15 (1997)
[44] Schwefel, H.-P.: Adaptive mechanismen in der biologischen evolution und ihr einfluss auf die evolutionsgeschwindigkeit. In: Interner Bericht der Arbeitsgruppe Bionik und Evolutionstechnik am Institut für Mess- und Regelungstechnik, TU Berlin (1974)
[45] Beyer, H.-G., Sendhoff, B.: Covariance matrix adaptation revisited – the CMSA evolution strategy –. In: Rudolph, G., Jansen, T., Lucas, S., Poloni, C., Beume, N. (eds.) PPSN 2008. LNCS, vol. 5199, pp. 123–132. Springer, Heidelberg (2008)
[46] Ostermeier, A., Gawelczyk, A., Hansen, N.: A derandomized approach to selfadaptation of evolution strategies. Evolutionary Computation 2(4), 369–380 (1994)
[47] Teytaud, F., Teytaud, O.: Why one must use reweighting in estimation of distributionalgorithms. In: Proceedings of the 11th Annual conference on Genetic and Evolutionary Computation (GECCO 2009), pp. 453–460 (2009)
[48] Grahl, J., Bosman, P.A.N., Rothlauf, F.: The correlation-triggered adaptive variance scaling idea. In: Proceedings of the 8th Annual conference on Genetic and Evolutionary Computation (GECCO 2006), pp. 397–404 (2006)
[49] Bosman, P.A.N., Grahl, J., Thierens, D.: Enhancing the performance of maximum–likelihood gaussian eDAs using anticipated mean shift. In: Rudolph, G., Jansen, T., Lucas, S., Poloni, C., Beume, N. (eds.) PPSN 2008. LNCS, vol. 5199, pp. 133–143. Springer, Heidelberg (2008)
[50] Kennedy, J., Eberhart, R.: Particle swarm optimization. In: Proceedings of the IEEE International Conference on Neural Networks, pp. 1942–1948 (1995)
[51] Storn, R., Price, K.: Differential evolution - a simple and efficient heuristic for global optimization over continuous spaces. Journal of Global Optimization 11, 341–359 (1997)
[52] Chakraborty, U.: Advances in Differential Evolution. SCI. Springer, Heidelberg (2008)
[53] Griffin, J.D., Kolda, T.G.: Nonlinearly-constrained optimization using asynchronous parallel generating set search. Technical report SAND2007-3257, Sandia National Laboratories (2007)
[54] Hestenes, M.R.: Multiplier and gradients methods. Journal of Optimization Theory and Applications 4(5), 303–320 (1969)
[55] Powell, M.J.D.: A method for nonlinear constraints in minimization problems. In: Fletcher, R. (ed.) Optimization, pp. 283–298. Academic Press, London (1969)
[56] Conn, A.R., Gould, N.I.M., Toint, P.L.: A globally convergent augmented Lagrangian algorithm for optimization with general constraints and simple bounds. SIAM Journal on Numerical Analysis 28(2), 545–572 (1991)
[57] Conn, A.R., Gould, N.I.M., Toint, P.L.: LANCELOT: A Fortran Package for Large-Scale Nonlinear Optimization (Release A). Computational Mathematics. Springer, Heidelberg (1992)
[58] Lewis, R.M., Torczon, V.: A direct search approach to nonlinear programming problems using an augmented Lagrangian method with explicit treatment of the linear constraints. Technical report WM-CS-2010-01, Dept. of Computer Science, College of William & Mary (2010)

[59] Fletcher, R., Leyffer, S., Toint, P.L.: A brief history of filter methods. Technical report ANL/MCS/JA-58300, Argonne National Laboratory (2006)
[60] Audet, C., Dennis Jr., J.E.: A pattern search filter method for nonlinear programming without derivatives. SIAM Journal on Optimization 14(4), 980–1010 (2004)
[61] Abramson, M.A.: NOMADm version 4.6 User's Guide. Dept. of Mathematics and Statistics, Air Force Institute of Technology (2007)
[62] Belur, S.V.: CORE: constrained optimization by random evolution. In: Koza, J.R. (ed.) Late Breaking Papers at the Genetic Programming 1997 Conference, pp. 280–286 (1997)
[63] Coello Coello, C.A.: Theoretical and numerical constraint handling techniques used with evolutionary algorithms: a survey of the state of the art. Computer Methods in Applied Mechanics and Engineering 191(11-12), 1245–1287 (2002)
[64] Parmee, I.C., Purchase, G.: The development of a directed genetic search technique for heavily constrained design spaces. In: Parmee, I.C. (ed.) Proceedings of the Conference on Adaptive Computing in Engineering Design and Control, pp. 97–102. University of Plymouth (1994)
[65] Surry, P.D., Radcliffe, N.J., Boyd, I.D.: A multi-objective approach to constrained optimisation of gas supply networks: the COMOGA method. In: Fogarty, T.C. (ed.) AISB-WS 1995. LNCS, vol. 993, pp. 166–180. Springer, Heidelberg (1995)
[66] Coello Coello, C.A.: Treating constraints as objectives for single-objective evolutionary optimization. Engineering Optimization 32(3), 275–308 (2000)
[67] Coello Coello, C.A.: Constraint handling through a multiobjective optimization technique. In: Proceedings of the 8th Annual conference on Genetic and Evolutionary Computation (GECCO 1999), pp. 117–118 (1999)
[68] Jimenez, F., Verdegay, J.L.: Evolutionary techniques for constrained optimization problems. In: Zimmermann, H.J. (ed.) 7th European Congress on Intelligent Techniques and Soft Computing (EUFIT 1999). Springer, Heidelberg (1999)
[69] Mezura-Montes, E., Coello Coello, C.A.: Constrained optimization via multiobjective evolutionary algorithms. In: Knowles, J., Corne, D., Deb, K., Deva, R. (eds.) Multiobjective Problem Solving from Nature. Natural Computing Series, pp. 53–75. Springer, Heidelberg (2008)
[70] Deb, K., Agrawal, S., Pratap, A., Meyarivan, T.: A fast and elitist multiobjective genetic algorithm: NSGA-II. IEEE Transactions on Evolutionary Computation 6(2), 182–197 (2002)
[71] Zitzler, E., Laumanns, M., Thiele, L.: SPEA2: improving the strength Pareto evolutionary algorithm for multiobjective optimization. In: Evolutionary Methods for Design, Optimisation and Control with Application to Industrial Problems (EUROGEN 2001), pp. 95–100 (2002)
[72] Schoenauer, M., Xanthakis, S.: Constrained GA optimization. In: Forrest, S. (ed.) Proceedings of the 5th International Conference on Genetic Algorithms (ICGA 1993), pp. 573–580. Morgan Kaufmann, San Francisco (1993)
[73] Montes, E.M., Coello Coello, C.A.: A simple multi-membered evolution strategy to solve constrained optimization problems. IEEE Transactions on Evolutionary Computation 9(1), 1–17 (2005)
[74] Liang, J., Suganthan, P.: Dynamic multi-swarm particle swarm optimizer with a novel constraint-handling mechanism. In: Yen, G.G., Lucas, S.M., Fogel, G., Kendall, G., Salomon, R., Zhang, B.-T., Coello Coello, C.A., Runarsson, T.P. (eds.) Proceedings of the 2006 IEEE Congress on Evolutionary Computation (CEC 2006), pp. 9–16. IEEE Press, Los Alamitos (2006)

[75] Raidl, G.R.: A unified view on hybrid metaheuristics. In: Almeida, F., Blesa Aguilera, M.J., Blum, C., Moreno Vega, J.M., Pérez Pérez, M., Roli, A., Sampels, M. (eds.) HM 2006. LNCS, vol. 4030, pp. 1–12. Springer, Heidelberg (2006)
[76] Talbi, E.G.: A taxonomy of hybrid metaheuristics. Journal of Heuristics 8(5), 541–564 (2002)
[77] Griewank, A.: Generalized descent for global optimization. Journal of Optimization Theory and Applications 34, 11–39 (1981)
[78] Duran Toksari, M., Güner, E.: Solving the unconstrained optimization problem by a variable neighborhood search. Journal of Mathematical Analysis and Applications 328(2), 1178–1187 (2007)

Chapter 5
Maximum Simulated Likelihood Estimation: Techniques and Applications in Economics

Ivan Jeliazkov and Alicia Lloro

Abstract. This chapter discusses maximum simulated likelihood estimation when construction of the likelihood function is carried out by recently proposed Markov chain Monte Carlo (MCMC) methods. The techniques are applicable to parameter estimation and Bayesian and frequentist model choice in a large class of multivariate econometric models for binary, ordinal, count, and censored data. We implement the methodology in a study of the joint behavior of four categories of U.S. technology patents using a copula model for multivariate count data. The results reveal interesting complementarities among several patent categories and support the case for joint modeling and estimation. Additionally, we find that the simulated likelihood algorithm performs well. Even with few MCMC draws, the precision of the likelihood estimate is sufficient for producing reliable parameter estimates and carrying out hypothesis tests.

5.1 Introduction

The econometric analysis of models for multivariate discrete data is often complicated by intractability of the likelihood function, which can rarely be evaluated directly and typically has to be estimated by simulation. In such settings, the efficiency of likelihood estimation plays a key role in determining the theoretical properties and practical appeal of standard optimization algorithms that rely on those estimates. For this reason, the development of fast and statistically efficient techniques for estimating the value of the likelihood function has been at the forefront of much of the research on maximum simulated likelihood estimation in econometrics.

In this paper we examine the performance of a method for estimating the ordinate of the likelihood function which was recently proposed in [8]. The method is rooted in Markov chain Monte Carlo (MCMC) theory and simulation [3, 4, 15, 18],

Ivan Jeliazkov · Alicia Lloro
Department of Economics, University of California,
Irvine, 3151 Social Science Plaza, Irvine, CA 92697
e-mail: ivan@uci.edu, alloro@uci.edu

S. Koziel & X.-S. Yang (Eds.): Comput. Optimization, Methods and Algorithms, SCI 356, pp. 85–100.
springerlink.com

and its ingredients have played a central role in Bayesian inference in econometrics and statistics. The current implementation of those methods, however, is intended to examine their applicability to purely frequentist problems such as maximum likelihood estimation and hypothesis testing.

We implement the methodology to study firm-level patent registration data in four patent categories in the "computers & instruments" industry during the 1980s. One goal of this application is to examine how patent counts in each category are affected by firm characteristics such as sales, workforce size, and research & development (R&D) capital. A second goal is to study the degree of complementarity or substitutability that emerges among patent categories due to a variety of unobserved factors, such as firms' internal R&D decisions, resource concentration, managerial dynamics, technological spillovers, and the relevance of innovations across category boundaries. These factors can affect multiple patent categories simultaneously and necessitate the specification of a joint empirical structure that can flexibly capture interdependence patterns.

We approach these tasks by considering a copula model for multivariate count data which enables us to pursue joint modeling and estimation. Because the outcome probabilities in the copula model are difficult to evaluate, we rely on MCMC simulation to evaluate the likelihood function. Moreover, to improve the performance of the optimization algorithm, we implement a quasi-Newton optimization method due to [1] that exploits a fundamental statistical relation to avoid direct computation of the Hessian matrix of the log-likelihood function. The application demonstrates that the simulated likelihood algorithm performs very well – even with few MCMC draws, the precision of the likelihood estimate is sufficient for producing reliable parameter estimates and hypothesis tests. The results support the case for joint modeling and estimation in our application and reveal interesting complementarities among several patent categories.

The remainder of this chapter is organized as follows. In Section 5.2, we present the copula model that we use in our application and the likelihood function that we use in estimation. The likelihood function is difficult to evaluate because it is given by a set of integrals with no closed-form solution. For this reason, in Section 5.3, we present the MCMC-based simulation algorithm for evaluating this function and discuss how it can be embedded in a standard optimization algorithm to maximize the log-likelihood function and yield parameter estimates and standard errors. Section 5.4 presents the results from our patent application and demonstrates the performance of the estimation algorithm. Section 5.5 offers concluding remarks.

5.2 Copula Model

In analyzing multiple data series, it is typically desirable to pursue joint modeling and estimation. Doing so allows researchers to investigate dependence structures among the individual variables of interest, leads to gains in estimation efficiency, and is also important for mitigating misspecification problems in nonlinear models.

In many applications, however, a suitable joint distribution may be unavailable or difficult to specify. This problem is particularly prevalent in multivariate discrete data settings, and in cases where the variables are of different types (e.g. some continuous, some discrete or censored). One area of research where incorporating a flexible and interpretable correlation structure has been difficult is the empirical analysis of multivariate count data [2, 21]. As a consequence, models for multivariate counts have typically sacrificed generality for the sake of retaining computational tractability. To deal with the aforementioned difficulties, we resort to a copula modeling approach whose origins can be traced back to [17].

Formally, a copula maps the unit hypercube $[0,1]^q$ to the unit interval $[0,1]$ and satisfies the following conditions:

1. $C(1,\ldots,1,a_p,1,\ldots,1) = a_p$ for every $p \in \{1,\ldots,q\}$ and all $a_p \in [0,1]$;
2. $C(a_1,\ldots,a_q) = 0$ if $a_p = 0$ for any $p \in \{1,\ldots,q\}$;
3. C is q-increasing, i.e. any hyperrectangle in $[0,1]^q$ has non-negative C-volume.

The generality of the approach rests on the recognition that a copula can be viewed as a q-dimensional distribution function with uniform marginals, each of which can be related to an arbitrary known cumulative distribution function (cdf) $F_j(\cdot)$, $j = 1,\ldots,q$. For example, if a random variable u_j is uniform $u_j \sim U(0,1)$, and $y_j = F_j^{-1}(u_j)$, then it is easy to show that $y_j \sim F_j(\cdot)$. As a consequence, if the variables $y_1,\ldots,y_q$ have corresponding univariate cdfs $F_1(y_1),\ldots,F_q(y_q)$ taking values in $[0,1]$, a copula is a function that can be used to link or "couple" those univariate marginal distributions to produce the joint distribution function $F(y_1,\ldots,y_q)$:

$$F(y_1,\ldots,y_q) = C(F_1(y_1),\ldots,F_q(y_q)). \tag{5.1}$$

A detailed overview of copulas is provided in [9], [13], and [20]. The key feature that will be of interest here is that they provide a way to model dependence among multiple random variables when their joint distribution is not easy to specify, including cases where the marginal distributions $\{F_j(\cdot)\}$ belong to entirely different parametric classes.

There are several families of copulas, but the Gaussian copula is a natural modeling choice when one is interested in extensions beyond the bivariate case. The Gaussian copula is given by

$$C(u|\Omega) = \Phi_q(\Phi^{-1}(u_1),\ldots,\Phi^{-1}(u_q)|\Omega), \tag{5.2}$$

where $u = (u_1,\ldots,u_q)'$, Φ represents the standard normal cdf, and Φ_q is the cdf for a multivariate normal vector $z = (z_1,\ldots,z_q)'$, $z \sim N(0,\Omega)$, where Ω is in correlation form with ones on the main diagonal. The data generating process implied by the Gaussian copula specification is given by

$$y_{ij} = F_{ij}^{-1}\{\Phi(z_{ij})\}, \quad z_i \sim N(0,\Omega), \quad i = 1,\ldots,n, \quad j = 1,\ldots,q, \tag{5.3}$$

where F_{ij} is a cdf specified in terms of a vector of parameters θ_j and covariates x_{ij}, q is the dimension of each vector $y_i = (y_{i1}, \ldots, y_{iq})'$, and n is the sample size. Note that the correlation matrix Ω for the latent z_i induces dependence among the elements of y_i and that the copula density will typically be analytically intractable.

The structures in 5.1, 5.2, and 5.3 are quite general and apply to both discrete and continuous outcomes. However, it is important to recognize that the inverse cdf mapping $F_{ij}^{-1}(\cdot)$ in 5.3 is one-to-one when y_{ij} is continuous and many-to-one when y_{ij} is discrete. Therefore, in the latter case it is necessary to integrate over the values of z_i that lead to the observed y_i in order to obtain their joint distribution. In our implementation, this integration is performed by MCMC methods.

In this chapter, we use the Gaussian copula framework to specify a joint model for multivariate count data, where each count variable $y_{ij} \in \{0, 1, 2, \ldots\}$ follows a variable-specific negative binomial distribution

$$y_{ij} \sim NB(\lambda_{ij}, \alpha_j), \tag{5.4}$$

with probability mass function (pmf) given by

$$\Pr(y_{ij}|\lambda_{ij}, \alpha_j) = \frac{\Gamma(\alpha_j + y_{ij}) r_{ij}^{\alpha_j} (1 - r_{ij})^{y_{ij}}}{\Gamma(1 + y_{ij}) \Gamma(\alpha_j)}, \quad \lambda_{ij} > 0, \quad \alpha_j > 0, \tag{5.5}$$

where $r_{ij} = \alpha_j / (\alpha_j + \lambda_{ij})$, and dependence on the covariates is modeled through $\lambda_{ij} = \exp(x_{ij}'\beta_j)$. Here, and in the remainder of this chapter, all vectors will be taken to be column vectors. The distribution in 5.5 has mean λ_{ij} and variance $\lambda_{ij}(1 + \lambda_{ij}/\alpha_j)$, so that, depending on α_j, it allows for varying degrees of over-dispersion. The variance can be much larger than the mean for small values of α_j, but in the limit (as $\alpha_j \to \infty$) the two are equal, as in the Poisson model where the conditional variance equals the conditional mean. Negative binomial models are carefully reviewed in [2], [6], and [21].

The cdf for the negative binomial distribution is obtained by summing the pmf in 5.5 for values less than or equal to y_{ij}:

$$F_j(y_{ij}|\lambda_{ij}, \alpha_j) = \sum_{k=0}^{y_{ij}} \Pr(k|\lambda_{ij}, \alpha_j). \tag{5.6}$$

To relate the negative binomial distribution to the Gaussian copula, the pmf and cdf computed in 5.5 and 5.6, respectively, can be used to find unique, recursively determined cutpoints

$$\begin{aligned} \gamma_{ij,U} &= \Phi^{-1}(F_j(y_{ij}|\beta_j, \alpha_j)) \\ \gamma_{ij,L} &= \Phi^{-1}(F_j(y_{ij}|\beta_j, \alpha_j) - \Pr(y_{ij}|\lambda_{ij}, \alpha_j)) \end{aligned} \tag{5.7}$$

that partition the standard normal distribution so that for $z_{ij} \sim N(0,1)$, we have $\Pr(z_{ij} \leq \gamma_{ij,U}) = F_j(y_{ij}|\lambda_{ij}, \alpha_j)$ and $\Pr(\gamma_{ij,L} < z_{ij} \leq \gamma_{ij,U}) = \Pr(y_{ij}|\lambda_{ij}, \alpha_j)$. Hence, the cutpoints in 5.7 provide the range $B_{ij} = (\gamma_{ij,L}, \gamma_{ij,U}]$ of z_{ij} that is consistent with each observed outcome y_{ij} in 5.3. In turn, because $z_i = (z_{i1}, \ldots, z_{iq})' \sim N(0, \Omega)$, the

Gaussian copula representation implies that the joint probability of observing the vector $y_i = (y_{i1}, \dots, y_{iq})'$ is given by

$$\Pr(y_i|\theta, \Omega) = \int_{B_{iq}} \cdots \int_{B_{i1}} f_N(z_i|0, \Omega) dz_i, \tag{5.8}$$

in which $f_N(\cdot)$ denotes the normal density and, for notational convenience, we let $\theta = (\theta_1', \dots, \theta_q')'$, where $\theta_j = (\beta_j', \alpha_j)'$ represents the parameters of the jth marginal model, which determine the regions of integration $B_{ij} = (\gamma_{ij,L}, \gamma_{ij,U}]$, $j = 1, \dots, q$. Figure 5.1 offers an example of how the region of integration is constructed in the simple bivariate case. Because of the dependence introduced by the correlation matrix Ω, the probabilities in 5.8 have no closed-form solution and will be estimated by MCMC simulation methods in this chapter. Once computed, the probabilities in 5.8, also called likelihood contributions, can be used to construct the likelihood function

$$f(y|\theta, \Omega) = \prod_{i=1}^{n} \Pr(y_i|\theta, \Omega). \tag{5.9}$$

The likelihood function is then used in obtaining maximum likelihood estimates $\hat{\theta}$ and $\hat{\Omega}$, standard errors, and in performing model comparisons and hypothesis tests. Because the likelihood contributions are obtained by simulation, $f(y|\theta, \Omega)$ in 5.9 is referred to as the simulated likelihood function, and the estimates $\hat{\theta}$ and $\hat{\Omega}$ are called maximum simulated likelihood estimates. We next discuss the simulation and optimization techniques that are used to obtain those estimates.

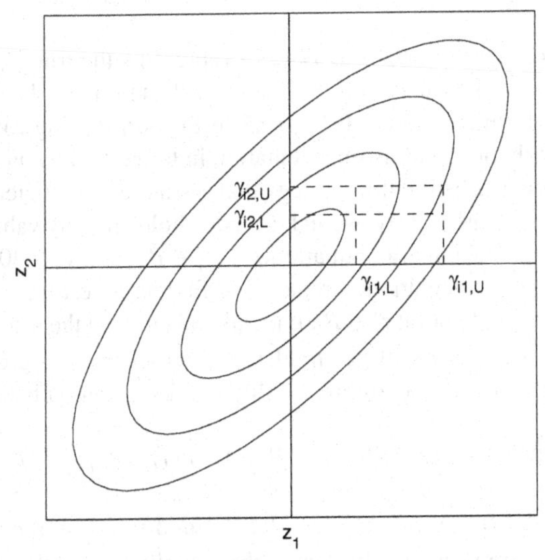

Fig. 5.1 An example of the region of integration implied by a bivariate Gaussian copula model

5.3 Estimation Methodology

Estimation by maximum simulated likelihood requires evaluation of the outcome probabilities for each observation vector y_i. Because each outcome probability in 5.8 is defined by an analytically intractable multidimensional integral, in Section 5.3.1 we describe a method for evaluating the outcome probabilities which was introduced in [8]. The method, called the Chib-Ritter-Tanner (CRT) method, stems from developments in MCMC simulation and Bayesian model choice and is well suited for evaluating outcome probabilities that comprise the likelihood of a variety of discrete data models. Because the CRT estimator produces continuous and differentiable estimates of 5.9, in Section 5.3.2 we describe how it can be applied in standard quasi-Newton gradient-based optimization using the Berndt-Hall-Hall-Hausman (BHHH) approach proposed in [1]. The BHHH approach exploits a fundamental statistical relation to avoid direct computation of the Hessian matrix of the log-likelihood function in the optimization algorithm.

5.3.1 *The CRT Method*

The CRT method, proposed in [8], is derived from theory and techniques in MCMC simulation and Bayesian model selection (see [3], [15]), where evaluation of multidimensional integrals with no analytical solution is routinely required. To understand the approach, note that the outcome probability in 5.8 can be rewritten as

$$\Pr(y_i|\theta,\Omega) = \int 1\{z_i \in B_i\} f_N(z_i|0,\Omega)dz_i = \frac{1\{z_i \in B_i\} f_N(z_i|0,\Omega)}{f_{TN_{B_i}}(z_i|0,\Omega)}, \tag{5.10}$$

where $B_i = B_{i1} \times B_{i2} \times \cdots \times B_{iq}$ and $f_{TN_{B_i}}(\cdot)$ represents the truncated normal density that accounts for the truncation constraints reflected in B_i. This representation follows by Bayes formula because $\Pr(y_i|\theta,\Omega)$ is the normalizing constant of a truncated normal distribution, and its representation in terms of the quantities in 5.10 is useful for developing an estimation strategy that is simple and efficient. As discussed in [3], this identity is particularly useful because it holds for any value of $z_i \in B_i$ and therefore, given that the numerator quantities $1\{z_i^* \in B_i\}$ and $f_N(z_i^*|0,\Omega)$ in 5.10 are directly available, the calculation is reduced to finding an estimate of the ordinate $f_{TN_{B_i}}(z_i^*|0,\Omega)$ at a single point $z_i^* \in B_i$, typically taken to be the sample mean of the draws $z_i \sim TN_{B_i}(0,\Omega)$ that will be simulated in the estimation procedure (details will be presented shortly). The log-probability is subsequently obtained as

$$\ln \widehat{\Pr}(y_i|\theta,\Omega) = \ln f_N(z_i^*|0,\Omega) - \ln \widehat{f_{TN_{B_i}}}(z_i^*|0,\Omega), \tag{5.11}$$

To estimate $\widehat{f_{TN_{B_i}}}(z_i^*|0,\Omega)$ in 5.11, the CRT method relies on random draws $z_i \sim TN_{B_i}(0,\Omega)$ which are produced by the Gibbs sampling algorithms of [5] or [16], where a new value for z_i is generated by iteratively simulating each element z_{ij} from its full-conditional density $z_{ij} \sim f(z_{ij}|y_{ij},\{z_{ik}\}_{k\neq j},\Omega) = TN_{B_{ij}}(\mu_{ij},\sigma_{ij}^2)$ for

$j = 1,\ldots,q$. In the preceding, μ_{ij} and σ_{ij}^2 are the conditional mean and variance of z_{ij} given $\{z_{ik}\}_{k\neq j}$ and are obtained by the usual formulas for a conditional Gaussian density. MCMC simulation of $z_i \sim TN_{B_i}(0,\Omega)$ is an important tool for drawing from this density, which is non-standard due to the multiple constraints defining the set B_i and the correlations in Ω.

The Gibbs transition kernel for moving from a point z_i to z_i^* is given by the product of univariate truncated normal full-conditional densities

$$K(z_i,z_i^*|y_i,\theta,\Omega) = \prod_{j=1}^{J} f(z_{ij}^*|y_i,\{z_{ik}^*\}_{k<j},\{z_{ik}\}_{k>j},\theta,\Omega). \tag{5.12}$$

Because the full-conditional densities represent the fundamental building blocks of the Gibbs sampler, the additional coding involved in evaluating 5.12 is minimized. By virtue of the fact that the Gibbs sampler satisfies Markov chain invariance (see [18, 4]), in our context we have that

$$f_{TN_{B_i}}(z_i^*|0,\Omega) = \int K(z_i,z_i^*|y_i,\theta,\Omega) f_{TN_{B_i}}(z_i|0,\Omega)dz_i, \tag{5.13}$$

a more general version of which was exploited for density estimation in [15]. Therefore, an estimate of $f_{TN_{B_i}}(z_i^*|0,\Omega)$ for use in 5.10 or 5.11 can be obtained by invoking 5.13 and averaging the transition kernel $K(z_i,z_i^*|y_i,\theta,\Omega)$ with respect to draws from the truncated normal distribution $z_i^{(g)} \sim TN_{B_i}(0,\Omega)$, $g = 1,\ldots,G$, i.e.

$$\widehat{f_{TN_{B_i}}}(z_i^*|0,\Omega) = \frac{1}{G}\sum_{g=1}^{G} K(z_i^{(g)},z_i^*|y_i,\theta,\Omega). \tag{5.14}$$

When repeated evaluation of 5.13 is required, e.g. in evaluating derivatives of $f(y|\theta,\Omega)$, one should remember to reset the random number generation seed used in the simulations. The CRT method produces continuous and differentiable estimates of $\Pr(y_i|\theta,\Omega)$ and can thus be applied directly in derivative-based optimization as discussed next.

5.3.2 Optimization Technique

Let ψ represent the vector of parameters that enter the log-likelihood function $\ln f(y|\psi)$. For the copula model that we considered in Section 5.2, ψ consists of the elements of θ and the unique off-diagonal entries of Ω (recall that Ω is symmetric positive definite matrix with ones on the main diagonal) and the likelihood function $f(y|\psi)$ is given in 5.9. Standard Newton-Raphson maximization of the log-likelihood function $\ln f(y|\psi)$ proceeds by updating the value of the parameter vector in iteration t, ψ_t, to a new value, ψ_{t+1}, using the formula

$$\psi_{t+1} = \psi_t - \lambda H_t^{-1} g_t, \tag{5.15}$$

where $g_t = \partial \ln f(y|\psi_t)/\partial \psi_t$ and $H_t = \partial^2 \ln f(y|\psi_t)/\partial \psi_t \partial \psi_t'$ are the gradient vector and Hessian matrix, respectively, of the log-likelihood function at ψ_t and λ is a step size. Gradient-based methods are widely used in log-likelihood optimization because many statistical models have well-behaved log-likelihood functions and gradients and Hessian matrices are often required for statistical inference, e.g. in obtaining standard errors or Lagrange multiplier test statistics. The standard Newton-Raphson method, however, has well-known drawbacks. One is that computation of the Hessian matrix can be quite computationally intensive. For a k dimensional parameter vector ψ, computing the Hessian requires $O(k^2)$ evaluations of the log-likelihood function. In the context of simulated likelihood estimation, where k can be very large and each likelihood evaluation can be very costly, evaluation of the Hessian presents a significant burden that adversely affects the computational efficiency of Newton-Raphson. Another problem is that $(-H)$ may fail to be positive definite. This may be due to purely numerical issues (e.g. the computed Hessian may be a poor approximation to the analytical one) or it may be caused by non-concavity of the log-likelihood function. In those instances, the Newton-Raphson iterations will fail to converge to a local maximum.

To deal with these difficulties, [1] noted that an application of a fundamental statistical relationship, known as the information identity, obviates the need for direct computation of the Hessian. Because we are interested in maximizing a statistical function given by the sum of the log-likelihood contributions over a sample of observations, it is possible to use statistical theory to speed up the iterations. In particular, by definition we have

$$\int f(y|\psi)dy = 1, \tag{5.16}$$

where it is assumed that if there are any limits of integration, they do not depend on the parameters ψ. With this assumption, an application of Leibniz's theorem implies that $\partial\{\int f(y|\psi)dy\}/\partial\psi = \int \partial f(y|\psi)/\partial\psi dy$. Moreover, because $\partial f(y|\psi)/\partial\psi = \{\partial \ln f(y|\psi)/\partial\psi\} f(y|\psi)$, upon differentiation of both sides of 5.16 with appropriate substitutions, we obtain

$$\int \frac{\partial \ln f(y|\psi)}{\partial \psi} f(y|\psi)dy = 0. \tag{5.17}$$

Differentiating 5.17 with respect to ψ once again (recalling that under our assumptions we can interchange integration and differentiation), we get

$$\int \left\{ \frac{\partial^2 \ln f(y|\psi)}{\partial \psi \partial \psi'} f(y|\psi) + \frac{\partial \ln f(y|\psi)}{\partial \psi} \frac{\partial f(y|\psi)}{\partial \psi'} \right\} dy = 0, \tag{5.18}$$

where, taking advantage of the equality $\partial f(y|\psi)/\partial\psi = \{\partial \ln f(y|\psi)/\partial\psi\} f(y|\psi)$ once again, we obtain the primary theoretical result underlying the BHHH approach

$$-\int \frac{\partial^2 \ln f(y|\psi)}{\partial \psi \partial \psi'} f(y|\psi)dy = \int \frac{\partial \ln f(y|\psi)}{\partial \psi} \frac{\partial \ln f(y|\psi)}{\partial \psi'} f(y|\psi)dy. \tag{5.19}$$

The left side of equation 5.19 gives $E(-H)$, whereas on the right side we have $E(gg')$ which also happens to be $Var(g)$ because from 5.17 we know that $E(g) = 0$. Now, because the log-likelihood is the sum of independent log-likelihood contributions, i.e. $\ln f(y|\psi) = \Sigma_{i=1}^{n} \ln f(y_i|\psi)$, it follows that

$$Var(g) = \sum_{i=1}^{n} Var(g_i) \approx \sum_{i=1}^{n} g_i g_i',$$

in which $g_i = \partial \ln f(y_i|\psi)/\partial \psi$. Therefore, the BHHH algorithm for maximizing the log-likelihood function relies on the recursions

$$\psi_{t+1} = \psi_t + \lambda B_t^{-1} g_t, \tag{5.20}$$

where $B_t = \Sigma_{i=1}^{n} \left[\frac{\partial \ln f(y_i|\psi_t)}{\partial \psi_t}\right] \left[\frac{\partial \ln f(y_i|\psi_t)}{\partial \psi_t}\right]'$ is used in lieu of $-H_t$ in 5.15.

Working with the outer product of gradients matrix, B_t, has several important advantages. First, computation of the gradients requires $O(k)$ likelihood evaluations and hence yields significant computational benefits relative to direct evaluation of H_t which requires $O(k^2)$ such evaluations. Note that $\{\partial \ln f(y_i|\psi_t)/\partial \psi_t\}$ are calculated anyway in computing g_t and that obtaining B_t only involves taking their outer product but requires no further evaluations of $\ln f(y|\psi)$. Second, B_t is necessarily positive definite, as long as the parameters are identified, even in regions where the log-likelihood is convex. Hence, the BHHH algorithm guarantees an increase in $\ln f(y|\psi)$ for a small enough step size λ. Finally, B_t is typically more computationally stable than H_t, thereby reducing numerical difficulties in practice (e.g. with inversion, matrix decomposition, etc.).

We make an important final remark about the interplay between simulation and optimization in maximum simulated likelihood estimation: precise estimation of the log-likelihood is essential for correct statistical inference. Specifically, it is crucial for computing likelihood ratio statistics, information criteria, marginal likelihoods and Bayes factors, and is also key to mitigating simulation biases in the maximum simulated likelihood estimation of parameters, standard errors, and confidence intervals (see [12], [19]). For instance, if the probabilities that enter $f(y|\psi)$ are estimated imprecisely, the maximum likelihood estimate will be biased (by Jensen's inequality) and the estimates of $\{\partial \ln f(y_i|\psi_t)/\partial \psi_t\}$ will be dominated by simulation noise. This adversely affects the estimated standard errors because B_t is inflated by simulation noise rather than capturing genuine log-likelihood curvature. Hence, relying on the modal value of B_t^{-1} as an estimate of the covariance matrix of $\hat{\psi}$ will produce standard errors and confidence bands that are too optimistic (too small). In extreme cases, parameters that are weakly identified may appear to be estimated well, due entirely to the simulation noise. Such problems can be recognized by examining the behavior of the estimated standard errors (square root of the diagonal of the modal value of B_t^{-1}) for different values of the simulation size G in 5.14 to determine whether they are stable or tend to decrease as G is increased.

Table 5.1 Descriptive statistics for the explanatory variables in the patent count application

Variable	Description	Mean	SD
$\ln(SALES)$	Log of real sales (millions)	6.830	1.703
$\ln(WF)$	Log of number of company employees	2.398	1.717
$\ln(RDC)$	Log of real R&D capital (millions)	5.593	1.815

5.4 Application

In this section, we implement the methodology developed earlier to study the joint behavior of firm-level patent registrations in four technology categories in the "computers & instruments" industry during the 1980s. We use the data sample of [10], which consists of $n = 498$ observations on 254 manufacturing firms from the U.S. Patent & Trademark Office data set discussed in [7] and [11]. The response variable is a 4×1 vector y_i ($i = 1, \ldots, 498$) containing firm-level counts of registered patents in communications (COM), computer hardware & software (CHS), computer peripherals (CP), and information storage (IS). The explanatory variables reflect the characteristics of individual firms and, in addition to a category specific intercept, include sales (SALES), workforce size (WF), and R&D capital (RDC). Sales are measured by the annual sales revenue of each firm, while the size of the workforce is given by the number of employees that the firm reports to stockholders. R&D capital is a variable constructed from the history of R&D investment using inventory and depreciation rate accounting standards discussed in [7]. All explanatory variables, except the intercept, are measured on the logarithmic scale. Table 5.1 contains variable explanations along with descriptive statistics.

To analyze these multivariate counts, we use a Gaussian copula model with negative binomial marginals which was presented in Section 5.2. The negative binomial specification is suitable for this application because patent counts exhibit a heavy right tail, and hence it is useful to specify a model that can account for the possible presence and extent of over-dispersion. In addition to examining how patents in each category are affected by firm characteristics, joint modeling allows us to study the interdependence of patent counts that emerges due to technological spillovers, managerial incentives, and internal R&D decisions. For instance, technological breakthroughs and know-how in one area may produce positive externalities and spill over to other areas. Moreover, significant discoveries may produce patents in multiple categories, resulting in positive correlation among patent counts. Alternatively, the advancement of a particular technology may cause a firm to re-focus and concentrate its resources to that area at the expense of others, thereby producing negative correlations. The dependence structure embodied in the correlation matrix Ω of the Gaussian copula model that we consider is intended to capture these and other factors that can affect multiple patent categories simultaneously.

We estimate the copula model by first estimating the parameters of each negative binomial model separately by maximum likelihood and then using those estimates as a starting point for maximizing the copula log-likelihood. The individual negative binomial models have well-behaved log-likelihood functions and are relatively fast and straightforward to estimate by standard optimization techniques such as those presented in Section 5.3.2. Parameter estimates for the independent negative binomial models and the joint Gaussian copula model are presented in Table 5.2.

Table 5.2 Maximum simulated likelihood estimates of independent negative binomial (NB) models and joint Gaussian copula model with standard errors in parentheses

	Independent NB Models				Gaussian Copula Model			
	COM	CHS	CP	IS	COM	CHS	CP	IS
Intercept	0.968	-1.712	-5.834	-2.105	0.917	-1.471	-6.099	-2.033
	(0.993)	(0.939)	(1.682)	(0.568)	(1.040)	(0.986)	(1.645)	(0.628)
$\ln(SALES)$	-0.297	-0.202	0.084	-0.190	-0.285	-0.194	0.242	-0.181
	(0.254)	(0.233)	(0.423)	(0.122)	(0.270)	(0.247)	(0.417)	(0.128)
$\ln(WF)$	0.763	0.353	0.273	0.218	0.759	0.319	0.085	0.219
	(0.210)	(0.194)	(0.378)	(0.140)	(0.222)	(0.203)	(0.376)	(0.147)
$\ln(RDC)$	0.081	0.611	0.717	0.631	0.078	0.580	0.665	0.608
	(0.122)	(0.091)	(0.120)	(0.080)	(0.128)	(0.105)	(0.148)	(0.089)
$\ln(\alpha_j)$	-0.174	-0.017	-0.564	-0.464	-0.184	0.026	-0.563	-0.451
	(0.090)	(0.091)	(0.131)	(0.110)	(0.101)	(0.098)	(0.150)	(0.119)
					1.000			
					(0.000)			
					0.072	1.000		
					(0.070)	(0.000)		
Ω					0.119	0.313	1.000	
					(0.075)	(0.053)	(0.000)	
					-0.080	0.225	0.115	1.000
					(0.074)	(0.063)	(0.080)	(0.000)

The results in Table 5.2 largely accord with economic theory. Of particular interest is the fact that in all cases the coefficients on $\ln(RDC)$ are positive, and for CHS, CP, and IS, they are also economically and statistically significant. Specifically, those point estimates are relatively large in magnitude and lie more than 1.96 standard errors away from zero, which is the 5% critical value for a two-sided test under asymptotic normality. This indicates that innovation in those categories is capital-intensive and the stock of R&D capital is a key determinant of patenting activity. The results also suggest that, all else being equal, the introduction of patents tends to be done by large firms, as measured by the size of the company workforce $\ln(WF)$. The coefficient on that variable in the communications category is large and statistically significant, whereas in the other three categories the estimates are positive but not significant at the customary significance levels. Interestingly, and perhaps counter-intuitively, the coefficients on $\ln(SALES)$ in these categories are predominantly negative (with the exception of computer peripherals), and none are statistically significant. To explain this puzzling finding, economists have proposed a rationalization that has to do with signaling in the presence of asymmetric information. In particular, firms that do not have steady sales revenue such as start-ups that have yet to establish a reliable customer base, are often cash constrained and may have to demonstrate their creditworthiness to potential lenders such as venture capitalists, banks, and individual investors in order to obtain loans. One way for such firms to exhibit their research innovations and overall productivity is to register patents. In this case patents serve a dual role – they protect the firm's innovations from infringement and also send a positive signal to potential outside stakeholders. In contrast, firms that have more reliable sources of revenue due to higher sales have lower incentives to patent their innovations and may instead opt to protect their intellectual property in other ways (e.g. by keeping trade secrets, entering into exclusive agreements with potential users of their technology, etc.). These considerations are especially relevant in the computers & instruments industry, where patents have short life cycles and can often be circumvented by competitors who "innovate around" registered research advances.

Table 5.2 also illustrates that over-dispersion is a common feature of all four data series, as demonstrated by the low estimates of $\{\ln(\alpha_j)\}$ across all categories in both the copula and univariate regression models. As a consequence, allowing for over-dispersion by considering a negative binomial specification, as opposed to estimating a Poisson model, appears well justified.

In Table 5.2, the estimated dependence matrix Ω in the Gaussian copula model reveals interesting complementarities among patent categories and supports the case for joint modeling and estimation. Specifically, the estimates suggest that patents in the computer hardware & software category are highly correlated with counts in the computer peripherals and information storage categories, while the correlation between patents in the communications category are relatively mildly correlated with those in the remaining categories. To test formally for the relevance of the copula correlation structure in this context, one can use the likelihood ratio and Lagrange multiplier tests. The log-likelihood for the restricted model (the independent negative binomial specification) is $L_R = -4050.06$ and for the unrestricted model (Gaussian copula model), it is $L_U = -4020.98$, leading to a likelihood ratio test statistic $-2(L_R - L_U) = 58.16$. This test statistic has a χ^2 distribution with 6 degrees of freedom (equal to the number of off-diagonal elements in Ω) and a 5% critical value of 12.59, suggesting that the data strongly reject the restricted (independent negative binomial) specification. The Lagrange multiplier test statistic is constructed from the gradient $g_R = \partial \ln f(y|\psi_R)/\partial\psi$ and curvature $B_R = \sum_{i=1}^{n} \frac{\partial \ln f(y_i|\psi_R)}{\partial\psi} \frac{\partial \ln f(y_i|\psi_R)}{\partial\psi'}$ of the log-likelihood function of the Gaussian copula model, both evaluated at the restricted maximum likelihood estimate ψ_R. Note that this corresponds to the case when Ω is an identity matrix and the Gaussian copula model is equivalent to fitting the four negative binomial models separately (in fact, these are the starting values we use in optimizing the copula log-likelihood). The Lagrange multiplier test statistic $LM = \{g_R'(B_R)^{-1}g_R\} = 64.58$ has the same asymptotic χ^2 distribution as the likelihood ratio statistic and also leads to strong rejection of the restricted model.

The parameter estimates and hypothesis tests presented above are based on maximizing an MCMC-based estimate of the log-likelihood function because that

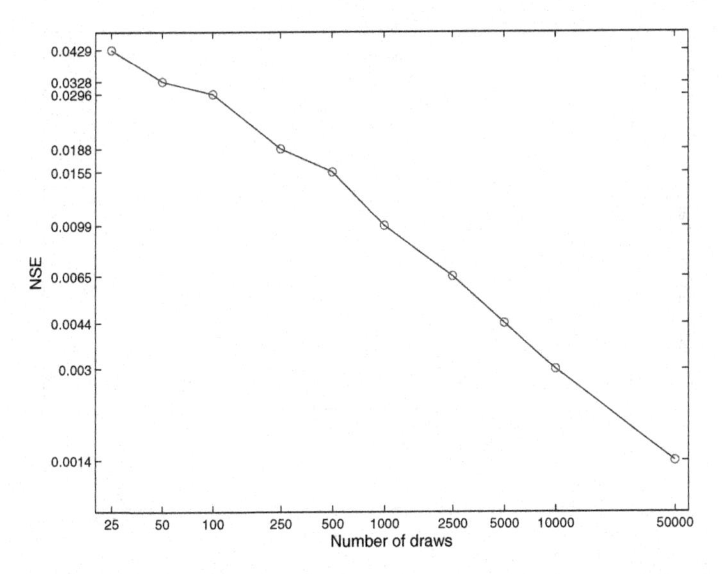

Fig. 5.2 Numerical standard errors (NSE) of the log-likelihood estimate as a function of the MCMC sample size in the CRT method (the axes, but not the values, are on the logarithmic scale)

function is analytically intractable. However, because the variability intrinsic in simulation-based estimation can affect the reliability of the results, it is important to examine the extent to which the point estimates, standard errors, and test statistics are affected by the performance of the simulated likelihood algorithm. In Figure 5.2, we have plotted the numerical standard error of the estimated log-likelihood ordinate $\ln \hat{f}(y|\psi)$ as a function of the simulation sample size G used in constructing the average in 5.14. The numerical standard error gives a measure of the variability of the estimated log-likelihood ordinate for fixed y and ψ if the simulation necessary to evaluate $\ln f(y|\psi)$ were to be repeated using a new Markov chain. The Figure demonstrates that the simulated likelihood algorithm performs very well – even with few MCMC draws, the precision of estimating $\ln \hat{f}(y|\psi)$ is sufficient for producing reliable parameter estimates and hypothesis tests. The low variability of $\ln \hat{f}(y|\psi)$ in our example is especially impressive because the numerical standard errors are obtained as the square root of the sum of variances of the $n = 498$ individual log-likelihood contributions. To be cautious, we have also verified the validity of the point estimates by initializing the algorithm at different starting values and also by estimating the model by Bayesian simulation methods similar to those proposed in [10] and [14], which do not rely on maximizing the log-likelihood.

At the end of Section 5.3.2, we discussed the possibility that in maximum simulated likelihood estimation the standard errors may be affected by simulation noise. To examine the extent to which variability in the log-likelihood estimate translates to downward biases in the standard errors of the parameter estimates, we compute the standard errors across several settings of the simulation size G, namely

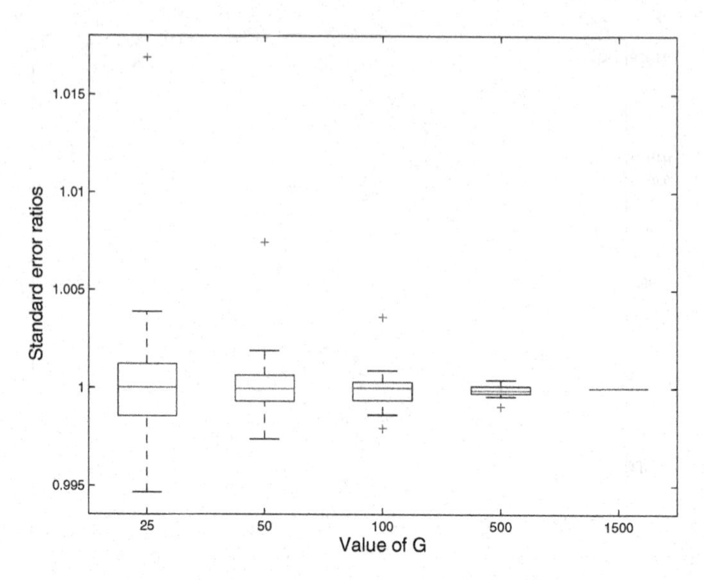

Fig. 5.3 Boxplots of the ratios of parameter standard errors estimated for each MCMC sample size setting G relative to those for $G = 1500$; the lines in the boxes mark the quartiles, the whiskers extend to values within 1.5 times the interquartile range, and outliers are displayed by "+"

$G \in \{25, 50, 100, 500, 1500\}$. We then compare the behavior of the standard errors for lower values of G relative to those for large G. Figure 5.3 presents boxplots of the ratios of the parameter standard errors estimated for each setting of G relative to those at the highest value $G = 1500$. The results suggest that while at lower values of G the standard error estimates are somewhat more volatile than at $G = 1500$, neither the volatility nor the possible downward bias in the estimates represents a significant concern. Because the CRT method produces very efficient estimates of the log-likelihood ordinate, such issues are not problematic even with small MCMC samples, although in practice G should be set conservatively high, subject to one's computational budget.

5.5 Concluding Remarks

This chapter has discussed techniques for obtaining maximum simulated likelihood estimates in the context of models for discrete data, where the likelihood function is obtained by MCMC simulation methods. These methods provide continuous and differentiable estimates that enable the application of widely used derivative-based techniques for obtaining parameter standard errors and test statistics. Because we are maximizing a log-likelihood function, we rely on the BHHH outer product of gradients method to simplify and speed up the computation of the Hessian matrix of the log-likelihood. The methodology is applied in a study of the joint behavior of four categories of U.S. technology patents using a Gaussian copula model for multivariate count data. The results support the case for joint modeling and estimation of the patent categories and suggest that the estimation techniques perform very well in practice. Additionally, the CRT estimates of the log-likelihood function are very efficient and produce reliable parameter estimates, standard errors, and hypothesis test statistics, mitigating any potential problems (discussed at the end of Section 5.3.2) that could arise due to maximizing a simulation-based estimate of the log-likelihood function.

We note that the simulated likelihood methods discussed here can be applied in optimization algorithms that do not require differentiation, for example in simulated annealing and metaheuristic algorithms which are carefully examined and summarized in [22]. At present, however, due to the computational intensity of evaluating the log-likelihood function at each value of the parameters, algorithms that require numerous evaluations of the objective function can be very time consuming, especially if standard errors have to be computed by bootstrapping. Nonetheless, the application of such algorithms is an important new frontier in maximum simulated likelihood estimation.

References

1. Berndt, E., Hall, B., Hall, R., Hausman, J.: Estimation and Inference in Nonlinear Structural Models. Annals of Economic and Social Measurement 3, 653–665 (1974)
2. Cameron, A.C., Trivedi, P.K.: Regression Analysis of Count Data. Cambridge University Press, Cambridge (1998)
3. Chib, S.: Marginal Likelihood from the Gibbs Output. Journal of the American Statistical Association 90, 1313–1321 (1995)
4. Chib, S., Greenberg, E.: Markov Chain Monte Carlo Simulation Methods in Econometrics. Econometric Theory 12, 409–431 (1996)
5. Geweke, J.: Efficient Simulation from the Multivariate Normal and Student-t Distributions Subject to Linear Constraints. In: Keramidas, E.M. (ed.) Proceedings of the Twenty-Third Symposium on the Interface. Computing Science and Statistics, pp. 571–578. Interface Foundation of North America, Inc, Fairfax (1991)
6. Greene, W.: Functional forms for the negative binomial model for count data. Economics Letters 99, 585–590 (2008)
7. Hall, B.H.: The Manufacturing Sector Master File: 1959–1987. NBER Working paper 3366 (1990)
8. Jeliazkov, I., Lee, E.H.: MCMC Perspectives on Simulated Likelihood Estimation. Advances in Econometrics: Maximum Simulated Likelihood 26, 3–39 (2010)
9. Joe, H.: Multivariate Models and Dependence Concepts. Chapman and Hall, London (1997)
10. Lee, E.H.: Essays on MCMC Estimation, Ph.D. Thesis, University of California, Irvine (2010)
11. Mairesse, J., Hall, B.H.: Estimating the Productivity of Research and Development: An Exploration of GMM Methods Using Data on French and United States Manufacturing Firms. NBER Working paper 5501 (1996)
12. McFadden, D., Train, K.: Mixed MNL Models for Discrete Response. Journal of Applied Econometrics 15, 447–470 (2000)

13. Nelsen, R.B.: An Introduction to Copulas. Springer, Heidelberg (1999)
14. Pitt, M., Chan, D., Kohn, R.: Efficient Bayesian Inference for Gaussian Copula Regression Models. Biometrika 93, 537–554 (2006)
15. Ritter, C., Tanner, M.A.: Facilitating the Gibbs Sampler: The Gibbs Stopper and the Griddy-Gibbs Sampler. Journal of the American Statistical Association 87, 861–868 (1992)
16. Robert, C.P.: Simulation of Truncated Normal Variables. Statistics and Computing 5, 121–125 (1995)
17. Sklar, A.: Fonctions de repartition 'a n dimensions et leurs marges. Publications de l'Institute de Statistique de l'Université de Paris 8, 229–231 (1959)
18. Tierney, L.: Markov Chains for Exploring Posterior Distributions. Annals of Statistics 22, 1701–1761 (1994)
19. Train, K.: Discrete Choice Methods with Simulation. Cambridge University Press, Cambridge (2003)
20. Trivedi, P.K., Zimmer, D.M.: Copula Modeling: An Introduction for Practitioners, Foundations and Trends in Econometrics, vol. 1, pp. 1–111 (2005)
21. Winkelmann, R.: Econometric Analysis of Count Data, 5th edn. Springer, Berlin (2008)
22. Yang, X.-S.: Engineering Optimization: An Introduction with Metaheuristic Application s Hoboken. John Wiley & Sons, New Jersey (2010)

Chapter 6
Optimizing Complex Multi-location Inventory Models Using Particle Swarm Optimization

Christian A. Hochmuth, Jörg Lässig, and Stefanie Thiem

Abstract. The efficient control of logistics systems is a complicated task. Analytical models allow to estimate the effect of certain policies. However, they necessitate the introduction of simplifying assumptions, and therefore, their scope is limited. To surmount these restrictions, we use Simulation Optimization by coupling a simulator that evaluates the performance of the system with an optimizer. This idea is illustrated for a very general class of multi-location inventory models with lateral transshipments. We discuss the characteristics of such models and introduce Particle Swarm Optimization for their optimization. Experimental studies show the applicability of this approach.

6.1 Introduction

Reducing cost and improving service is the key to success in a competitive economic climate. Although these objectives seem contradictory, there is a way to achieve them. Spreading of service locations improves service and pooling of resources can decrease cost if lateral transshipments are allowed between the locations. The design and control of such multi-location systems is an important non-trivial

Christian A. Hochmuth
Manufacturing Coordination and Technology,
Bosch Rexroth AG, 97816 Lohr am Main, Germany
e-mail: christian.hochmuth@boschrexroth.de

Jörg Lässig
Institute of Computational Science, University of Lugano, Via Giuseppe Buffi 13,
6906 Lugano, Switzerland
e-mail: joerg.laessig@usi.ch

Stefanie Thiem
Institute of Physics, Chemnitz University of Technology,
09107 Chemnitz, Germany
e-mail: stefanie.thiem@cs.tu-chemnitz.de

S. Koziel & X.-S. Yang (Eds.): Comput. Optimization, Methods and Algorithms, SCI 356, pp. 101–124.
springerlink.com

task. Therefore, we need suitable mathematical models – Multi-Location Inventory Models with lateral Transshipments (MLIMT) – to describe the following situation [14]. A given number of locations has to meet a demand for some products during a defined planning horizon. Each location can replenish its stock either by ordering from an outside supplier or by transshipments from other locations. The problem arises to define such ordering and transshipment decisions (OD and TD) that optimize defined performance measures for the whole system.

The MLIMT presented here has been developed with respect to three aspects. First, we assume discrete review for ordering in conjunction with continuous review for transshipments. Second, we propose an MLIMT simulator as general as possible to abandon restrictions of existing studies (see Section 6.2) and to ensure broad applicability. Third, we follow a Simulation Optimization (SO) approach by iteratively connecting the MLIMT simulator with a Particle Swarm Optimization (PSO) algorithm to investigate the search space. Hence, we contribute to the application of SO by describing a multi-location inventory model that is far more general, and thus, more complex than preceding models. We show that it is possible to evaluate, and thus, to optimize policies for arbitrary demand processes as well as arbitrary ordering, demand satisfaction, pooling and transshipment modes. In fact, SO provides the solution for the optimal control of complex logistics networks. Moreover, we contribute to the methodology of SO by integrating a PSO algorithm for this specific application. So far, PSO has been implemented for a Single-Warehouse, Multi-Retailer system by Köchel and Thiem [23], and we aim to show its applicability for even more complex logistics networks.

After a discussion of related work in Section 6.2 and a brief introduction of the SO approach in Section 6.3, we present the characteristics of a general MLIMT and delve into the implemented simulation model in Section 6.4. In Section 6.5 we describe the applied PSO in detail. Experimental studies are discussed in Section 6.6, followed by concluding remarks in Section 6.7.

6.2 Related Work

At present a great variety of models and approaches exists dealing with this decision problem. The most common and broadest investigated class of models assumes a single product, discrete review, independent and identically distributed demand, backlogging, complete pooling, emergency lateral transshipments at the end of a period, zero lead times, linear cost functions, and the total expected cost criterion as performance measure (see Köchel [19], Chiou [3] for a review). However, MLIMTs generally do not allow analytical solutions due to transshipments. TDs change the state of the system and thereby influence the OD. Thus it is impossible to define the total consequences of an OD. Approximate models and simulations are alternatives, see e.g., Köchel [18, 19], Robinson [30]. Additional problems connected with TDs arise for continuous review models. One is to prevent undesirable forth-and-back transshipments. This is narrowly connected with the problem to forecast the demand during the transshipment time and the time interval elapsing from the release moment of a TD until the next order quantity will arrive. Therefore, continuous review MLIMTs are usually investigated under several simplifying assumptions, e.g., two locations [7, 33], Poisson demand [25], a fixed ordering policy not considering future transshipments [27], restriction to simple rules such as a one-for-one ordering policy [25] and an all-or-nothing transshipment policy [7], or the limitation that at most one transshipment with negligible time and a single shipping point during an order cycle is possible [33]. Nowhere the question for optimal ordering and transshipment policies has been answered. All models work with a given ordering policy and heuristic transshipment rules. In few cases simulation is used either for testing approximate analytical models [27, 33] or for the definition of the best reorder point s for a (s,S)-ordering policy [24] by linear search and simulation. Thus, the investigations are restricted to small-size models. Herer et al. [12] calculates optimal order-up-to levels S using a sample-path-based optimization procedure and subsequently finds optimal transshipment policies for given ordering policies applying linear programming to a network flow model. Extensions include finite transportation capacities [28] and positive replenishment lead times [10]. Furthermore, we investigate the effect of non-stationary transshipment policies under continuous review. Thus, the complexity of this general model motivates the application of simulation-based optimization with PSO instead of gradient-based methods. In this regard, we follow an approach similar to Arnold et al. [1], and recently Belgasmi et al. [2], who analyze the effect of different parameters using evolutionary optimization.

6.3 Simulation Optimization

The Simulation Optimization (SO) approach is well known in the field of Industrial Engineering. Its key advantage is that various performance measures can be optimized for in fact arbitrary models. Among many others, Guariso et al. [11], Willis and Jones [32], Iassinovski et al. [15] introduce comprehensive SO frameworks. Some notable examples are finding optimal order policies [8, 1], sequencing and lot-sizing in production [16], production planning and control in remanufacturing [26], and optimizing multi-echelon inventory systems [22]. For a review of SO in general and with regard to inventory problems see Köchel [20].

In general SO comes in two distinguishable flavors. Non-iterative (non-recursive or retrospective) SO decouples simulation and optimization, while iterative (recursive or prospective) SO integrates both functional components into a self-adapting search method. For an overview on SO approaches we refer to Fu et al. [9] and Köchel [20]. In case of non-iterative SO, the objective function of the model is estimated by simulation prior to the optimization. Thus, to cover the search space in sufficient accuracy, extensive simulation is necessary, especially if the objective function is unknown. In contrast, in case of iterative SO, simulation is used to evaluate actual solution candidates, and therefore, simulation is adapted to the current state of the search. The general idea of iterative SO is outlined in Figure 6.1. For a given decision problem an optimizer proposes candidate solutions. Using the results of simulation experiments, the performance of these candidate solutions is estimated. On the basis of the estimated performance, the optimizer decides to accept or reject the current decisions. Acceptance stops the search process whereas rejection continues it.

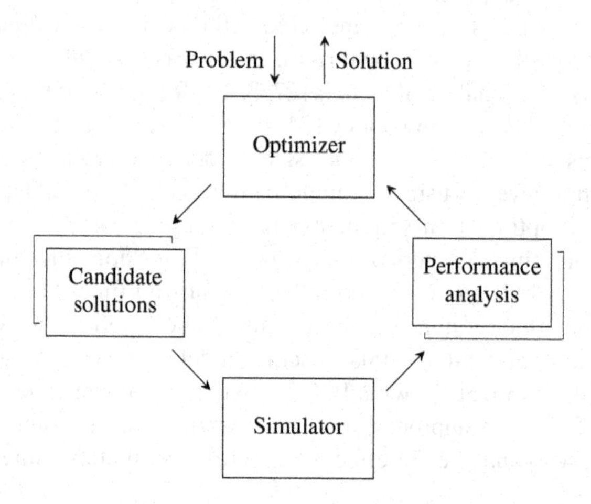

Fig. 6.1 Scheme for the iterative Simulation Optimization approach. The optimizer proposes new candidate solutions for the given problem, whose performance are estimated by simulation experiments. Depending on the estimated performance, the optimizer decides to either accept the current decisions and stop the search process or to reject it and to continue

As seen from Figure 6.1, iterative SO is based upon two main elements – a simulator for the system to be investigated and an optimizer that finds acceptable solutions. Generic simulators and optimizers are compatible, and thus SO is suited for the solution of arbitrary complex optimization problems. In the past different applications of the outlined approach especially to inventory problems [21, 22, 16, 20] have been implemented. In most cases Genetic Algorithms (GA) have been applied so far. But just as GAs also Particle Swarm Optimization (PSO) is in fact suitable for very general optimization problems. Contrary to gradient-based approaches, local optima can be left. Hence, these methods are predestined for unknown or complicated fitness landscapes. However, it is not guaranteed to find the global optimum, but a very good solution is usually returned in reasonable time. Furthermore, they rely only on a small amount of information and can be designed independently from the application domain. It will be interesting to see if PSO deals as excellently with the random output of stochastic simulation as GAs.

6.4 Multi-Location Inventory Models with Lateral Transshipments

6.4.1 Features of a General Model

A simulation model can in principle represent any real system with arbitrary accuracy. However, our objective is not to over-size a simulation model for all possible inventory systems. Instead we develop a simulation model capable of evaluating solutions for an important class of systems in reasonable time. First, we describe such a general Multi-Location Inventory Model with Lateral Transshipments (MLIMT).

The visible complexity of an MLIMT is of course determined by its *number of locations* N. With respect to the analytical tractability the cases $N = 2$ and $N > 2$ can be distinguished. Thus, the limitations of analytical models are obvious, and in order to solve real-world problems, it is crucial to surmount this restriction. The general case is illustrated in Figure 6.2.

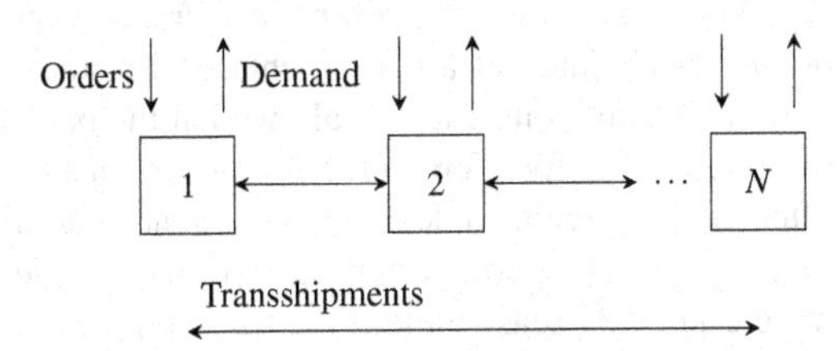

Fig. 6.2 Logical view of a general Multi-Location Inventory Model with Lateral Transshipments (MLIMT). Each of the N locations may refill its stock either by ordering from an outside supplier or by transshipments from other locations in order to meet the demand. The locations are on an equal level without any predefined structure

Each of the N locations faces a certain demand for a single *product* or a finite number of products. In the latter case a substitution order between products may be defined. Most approaches assume a single product. For the consideration of multiple products, sequential simulation and optimization is feasible, unless fixed cost or finite resources are shared among products. However, this limitation is negligible provided that shared fixed cost is insignificant relative to total fixed and variable cost, and capacities for storage and transportation are considered to be infinite.

The *ordering mode* defines when to order, i.e., the *review scheme*, and what *ordering policy* to use. The review scheme defines the time moments for ordering. Discrete and continuous review are the alternatives. Under the discrete review scheme the planning horizon is divided into periods. Usually the ordering policy is defined by its type and corresponding parameters (e.g., order-up-to, one-for-one, (s,S), (R,Q)).

Central to the model specification is the definition of the *demand process*. It may be deterministic or random, identical or different for all locations, stationary or non-stationary in time, independent or dependent across locations and time, with complete or incomplete information. To draw a reliable picture of the system performance, it is crucial to assume a realistic demand process, and thus, to track and to extrapolate orders in a real-word system.

Arriving demand is handled at each individual location according to the *demand satisfaction mode*. It is common to assume a queue for waiting demand. By defining an infinite, finite or zero queueing capacity, the backlogging, intermediate and lost-sales cases are distinguished. The service policy defines at what position an arriving client is enqueued. Eventually, clients may leave the location after a random impatient time if their demand is not or only partially served.

Clearly, it may be advantageous to balance excess and shortage among locations to serve demand that would otherwise be lost. The *pooling mode* comprises all rules by which the on-hand inventory is used to respond to shortages in the MLIMT. Pooling may be complete or partial and defines which locations and what amount of available product units is pooled. Furthermore, we distinguish between pooling of stocks and pooling of residuals.

Still, it must be defined by the *transshipment mode* when to transship and what transshipment policy to use. There may be *preventive* lateral transshipments to anticipate a stock-out or *emergency* lateral transshipments after a stock-out is observed. E.g., preventive lateral transshipments may be allowed at the beginning of a period in a discrete review model, i.e., before demand realization, or at a given moment during a period, i.e., after partially realized demand. Emergency lateral transshipments are usually allowed at the end of a period after realization of the demand.

Transshipments are especially reasonable if the transshipment *lead time* is practically negligible. In general, lead times for orders and transshipments of product units may be positive constants or random. Again, a distinction with respect to analytical tractability can be made. Although the effect of lead times for real-world systems may be pronounced, many analytical models assume zero lead times.

To measure the system performance, *cost and gain functions* are defined. There may incur cost for ordering, storing and transhipping product units as well as for waiting and lost demand. These functions may be linear, linear with set-up part, or generally non-linear. A location may also earn gain from sold units. The cost is tracked in a certain *planning horizon*, which may be finite or infinite. In case of periodic review it may consist of a single period.

Finally, as *optimization criterion* various cost criteria can be used such as total expected cost, total expected discounted cost, long-run average cost, and non-cost criteria such as service rates or expected waiting times. Both criteria types can be combined to formulate a multi-objective problem. Alternatively, one criterion is optimized while given restrictions must be satisfied for others, or different aspects such as service are represented by cost functions, e.g., out-of-stock cost.

The most common class of models defines a single product, discrete review, demand independent and identically distributed across time, backlogging, complete pooling, emergency lateral transshipments at the end of a period, zero lead times, linear cost functions, and a total expected cost criterion. However, even for that simple type of an MLIMT an analytical solution is not possible in general due to the transshipments at the end of a period. Potential transshipments had to be taken into account for the ordering decision at the beginning of a period. But after the demand realization at the end of a period the optimal transshipment decision results from an open (linear) transportation problem. Such problems do not have closed form solutions. Therefore, prior to the demand realization no expression is available for the cost savings from transshipments. Both approximate models and simulation are potential solutions, e.g., Köchel [18, 19] and Robinson [30].

6.4.2 Features of the Simulation Model

The simulation model offers features that allow the mapping of very general situations. The simulator is in principle suited for models with an arbitrary number of independent non-homogeneous locations, a single product, constant location-dependent delivery and transshipment lead times, and unlimited transportation resources. The most important extensions of existing models are the following ones.

With regard to the *ordering mode*, we assume a periodic review scheme with fixed length $t_{P,i}$ of the review period for orders at location i. In principle arbitrary ordering policies can be realized within the simulation model and so far (s_i,S_i)- and (s_i,nQ_i)- ordering policies have been implemented.

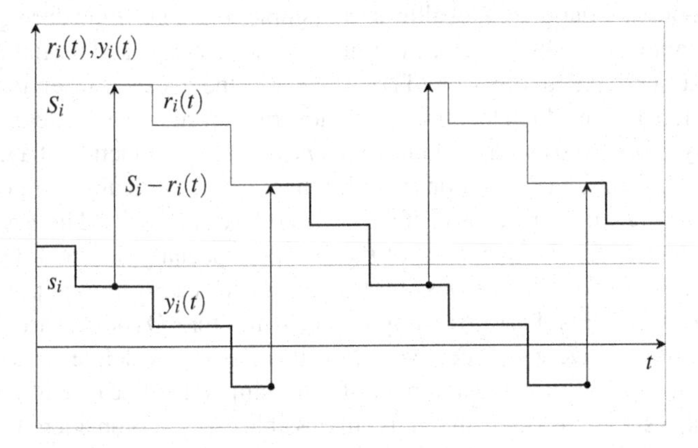

Fig. 6.3 (s_i,S_i)-ordering policy. If the inventory position r_i of location i drops below the reorder point s_i at the end of an order period, an order is released to the order-up-to level S_i, i.e., $S_i - r_i$ product units are ordered. Analogously, but under continuous review, a transshipment order (TO) of $H_i - f_{TO,i}(t)$ product units is released, if the state function $f_{TO,i}(t)$ falls below h_i

Clients arrive at the locations according to a compound renewal *demand process*. Such a process is described by two independent random variables T_i and B_i for the inter-arrival time of clients at location i and their demand, $i = 1\ldots N$, respectively. Thus, exact holding and penalty cost can be calculated, which is not the case for models with discrete review, where the whole demand of a period is transferred to the end of a period. That disadvantage does not exist for models with continuous review. However, in almost all such models a Poisson demand process is assumed – a strong restriction as well.

Concerning the *demand satisfaction mode*, most models assume the back-order or the lost-sales cases. An arriving client is enqueued according to a specific service policy, such as First-In-First-Out (FIFO) and Last-In-First-Out (LIFO), sorting clients by their arrival time, Smallest-Amount-Next (SAN) and Biggest-Amount-Next (BAN), sorting clients by their unserved demand, and Earliest-Deadline-First (EDF). In addition, a random impatient time is realized for each client.

To balance excess and shortage, the simulation model permits all pooling modes from complete to time-dependent partial pooling. A symmetric $N \times N$ matrix $\mathbf{P} = (p_{ij})$ defines pooling groups in such a way that two locations i and j belong to the same group if and only if $p_{ij} = 1$, $p_{ij} = 0$ otherwise. The following reflection is crucial. Transshipments allow the fast elimination of shortages, but near to the end of an order period transshipments may be less advantageous. Therefore, a parameter $t_{\mathrm{pool},i} \in [0, t_{\mathrm{P},i}]$ is defined for each location i. After the kth order request, location i can get transshipments from all other locations as long as for the actual time $t \leq k t_{\mathrm{P},i} + t_{\mathrm{pool},i}$ holds. For all other times location i can receive transshipments only from locations that are in the same pooling group. Thus, the transshipment policies become non-stationary in time.

Transshipments are in fact in the spotlight of this chapter. Regarding the *transshipment mode*, our simulation model allows transshipments at any time during an order cycle (continuous review) as well as multiple shipping points and partial deliveries to realize a transshipment decision (TD). To answer the question when to transship what amount between which locations, a great variety of rules can be defined. Broad applicability is achieved by three main ideas – *priorities*, introduction of a *state function* and generalization of common transshipment *rules*. Difficulties are caused by the problem to calculate the effects of a TD. Therefore, TDs should be based on appropriate forecasts for the dynamics of the model, especially the stock levels. The MLIMT simulator offers several possibilities. For each location transshipment orders (TO) and product offers (PO) are distinguished. Times for TOs or POs are the arrival times of clients or deliveries, respectively. Priorities are used to define the sequence of transshipments in one-to-many and many-to-one situations. Because of continuous time only such situations occur, and thus, all possible cases are considered. The three rules, Biggest-Amount-Next (BAN), Minimal-Transshipment-Cost per unit (MTC) and Minimal-Transshipment-Time (MTT) may be combined arbitrarily. State functions are used to decide when to release a TO or PO. The following variables for each location i and time $t \geq 0$ are used in further statements:

$y_i(t)$	Inventory level
$y_i^{\pm}(t) = \max(\pm y_i(t), 0)$	On-hand stock (+) and shortage (−), respectively
$b_{\mathrm{ord},i}(t)$	Product units ordered but not yet delivered
$b_{\mathrm{ord},ki}$	Product units ordered in the k-th request
$b_{\mathrm{tr},i}(t)$	Transshipments on the way to location i
$r_i(t) = y_i(t) + b_{\mathrm{ord},i}(t) + b_{\mathrm{tr},i}(t)$	Inventory position

$t_{\mathrm{P},i}$	Order period time
$t_{\mathrm{A},i}$	Delivery lead time of an order
$n_{\mathrm{ord},i} = \lfloor t_{\mathrm{A},i}/t_{\mathrm{P},i} \rfloor$	Number of periods to deliver an order

To decide at time t in location i about a TO or PO, the state functions $f_{\mathrm{TO},i}(t)$ and $f_{\mathrm{PO},i}(t)$ are defined based on the available stock plus expected transshipments $f_{\mathrm{TO},i}(t) = y_i(t) + b_{\mathrm{tr},i}(t)$ and the on-hand stock $f_{\mathrm{PO},i}(t) = y_i^+(t)$, respectively. Since fixed cost components for transshipments are feasible, a heuristic (h_i, H_i)-rule for TOs is suggested in the following way, which is inspired by the (s_i, S_i)-rule for order requests $(h_i \leq H_i)$.

If $f_{\mathrm{TO},i}(t) < h_i$
    release a TO for $H_i - f_{\mathrm{TO},i}(t)$ product units.

However, in case of positive transshipment times it may be advantageous to take future demand into account. Thus, a TO is released on the basis of a forecast of the state function $f_{\mathrm{TO},i}(t')$ for a time moment $t' \geq t$, and the transshipment policies become non-stationary in time. The MLIMT simulator offers three such time moments: $t' = t$, the current time (i.e., no forecast), $t' = t_1$, the next order review moment, and $t' = t_2$, the next potential moment of an order supply. For instance, the state function $f_{\mathrm{TO},i}(t) = y_i(t) + b_{\mathrm{tr},i}(t), t \geq 0$ is considered. Let $kt_{\mathrm{P},i} \leq t < (k+1)t_{\mathrm{P},i}$, i.e., we assume that we are in the review period after the kth order request. Then t_1 is defined as follows.

$$t_1 = (k+1)t_{\mathrm{P},i} \,. \tag{6.1}$$

For t_2 we introduce two events $ev(t) \leftrightarrow \{$in the actual period there has not been an order supply until $t\}$ and $\overline{ev}(t) \leftrightarrow \{$there has been an order supply until $t\}$.

$$t_2 = (k - n_{\mathrm{ord},i})t_{\mathrm{P},i} + t_{\mathrm{A},i} + \begin{cases} 0 & ev(t) \leftrightarrow t < (k - n_{\mathrm{ord},i})t_{\mathrm{P},i} + t_{\mathrm{A},i} \\ t_{\mathrm{P},i} & \overline{ev}(t) \leftrightarrow t \geq (k - n_{\mathrm{ord},i})t_{\mathrm{P},i} + t_{\mathrm{A},i} \end{cases} . \tag{6.2}$$

Using $m_i = \langle B_i \rangle / \langle T_i \rangle$ as long-run demand per time unit at location i, the following forecasts are used, illustrated in Figures 6.4 and 6.5:

$$\hat{f}_{\mathrm{TO},i}(t) = f_{\mathrm{TO},i}(t) = y_i(t) + b_{\mathrm{tr},i}(t) \,, \tag{6.3}$$

$$\hat{f}_{\mathrm{TO},i}(t_1) = f_{\mathrm{TO},i}(t) - m_i(t_1 - t) + \begin{cases} b_{\mathrm{ord},k'i}, \; k' = k - n_{\mathrm{ord},i} & ev(t) \\ 0 & \overline{ev}(t) \end{cases} , \tag{6.4}$$

$$\hat{f}_{\mathrm{TO},i}(t_2) = f_{\mathrm{TO},i}(t) - m_i(t_2 - t) \,. \tag{6.5}$$

Thus, replacing function $f_{\mathrm{TO},i}(t)$ by various forecast functions, a great variety of rules can be described to control the release of TOs. We remark that in case of linear transshipment cost functions without set-up part the (h_i, H_i)-rule degenerates to (H_i, H_i). A well-designed optimization algorithm will approximate that solution. Therefore, we work generally with the (h_i, H_i)-rule. To serve a TO, at least one location has to offer some product quantity. To decide when to offer what amount, an additional control parameter is introduced – the offering level o_i, corresponding to the

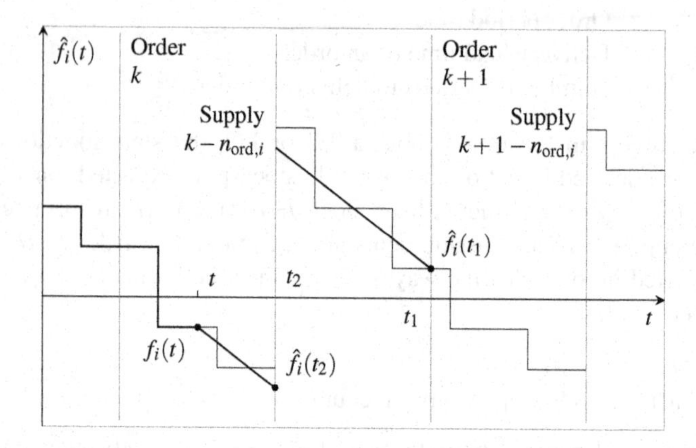

Fig. 6.4 Forecast functions for $ev(t) \leftrightarrow t < (k - n_{\mathrm{ord},i})t_{\mathrm{P},i} + t_{\mathrm{A},i}$. In the actual period there has not been an order supply until t. Thus, the time moment t_2 of the next order supply $k - n_{\mathrm{ord},i}$ is in the current period, and the supplied amount must be considered to forecast $\hat{f}_i(t_1)$

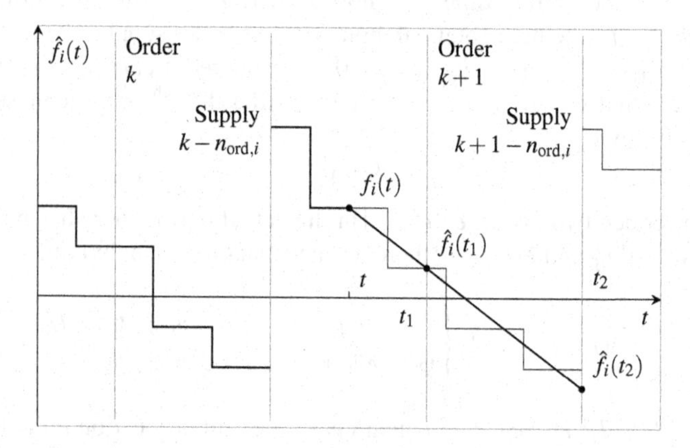

Fig. 6.5 Forecast functions for $\overline{ev}(t) \leftrightarrow t \geq (k - n_{\mathrm{ord},i})t_{\mathrm{P},i} + t_{\mathrm{A},i}$. In the actual period there has been an order supply until t, and thus, the time moment t_2 of the next order supply $k+1-n_{\mathrm{ord},i}$ is in the next period, not affecting $\hat{f}_i(t_1)$

hold-back level introduced by Xu et al. [33]. Since only on-hand stock can be transshipped, the state function $f_{\mathrm{PO},i}(t) = y_i^+(t)$ is defined. The offered amount $y_i^+(t) - o_i$ must not be smaller than a certain value $\Delta o_{\mathrm{min},i}$ to prevent undesirably small and frequent transshipments. Similar forecasts are applied to take future demand into account with forecast moments t, t_1, and t_2. For details we refer to Hochmuth [13]. Thus, the PO rule is as follows.

If $\hat{f}_{\mathrm{PO},i}(t) - o_i \geq \Delta o_{\mathrm{min},i}$
release a PO for $\hat{f}_{\mathrm{PO},i}(t) - o_i$ product units.

Thus, the set of available transshipment policies is extended, including all commonly used policies, and allowing multiple shipping points with partial deliveries.

In order to measure the system performance by *cost and gain functions*, order, holding, shortage (waiting) and transshipment cost functions may consist of fixed values, components linear in time, and components linear in time and units. Fixed cost arises from each non-served demand unit. All cost values are location-related. The gain from a unit, sold by any location, is a constant. To track cost for infinite *planning horizons*, appropriate approximations must be used. The only problem with respect to finite horizons is the increase in computing time to get a sufficiently accurate estimate, although the extent can be limited using parallelization.

Choosing cost function components in a specific way, cost *criteria* as well as non-cost criteria can be used, e.g., the average ratio of customers experiencing a stock-out or the average queue time measured by out-of-stock cost, or the efficiency of logistics, indicated by order and transshipment cost.

6.5 Particle Swarm Optimization

Particle Swarm Optimization (PSO), originally proposed by Kennedy and Eberhart [17], has been successfully applied to many real-world optimization problems in recent years [34, 6, 29]. PSO uses the dynamics of swarms to find solutions to optimization problems with continuous solution space. Meanwhile, many different versions and additional heuristics were introduced, where we restrict our considerations here to the Standard PSO 2007 algorithm by Clerc [4].

PSO is, similar to Genetic Algorithms or ensemble-based approaches [31], an iterative population-based approach, i.e., PSO works with a set of feasible solutions, the swarm. Let N denote the number of swarm individuals (particles) or the swarm size, respectively. The basic idea of PSO is that all swarm individuals move partly randomly through the solution space $\mathscr{S}$. Thereby individuals can share information about their so far best previous position $\mathbf{r}^{\text{bsf}}$, where each particle has a number of K informants. Additionally, each individual i has an internal memory to store its best so far (locally best) solution $\mathbf{r}_i^{\text{lbsf}}$. In every iteration the movement of each individual beginning from its actual position $\mathbf{r}_i$ is then given by a trade off between its current velocity $\mathbf{v}_i$, a movement in the direction of its locally best solution $\mathbf{r}_i^{\text{lbsf}}$ (cognitive component) and of its so far best known solution $\mathbf{r}_i^{\text{bsf}}$ of its informants in the swarm (social component). Thus, the equations of motion for one individual i and a discrete time parameter t are given by

$$\mathbf{v}_i^{t+1} = w \cdot \mathbf{v}_i^t + c_1 \cdot \mathscr{R}_1 \cdot \left(\mathbf{r}_i^{\text{lbsf},t} - \mathbf{r}_i^t\right) + c_2 \cdot \mathscr{R}_2 \cdot \left(\mathbf{r}_i^{\text{bsf},t} - \mathbf{r}_i^t\right) \tag{6.6}$$

$$\mathbf{r}_i^{t+1} = \mathbf{r}_i^t + \mathbf{v}_i^{t+1} \,. \tag{6.7}$$

The diagonal matrices $\mathscr{R}_1$ and $\mathscr{R}_2$ contain uniform random numbers in $[0,1)$ and thus randomly weight each component of the connecting vector $(\mathbf{r}_i^{\text{lbsf}} - \mathbf{r}_i)$ from the current position $\mathbf{r}_i$ to the locally best solutions $\mathbf{r}_i^{\text{lbsf}}$. The vector $(\mathbf{r}_i^{\text{bsf}} - \mathbf{r}_i)$ is treated analogously. Since every component is multiplied with a different random number, the vectors are not only changed in length but also perturbed from their original direction. The new position follows then from the superposition of the three vectors. By choosing the cognitive parameter c_1 and the social parameter c_2, the influence of these two components can be adjusted. The Standard PSO 2007 setup uses $c_1 = c_2 = 0.5 + \ln(2) \simeq 1.193$ and $w = 1/(2\ln 2) \simeq 0.721$ as proposed by Clerc and Kennedy [5]. A number of $N = 100$ particles is chosen with a number of $K = 3$ informants.

The pseudo-code of the solution update for the swarm is shown in Algorithm 1. The position and the velocity components for the different particles i and dimension d are written as subscripts, i.e., v_{id} is the d-th component of the velocity vector $\mathbf{v}_i$ of particle i. The iterative solution update of the vector $\mathbf{v}_i$ is visualized in Figure 6.6.

Algorithm 1: Position and velocity update rule in PSO

Data: position $\mathbf{r}_i$, velocity $\mathbf{v}_i$, locally best position $\mathbf{r}_i^{\text{lbsf}}$ and globally best position $\mathbf{r}_i^{\text{bsf}}$ for each particle i, cognitive parameter c_1 and social parameter c_2.
Result: $\{\mathbf{r}_i\}$ with $\mathbf{r}_i \in \mathscr{S}$ and updated velocities $\{\mathbf{v}_i\}$.

```
1  begin
2    forall particles i do
3      forall dimensions d do
4        R_1 ← get_uniform_random_number(0,1);
5        R_2 ← get_uniform_random_number(0,1);
6        v_id ← w·v_id + c_1·R_1·(r_id^lbsf − r_id) + c_2·R_2·(r_id^bsf − r_id);
7        r_id ← r_id + v_id;
8
9    return {r_i},{v_i};
10 end
```

6.6 Experimentation

6.6.1 System Setup

Using the MLIMT simulator in combination with PSO, different scenarios are optimized for the five-location model shown in Figure 6.7. We are particularly interested in the effect of the following three factors that are tested in combination.

First, we test the two service policies First-In-First-Out (FIFO) and Earliest-Deadline-First (EDF). Second, for the transshipment orders (TOs) we either monitor the current time, i.e., $t' = t$, or we use forecasting with $t' = t_1$, the next order review moment. Third, for the pooling strategy two different policies are applied. For the first option, the stocks of all locations i are completely pooled, i.e., $t_{\text{pool},i} = t_{\text{P},i}$. Thus, only ordering and transshipment policies are optimized, not pooling. For the second option, only locations next to each other in Figure 6.7 are in one pooling group, i.e., there are four different pooling groups. Therefore, lateral transshipments are limited to adjacent locations at the end of an order period.

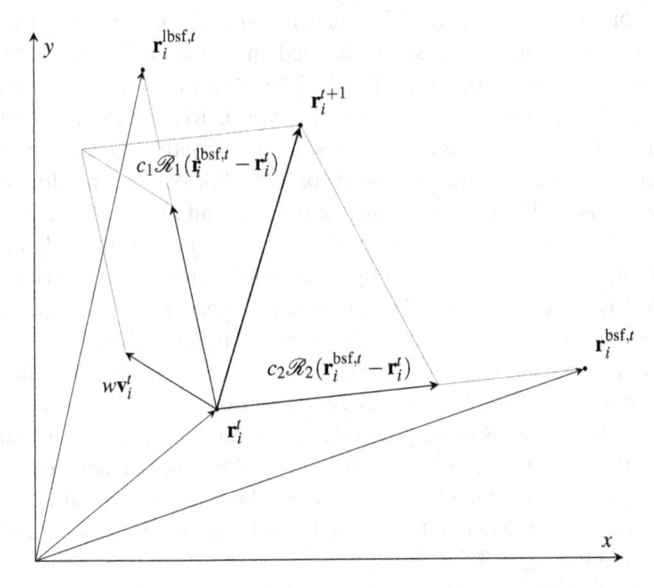

Fig. 6.6 Iterative solution update of PSO in two dimensions. From the current particle position $\mathbf{r}_i^t$ the new position $\mathbf{r}_i^{t+1}$ is obtained by vector addition of the velocity, the cognitive and the social component

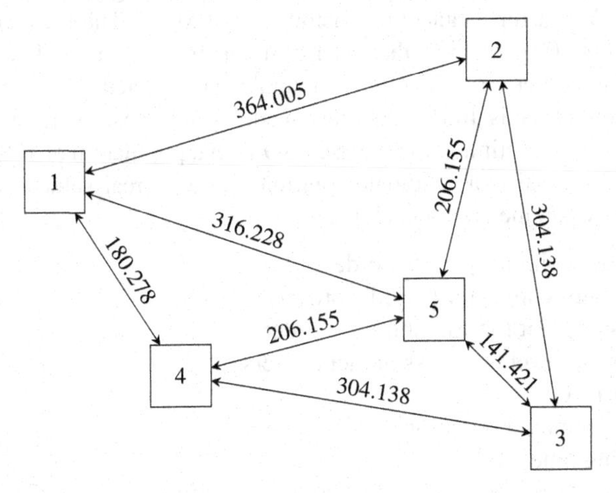

Fig. 6.7 Topology of the five-location model. An edge between two locations indicates that these belong to the same pooling group, i.e., lateral transshipments are feasible at all times. Along the edges distances in kilometers are shown

The other parameters are identical for all optimization runs. All locations i use an (s_i, S_i)-ordering policy, where the initial reorder points are $s_i = 0$ and the initial order-up-to levels are $S_1 = 600$, $S_2 = 900$, $S_3 = 1{,}200$, $S_4 = 1{,}500$ and $S_5 = 1{,}800$.

For all locations i an order period is equal to 10 days, i.e., $t_{P,i} = 10$ days. The distances between all locations are visualized in Figure 6.7, and the transshipment velocity is chosen to be 50.00 km/h. The state function chosen for TOs is $f_{TO,i}(t) = y_i(t) + b_{tr,i}(t)$ and for POs $f_{PO,i}(t) = y_i^+(t)$. To analyze the effect of forecasting demand for ordering transshipments, all combinations of the current time t and the forecast moment t_1 are compared for TOs. For offering product units, the current time t is used. The priority sequence for TOs and POs is MTC, BAN, MTT.

The inter-arrival time of customers to a location i is an exponentially distributed random variable with $\langle T_i \rangle = 2\,\text{h}$. The impatient time is triangularly distributed in the interval $(0\,\text{h}, 8\,\text{h})$, i.e., $\langle W_i \rangle = 4\,\text{h}$. The customer demand is for all locations i uniformly distributed in $[0, B_{\text{max}}]$ but with different maximum values, i.e., $B_{\text{max},1} = 10$, $B_{\text{max},2} = 15$, $B_{\text{max},3} = 20$, $B_{\text{max},4} = 25$, $B_{\text{max},5} = 30$, respectively. The initial inventory of the five locations i is chosen to be $I_{\text{start},1} = 600$, $I_{\text{start},2} = 900$, $I_{\text{start},3} = 1{,}200$, $I_{\text{start},4} = 1{,}500$, $I_{\text{start},5} = 1{,}800$, respectively. However, the initialization values will not have an influence after the transition time. The maximum capacity of the storage is 10,000 product units for each location. The regular order delivery times at the end of each period are location dependent as well. For location i the times are $t_1 = 2.0$ d, $t_2 = 2.5$ d, $t_3 = 3.0$ d, $d_4 = 3.5$ d and $t_5 = 4.0$ d.

The cost for storing product units is 1.00 € per unit and day, whereas the order and transshipment cost is 1.00 € per unit and per day transportation time. The fixed transshipment cost is 10.00 € for each location and the gain per unit sold is 100.00 €. The out-of-stock cost per product unit and waiting time are 1.00 €/h and the out-of-stock cost for a canceling customer is 50.00 €. The fixed cost for each periodic order is 500.00 €, and the order cost per product unit and day is 1.00 €. The optimization criterion is the minimum total cost expected.

The simulation time is 468 weeks plus an additional transition time of 52 weeks in the beginning. For optimization we use PSO with a population of 100 individuals i, where an individual is a candidate solution $\mathbf{r}_i$, i.e., a real-valued vector of the following policy parameters for each location i.

s_i	Reorder point for periodic orders
S_i	Order-up-to level for periodic orders
h_i	Reorder point for transshipment orders
H_i	Order-up-level for transshipment orders
o_i	Offer level
$\Delta o_{\text{min},i}$	Minimum offer quantity
$t_{\text{pool},i}$	Pooling time

The optimization stops if a new optimum has not occurred for the last 2,000 cycles, but at least 10,000 cycles must be realized to prevent early convergence in a local optimum. On machines with two dual-core Opteron 270 2GHz processors, one iteration consumes about 15 seconds runtime. For all experiments total results, optimized parameter values and cost function values are determined. After the optimization the minimum absolute values of all parameters not changing the cost function values are determined using a binary search.

6.6.2 Results and Discussion

Tables 6.1 and 6.2 show the total results for the service policies FIFO and EDF, respectively. For each service policy all combinations of forecasting demand for transshipment requests and assigning pooling groups are evaluated. Solutions 1 and 2 monitor the current net inventory level for deciding on transshipment orders, while solutions 3 and 4 forecast demand to the time of the next order. Solutions 1 and 3 allow transshipments between all locations, while solutions 2 and 4 confine transshipments to adjacent locations at the end of an order period. For each solution total cost per annum and the number of optimization cycles are shown. The rank sorts the eight systems from Tables 6.1 and 6.2 with respect to their solution quality.

Table 6.1 Overall results for the service policy First-In-First-Out (FIFO). Monitoring the net-inventory level at the current time t while limiting transshipments to adjacent locations at the end of the order period is the optimal policy

	Transshipment order	Pooling	Result in € p.a.	Cycle optimum (total)	Rank
1	current time t	all	$-19,666,901.01$	9,562 (11,562)	3
2	current time t	adjacent	$\mathbf{-19,758,378.05}$	9,562 (11,562)	**2**
3	time of next order t_1	all	$-19,530,460.22$	9,562 (11,562)	8
4	time of next order t_1	adjacent	$-19,562,191.79$	9,562 (11,562)	7

Table 6.2 Total results for service policy Earliest-Deadline-First (EDF). Observing the current net inventory level and restricting pooling dominates the other choices, while EDF is even slightly better than FIFO

	Transshipment order	Pooling	Result in € p.a.	Cycle optimum (total)	Rank
1	current time t	all	$-19,652,741.48$	8,259 (10,259)	4
2	current time t	next	$\mathbf{-19,762,585.05}$	8,259 (10,259)	**1**
3	time of next order t_1	all	$-19,565,348.10$	9,562 (11,562)	6
4	time of next order t_1	next	$-19,587,244.27$	9,562 (11,562)	5

Looking at the total results, there exists a lower bound regardless of the individual policies. However, for both service policies solution 2 yields the best performance. It is advantageous for this system to order transshipments based on the current net-inventory level and to limit transshipments to adjacent locations at the end of an order period. But even though the total results are similar, the optimal model structure varies significantly. Therefore, all solutions are investigated in detail. The optimal parameter values of the four considered systems are listed in Tables 6.3 and 6.4 for FIFO and EDF, respectively. Pooling times $t_{\text{pool},i}$ are optimized if transshipments are restricted to adjacent locations, and set to the order period time $t_{\text{P},i}$ otherwise. The resulting flows are visualized in Figures 6.8–6.10 for the three best solutions.

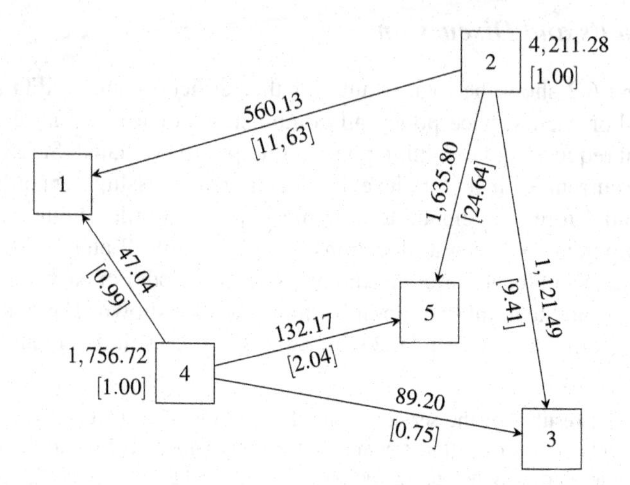

Fig. 6.8 Flows for solution 2 (rank 1) using the service policy Earliest-Deadline-First (EDF). Locations 2 and 4 act as hubs. The volumes ordered per period are listed next to these locations, as well as the order frequency in square brackets. Transshipments are indicated by directed edges, in conjunction with transshipment volumes and frequencies in square-brackets

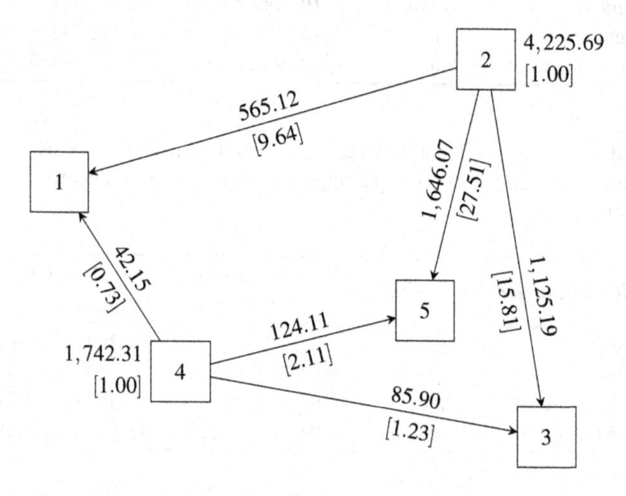

Fig. 6.9 Flows for solution 2 (rank 2) using the service policy First-In-First-Out (FIFO). This solution is similar to EDF solution 2. Locations 2 and 4 act as hubs, while locations 1, 3 and 5 just receive transshipments, and thereby, act as spokes. Periodic order volumes are listed next to the hubs, as well as transshipment volumes along the edges, and frequencies in square brackets

The figures illustrate that the solutions 2 for FIFO and EDF, respectively, are very similar. Moreover, there are two observations. First, there are locations periodically receiving and offering product units. These locations act as hubs in a hub-and-spoke structure. Second, there are locations just receiving transshipments from other

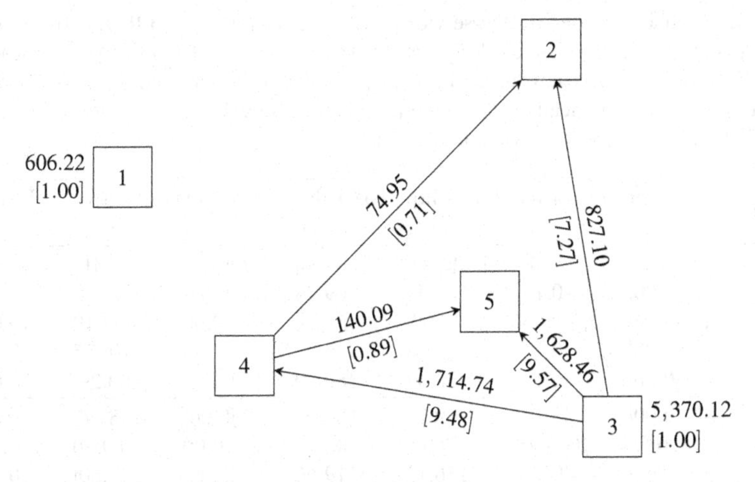

Fig. 6.10 Flows for (First-In-First-Out) FIFO solution 1 (rank 3). Location 1 is isolated, ordering products as indicated by the periodic order volume and frequency in square brackets next to the location, but not exchanging transshipments. The offcut network is integrated with location 3 as a hub. Transshipments are visualized by directed edges along with volumes per period and frequencies in square brackets.

locations, and thus, never receiving periodic orders. Thereby, these locations act as spokes. In EDF solution 2 – the best solution – locations 2 and 4 are considered as hubs, while locations 1, 3 and 5 are spokes, see Figure 6.8. Thus, transshipments take the role of periodic orders rather than just eliminating shortages due to stochastic demand. This is a consequence of the general definition of transshipments under continuous review, fixed order cost, and order lead times.

Furthermore, some solutions show a specific characteristic. A particular location is isolated, receiving periodic orders but never exchanging transshipments, e.g., location 1 in FIFO solution 1, cp. Figure 6.10. That points to the limitations of the proposed heuristic. Ordering and offering decisions are based upon the inventory level, not differentiating between target locations. Therefore, in specific situations it may be more economical not to exchange transshipments at all. Setting up pooling groups is a potential way to limit the complexity and to guide the optimization process in this case. Of course, complete linear optimization of the transport problem would be feasible, too, but at the cost of continuous review.

After studying elaborate model structures, which solution should a user implement? Tables 6.1 and 6.2 show the individual overall cost function values of all solutions for FIFO and EDF, respectively. By further evaluating specific cost functions as presented in Tables 6.5 and 6.6, decisions are better informed. Low out-of-stock cost corresponds to high service quality, and low order and transshipment cost indicates efficient logistics, if the total results are comparable. FIFO solution 1 in Table 6.5 leads to the least out-of-stock cost for all considered systems. A case in point for contradictive objectives is FIFO solution 4. Product units are constantly

Table 6.3 Parameter values for the service policy First-In-First-Out (FIFO) corresponding to the systems specified in Table 6.1. Prohibitive values, e.g., reorder points s_i never leading to a positive ordering decision, are enclosed in square brackets. Thus, hubs can be identified as locations which periodically order and offer product units. Spokes never receive periodic orders but replenish their stock via transshipments

	i	Periodic order s_i	S_i	Transshipment order h_i	H_i	Product offer o_i	$\Delta o_{\min,i}$	Pooling $t_{\text{pool},i}$
1	1	457.52	849.45	[−22.49]	[0.00]	[799.91]	[0.10]	10.00 d
	2	[0.00]	[0.10]	169.40	269.74	[0.00]	[276.52]	10.00 d
	3	1,980.59	6,305.89	[0.00]	[0.10]	0.00	447.19	10.00 d
	4	[0.00]	[0.10]	822.94	948.94	0.00	616.75	10.00 d
	5	[0.00]	[0.10]	332.47	480.29	[0.00]	[480.29]	10.00 d
2	1	[0.00]	[0.10]	69.21	123.65	[0.00]	[128.98]	0.00 d
	2	1,896.58	5,075.77	[0.00]	[0.10]	0.00	169.30	0.00 d
	3	[0.00]	[0.10]	156.17	219.60	[0.00]	[219.60]	0.00 d
	4	1,538.05	2,737.03	[−109.73]	[0.00]	0.00	371.50	0.00 d
	5	[0.00]	[0.10]	173.15	222.02	[0.00]	[222.02]	0.00 d
3	1	1,944.91	3,606.09	[0.00]	[0.10]	0.00	580.69	10.00 d
	2	[0.00]	[0.10]	172.74	229.51	[0.00]	[1,111.98]	10.00 d
	3	835.21	1,739.77	[0.00]	[0.10]	[0.00]	[1,696.18]	10.00 d
	4	1,247.85	2,243.85	[−56.56]	[0.00]	[0.00]	[2,063.09]	10.00 d
	5	[0.00]	[0.10]	693.24	693.35	[2,124.58]	[36.83]	10.00 d
4	1	[−24.46]	[0.00]	−123.47	85.79	[0.00]	[706.60]	0.00 d
	2	2,218.16	7,367.75	936.48	4,158.19	252.98	6.99	0.00 d
	3	[0.00]	[0.10]	71.52	166.59	[0.00]	[1,359.51]	0.00 d
	4	[0.00]	[0.10]	502.27	518.07	[1,624.89]	[327.09]	5.89 d
	5	[0.00]	[0.10]	3,842.22	5,958.31	599.10	517.20	0.00 d

Table 6.4 Parameter values for the service policy Earliest-Deadline-First (EDF) corresponding to the systems specified in Table 6.2. Square brackets indicate prohibitive values, never leading to a positive ordering or transshipment decision. Hence, analogous to FIFO, there are hubs, that order periodically, and spokes, that receive transshipments but never periodically orders

	i	Periodic order		Transshipment order		Product offer		Pooling
		s_i	S_i	h_i	H_i	o_i	$\Delta o_{\min,i}$	$t_{\text{pool},i}$
1	1	[0.00]	[0.10]	1,119.30	2,636.55	0.00	356.60	10.00 d
	2	839.26	4,918.42	[0.00]	[0.10]	943.20	0.10	10.00 d
	3	[0.00]	[0.10]	513.79	601.69	[0.00]	[636.77]	10.00 d
	4	1,276.91	2,272.90	[−133.33]	[0.00]	[1,462.34]	[170.68]	10.00 d
	5	[0.00]	[0.10]	740.30	872.29	[0.00]	[872.29]	10.00 d
2	1	[0.00]	[0.10]	82.35	126.48	[0.00]	[128.99]	0.00 d
	2	1,826.91	5,069.91	[0.00]	[0.10]	0.00	229.32	0.00 d
	3	[0.00]	[0.10]	128.96	240.01	[0.00]	[240.02]	0.00 d
	4	1,610.11	2,738.23	[−133.32]	[0.00]	0.00	361.77	0.00 d
	5	[0.00]	[0.10]	171.41	226.53	[0.00]	[226.54]	0.00 d
3	1	457.03	848.96	[−66.15]	[−66.05]	0.00	834.33	10.00 d
	2	[0.00]	[0.10]	345.71	348.45	295.33	901.85	10.00 d
	3	1,200.00	4,543.22	[0.00]	[0.10]	461.66	411.10	10.00 d
	4	1,242.79	2,238.78	[−219.85]	[0.00]	270.82	1,251.48	10.00 d
	5	[0.00]	[0.10]	618.03	642.67	0.00	1,580.89	10.00 d
4	1	[0.00]	[0.10]	239.10	239.46	[0.00]	[697.94]	0.00 d
	2	900.00	5,112.43	[0.00]	[0.10]	919.15	145.43	0.00 d
	3	756.54	1,658.42	[−122.57]	[0.00]	[1,125.11]	[349.98]	0.00 d
	4	[−26.06]	[0.00]	522.87	623.86	893.38	485.86	6.01 d
	5	[−55.03]	[0.00]	1,015.13	1,015.24	0.00	1,679.06	0.00 d

Table 6.5 Cost function values for service policy First-In-First-Out (FIFO) corresponding to the systems specified in Table 6.1. Different performance aspects correlate with individual cost functions, e.g., high service quality with low out-of-stock cost, and efficient logistics with low order and transshipment cost

	i	**Inventory cost** in € p.a.	**Out-of-stock cost** in € p.a.	**Periodic order cost** in € p.a.	**Transshipment cost** in € p.a.	**Gain** in € p.a.
1	1	155,087.59	411.44	62,299.55	0.00	−2,208,904.47
	2	69,562.63	1,255.94	0.00	0.00	−3,283,135.39
	3	762,445.29	1,022.29	604,583.39	40,015.55	−4,400,362.71
	4	283,973.96	28.74	0.00	2,260.93	−5,459,167.75
	5	136,758.42	1,772.39	0.00	0.00	−6,436,808.79
	Σ	1,407,827.89	4,490.79	666,882.94	42,276.48	−21,788,379.12
2	1	29,054.96	231.05	0.00	0.00	−2,210,119.90
	2	725,377.76	3,427.50	402,704.69	46,190.34	−3,263,712.13
	3	57,920.84	132.53	0.00	0.00	−4,408,758.96
	4	426,709.86	2,336.54	240,136.41	3,281.27	−5,432,982.09
	5	61,878.88	1,199.26	0.00	0.00	−6,443,386.84
	Σ	1,300,942.30	7,326.86	642,841.09	49,471.61	−21,758,959.92
3	1	195,878.78	7,102.64	255,016.95	30,223.49	−2,160,377.96
	2	181,524.08	5,197.14	0.00	0.00	−3,248,107.96
	3	282,963.26	2,473.43	149,532.34	0.00	−4,385,451.48
	4	356,317.34	2,970.94	207,771.43	0.00	−5,426,028.47
	5	462,172.90	640.30	0.00	0.00	−6,450,279.37
	Σ	1,478,856.36	18,384.46	612,320.71	30,223.49	−21,670,245.23
4	1	131,398.36	3,234.10	0.00	0.00	−2,189,556.93
	2	77,948.69	2,580.25	560,777.79	356,487.74	−3,271,890.82
	3	213,627.07	2,851.20	0.00	0.00	−4,379,903.26
	4	328,564.02	527.37	0.00	0.00	−5,454,538.59
	5	185,321.49	1,562.78	0.00	309,517.80	−6,440,700.86
	Σ	936,859.63	10,755.70	560,777.79	666,005.55	−21,736,590.46

being shipped, and thus, inventory cost is low, while transshipment cost is excessive. Therefore, it is reasonable to evaluate the comparative effects of all solutions in certain aspects, if the total results are inconclusive. To emphasize the importance of these aspects, the cost function coefficients are adjusted accordingly.

Table 6.6 Cost function values for service policy Earliest-Deadline-First (EDF) corresponding to the systems specified in Table 6.2. Out-of-stock cost monitor service quality, while order and transshipment cost highlight logistics efficiency

	i	**Inventory cost** in € p.a.	**Out-of-stock cost** in € p.a.	**Periodic order cost** in € p.a.	**Transshipment cost** in € p.a.	**Gain** in € p.a.
1	1	365,747.27	1,146.62	0.00	27,696.44	−2,203,858.57
	2	244,008.31	2,268.36	425,373.39	37,028.38	−3,274,103.38
	3	165,081.55	1,631.23	0.00	0.00	−4,393,509.79
	4	366,053.23	2,325.96	208,040.62	0.00	−5,433,807.75
	5	245,453.35	1,407.04	0.00	0.00	−6,440,723.73
	Σ	1,386,343.71	8,779.21	633,414.00	64,724.82	−21,746,003.22
2	1	31,999.32	175.38	0.00	0.00	−2,210,516.80
	2	728,726.45	1,614.28	401,393.38	43,385.79	−3,279,887.25
	3	56,645.58	284.35	0.00	0.00	−4,407,542.41
	4	420,698.60	2,864.40	241,972.40	3,284.45	−5,425,756.66
	5	62,273.66	1,816.12	0.00	0.00	−6,436,016.10
	Σ	1,300,343.62	6,754.53	643,365.78	46,670.24	−21,759,719.22
3	1	154,912.34	419.23	62,298.78	0.00	−2,208,866.14
	2	203,470.87	765.83	0.00	0.00	−3,287,028.57
	3	332,156.91	2,131.02	440,773.80	20,136.40	−4,387,147.81
	4	354,632.10	3,077.18	207,722.22	0.00	−5,424,566.47
	5	391,085.60	2,037.10	0.00	0.00	−6,433,358.52
	Σ	1,436,257.83	8,430.37	710,794.80	20,136.40	−21,740,967.51
4	1	142,997.01	314.92	0.00	0.00	−2,209,536.26
	2	209,395.76	5,015.98	450,905.39	35,723.49	−3,249,231.08
	3	257,170.43	5,751.66	148,471.55	0.00	−4,350,342.94
	4	336,575.82	1,066.56	0.00	13,377.00	−5,448,040.82
	5	478,784.82	2,159.82	0.00	13,033.75	−6,430,837.13
	Σ	1,424,923.84	14,308.94	599,376.94	62,134.24	−21,687,988.22

6.7 Conclusion and Future Work

The Simulation Optimization approach is applicable to very general multi-location inventory systems. The concept presented in this chapter iteratively combines a simulator with Particle Swarm Optimization. This concept allows the investigation of complex models with few assumptions and is theoretically not limited to a location count contrary to analytical approaches. Due to the complexity of the model, it is difficult to understand the effect of certain policies. Therefore, valuable insights regarding the dynamics of the system are obtained through simulation in addition to the optimal parameter set. However, applying global optimization to complex models still involves a certain risk to end in a local optimum. That risk is confined by extending the simulation time and the optimization cycle count. The optimum depends on the model specification and shows a specific structure. The development of such a structure is one of the most intriguing aspects, and the question arises, what conditions have a promoting effect.

As aforementioned an advantage of the Simulation Optimization of multi-location inventory systems with lateral transshipments is that the model itself is straightforward extendable. Functional extensions are, e.g., policies for periodic orders, transshipment orders and product offers. Extending the parameter set itself, the capacity of the locations can be optimized by introducing estate and energy cost for unused storage. Thus, not only the flows of transshipments are optimized, but also the allocation of capacities. In addition to static aspects of the model, the parameter set may be extended by dynamic properties such as the location-specific order period time. Besides these extensions there is an idea regarding orders from more than one location at a time. Under specific circumstances one location evolves as a supplier, ordering and redistributing product units. Therefore, the basic idea is to release an order by several locations and to solve the Traveling Salesman Problem with minimal cost. However, the existing heuristics already seem to approximate such a transportation logic well, and thus, the inclusion of more elaborate policies is expected just to increase complexity. Further research may also concentrate on characteristics favoring demand forecast and promoting certain flows through a location network leading to a structure.

Acknowledgment. The authors would like to thank the Robert Bosch doctoral program, the German Academic Exchange Service and the Foundation of the German Business for funding their research.

References

[1] Arnold, J., Kochel, P., Uhlig, H.: With Parallel Evolution towards the Optimal Order Policy of a Multi-Location Inventory with Lateral Transshipments. In: Papachristos, S., Ganas, I. (eds.) Research Papers of the 3rd ISIR Summer School, pp. 1–14 (1997)

[2] Belgasmi, N., Saïd, L.B., Ghédira, K.: Evolutionary Multiobjective Optimization of the ulti-Location Transshipment Problem. Operational Research 8(2), 167–183 (2008)

[3] Chiou, C.-C.: Transshipment Problems in Supply Chain Systems: Review and Extensions. Supply Chain, Theory and Applications, 558–579 (2008)

[4] Clerc, M.: Standard PSO (2007), `http://www.particleswarm.info/Programs.html` Online: accessed July 31, 2010

[5] Clerc, M., Kennedy, J.: The Particle Swarm – Explosion, Stability, and Convergence in a Multidimensional Complex Space. IEEE Transactions on Evolutionary Computation 6(1), 58–73 (2002)

[6] Dye, C.-Y., Hsieh, T.-P.: A Particle Swarm Optimization for Solving Joint Pricing and Lot-Sizing Problem with Fluctuating Demand and Unit Purchasing Cost. Computers & Mathematics with Applications 60, 1895–1907 (2010)

[7] Evers, P.T.: Heuristics for Assessing Emergency Transshipments. European Journal of Operational Research 129, 311–316 (2001)

[8] Fu, M.C., Healy, K.J.: Techniques for Optimization via Simulation: An Experimental Study on an (s;S) Inventory System. IIE Transactions 29, 191–199 (1997)
[9] Fu, M.C., Glover, F.W., April, J.: Simulation Optimization: A Review, New Developments and Applications. In: Kuhl, M.E., Steiger, N.M., Armstrong, F.P., Joines, J.A. (eds.) Proceedings of the 2005 Winter Simulation Conference, pp. 83–95 (2005)
[10] Gong, Y., Yücesan, E.: Stochastic Optimization for Transshipment Problems with Positive Replenishment Lead Times. International Journal of Production Economics (2010) (in Press, Corrected Proof)
[11] Guariso, G., Hitz, M., Werthner, H.: An Integrated Simulation and Optimization Modelling Environment for Decision Support. Decision Support Systems 16(2), 103–117 (1996)
[12] Herer, Y.T., Tzur, M., Yücesan, E.: The Multilocation Transshipment Problem. IIE Transactions 38, 185–200 (2006)
[13] Hochmuth, C.A.: Design and Implementation of a Software Tool for Simulation Optimization of Multi-Location Inventory Systems with Transshipments. Master's thesis, Chemnitz University of Technology, In German (2008)
[14] Hochmuth, C.A., Lássig, J., Thiem, S.: Simulation-Based Evolutionary Optimization of Complex Multi-Location Inventory Models. In: 3rd IEEE International Conference on Computer Science and Information Technology (ICCSIT), vol. 5, pp. 703–708 (2010)
[15] Iassinovski, S., Artiba, A., Bachelet, V., Riane, F.: Integration of Simulation and Optimization for Solving Complex Decision Making Problems. International Journal of Production Economics 85(1), 3–10 (2003)
[16] Kämpf, M., Köchel, P.: Simulation-Based Sequencing and Lot Size Optimisation for a Production-and-Inventory System with Multiple Items. International Journal of Production Economics 104, 191–200 (2006)
[17] Kennedy, J., Eberhart, R.: Particle Swarm Optimization. In: Proceedings of IEEE International Conference on Neural Networks, vol. 4, pp. 1942–1948 (1995)
[18] Köchel, P.: About the Optimal Inventory Control in a System of Locations: An Approximate Solution. Mathematische Operationsforschung und Statistik, Serie Optimisation 8, 105–118 (1977)
[19] Köchel, P.: A Survey on Multi-Location Inventory Models with Lateral Transshipments. In: Papachristos, S., Ganas, I. (eds.) Inventory Modelling in Production and Supply Chains, Research Papers of the 3rd ISIR Summer School, Ioannina, Greece, pp. 183–207 (1998)
[20] Köchel, P.: Simulation Optimisation: Approaches, Examples, and Experiences. Technical Report CSR-09-03, Department of Computer Science, Chemnitz University of Technology (2009)
[21] Kochel, P., Arnold, J.: Evolutionary Algorithms for the Optimization of Multi-Location Systems with Transport. In: Simulationstechnik, Proceedings of the 10th Symposium in Dresden, pp. 461–464. Vieweg (1996)
[22] Köchel, P., Nieländer, U.: Simulation-Based Optimisation of Multi-Echelon Inventory Systems. International Journal of Production Economics 93-94, 505–513 (2005)
[23] Köchel, P., Thiem, S.: Search for Good Policies in a Single-Warehouse, Multi-Retailer System by Particle Swarm Optimisation. International Journal of Production Economics (2010) (in press, corrected proof)
[24] Kukreja, A., Schmidt, C.P.: A Model for Lumpy Parts in a Multi-Location Inventory System with Transshipments. Computers & Operations Research 32, 2059–2075 (2005)
[25] Kukreja, A., Schmidt, C.P., Miller, D.M.: Stocking Decisions for Low- Usage Items in a Multilocation Inventory System. Management Science 47, 1371–1383 (2001)

[26] Li, J., González, M., Zhu, Y.: A Hybrid Simulation Optimization Method for Production Planning of Dedicated Remanufacturing. International Journal of Production Economics 117(2), 286–301 (2009)
[27] Minner, S., Silver, E.A., Robb, D.J.: An Improved Heuristic for Deciding on Emergency Transshipments. European Journal of Operational Research 148, 384–400 (2003)
[28] Özdemir, D., Yücesan, E., Herer, Y.T.: Multi-Location Transshipment Problem with Capacitated Transportation. European Journal of Operational Research 175(1), 602–621 (2006)
[29] Parsopoulos, K.E., Skouri, K., Vrahatis, M.N.: Particle swarm optimization for tackling continuous review inventory models. In: Giacobini, M., Brabazon, A., Cagnoni, S., Di Caro, G.A., Drechsler, R., Ekárt, A., Esparcia-Alcázar, A.I., Farooq, M., Fink, A., McCormack, J., O'Neill, M., Romero, J., Rothlauf, F., Squillero, G., Uyar, A.Ş., Yang, S. (eds.) EvoWorkshops 2008. LNCS, vol. 4974, pp. 103–112. Springer, Heidelberg (2008)
[30] Robinson, L.W.: Optimal and Approximate Policies in Multi-Period Multi- Location Inventory Models with Transshipments. Operations Research 38, 278–295 (1990)
[31] Ruppeiner, G., Pedersen, J.M., Salamon, P.: Ensemble Approach to Simulated annealing. Jounal de Physique I 1(4), 455–470 (1991)
[32] Willis, K.O., Jones, D.F.: Multi-Objective Simulation Optimization through Search Heuristics and Relational Database Analysis. Decision Support Systems 46(1), 277–286 (2008)
[33] Xu, K., Evers, P.T., Fu, M.C.: Estimating Customer Service in a Two- Location Continuous Review Inventory Model with Emergency Transshipments. European Journal of Operational Research 145, 569–584 (2003)
[34] Zhan, Z.-H., Feng, X.-L., Gong, Y.-J., Zhang, J.: Solving the Flight Frequency Programming Problem with Particle Swarm Optimization. In: Proceedings of the 11th Congress on Evolutionary Computation, CEC 2009, pp. 1383–1390. IEEE Press, Los Alamitos (2009)

Chapter 7
Traditional and Hybrid Derivative-Free Optimization Approaches for Black Box Functions

Genetha Anne Gray and Kathleen R. Fowler

Abstract. Picking a suitable optimization solver for any optimization problem is quite challenging and has been the subject of many studies and much debate. This is due in part to each solver having its own inherent strengths and weaknesses. For example, one approach may be global but have slow local convergence properties, while another may have fast local convergence but is unable to globally search the entire feasible region. In order to take advantage of the benefits of more than one solver and to overcome any shortcomings, two or more methods may be combined, forming a hybrid. Hybrid optimization is a popular approach in the combinatorial optimization community, where metaheuristics (such as genetic algorithms, tabu search, ant colony, variable neighborhood search, etc.) are combined to improve robustness and blend the distinct strengths of different approaches. More recently, metaheuristics have been combined with deterministic methods to form hybrids that simultaneously perform global and local searches. In this Chapter, we will examine the hybridization of derivative-free methods to address black box, simulation-based optimization problems. In these applications, the optimization is guided solely by function values (*i.e.* not by derivative information), and the function values require the output of a computational model. Specifically, we will focus on improving derivative-free sampling methods through hybridization. We will review derivative-free optimization methods, discuss possible hybrids, describe intelligent hybrid approaches that properly utilize both methods, and give an examples of the successful application of hybrid optimization to a problem from the hydrological sciences.

Genetha Anne Gray
Department of Quantitative Modeling & Analysis,
Sandia National Laboratories,
P.O. Box 969, MS 9159, Livermore, CA 94551-0969 USA
e-mail: `gagray@sandia.gov`

Kathleen R. Fowler
Department of Mathematics & Computer Science,
Clarkson University, P.O. Box 5815, Potsdam, NY, 13699-5815 USA
e-mail: `kfowler@clarkson.edu`

S. Koziel & X.-S. Yang (Eds.): Comput. Optimization, Methods and Algorithms, SCI 356, pp. 125–151.
springerlink.com

7.1 Introduction and Motivation

Computer simulation is an important tool that is often used to reduce the costs associated with the study of complex systems in science and engineering. In recent years, simulation has been paired with optimization in order to design and control such systems. The resulting *simulation-based* problems have objective functions and/or constraints which rely on the output from sophisticated simulation programs. Often, a simulator is referred to as a *black box* since it is defined solely by its input and output and not by the actual program being executed. In other words, the underlying structure of simulation is unknown. In these applications, the problem characteristics are mathematically challenging in the optimization landscapes may be disconnected, nonconvex, nonsmooth, or contain undesirable, multiple local minima. Gradient-based optimization methods are well known to perform poorly on problems with these characteristics as derivatives are often unavailable and approximations may be insufficient [14]. Moreover, it has been shown that derivative approximations of functions that incorporate noisy data may contain too much error to be useful [56]. Instead, *derivative-free optimization (DFO)* methods, which advance using only function values, are applied. A variety of DFO methods have emerged and matured over the years to address simulation-based problems, and many are supported theoretically with convergence criteria established. In this Chapter, we will review some such methods and demonstrate their utility on a water management application proposed specifically in the literature as a simulation-based optimization benchmarking problem.

To obtain a solution to a simulation-based problem, one seeks an optimization algorithm that is (*i*) reliable in the sense that similar solutions can be obtained using different initial points or optimization parameters, (*ii*) accurate in that a reasonable approximation to the global minimum is obtained, and (*iii*) efficient with respect to finding a solution using as few function calls as possible. The role of efficiency is particularly important in simulation-based applications because the optimization is guided solely by function values defined in terms of output from a black box. In practice, the computational time required to complete these simulations can range from a few seconds to a few days depending on the application and problem size. Thus, parallel implementations of both the simulator and the optimization methods are often essential for computational tractability of black-box problems.

Picking a suitable optimization algorithm that meets these criteria is quite challenging and has been the subject of many studies and much debate. This is because every optimization technique has inherent strengths and weaknesses. Moreover, some optimization algorithms contain characteristics which make them better suited to solve particular kinds of problems. Hybridization, or the combining of two or more complementary, but distinct methods, allows the user to take advantage of the beneficial elements of multiple methods. For example, consider two methods A and B where method A is capable of handling noise and undefined points and method B excels in smooth regions with small amounts of noise. In this case, method A may be unacceptably slow to find a solution while method B may fail in noisy or discontinuous regions of the domain. By forming a hybrid, method A can help overcome difficult regions of the domain and method B can be applied for fast convergence and efficiency. This Chapter also explores the promise of hybrid approaches and demonstrates some results for the water management problem.

Throughout this Chapter, the problem of interest is

$$\min_{x \in \Omega} f(x), \tag{7.1}$$

where the objective function is $f : \mathbb{R}^n \to \mathbb{R}$ and Ω defines the feasible search space. In practice, Ω may be comprised of component-wise bound constraints on the decision variable x in combination with linear and nonlinear equality or inequality constraints. Often, Ω may be further defined in terms of state variables determined by simulation output. The example in this Chapter includes such constraints. In addition, integer and categorical variables (for example those which require a 'yes' or 'no') are present in many engineering applications. There are a variety of DFO methods equipped to handle these classes of problems and several are discussed later. For the application in this work, we consider both real-valued and mixed integer problem formulations.

The rest of this Chapter is outlined as follows: In Section 2, an example of a black box optimization problem from hydrology is introduced. Then, in Section 3, some DFO approaches are introduced including a genetic algorithm (GA), DIRECT, asynchronous parallel pattern search (APPS) and implicit filtering. In addition, these methods are demonstrated on the example introduced in Section 2. Section 4 describes some hybrid methods created using the classical DFO methods from Section 3, and describes their performance on the example problem. Finally, Section 5 summarizes all the information given in this Chapter and gives some ideas regarding future research directions for hybrid optimization.

7.2 A Motivating Example

To demonstrate the strengths and weaknesses of some DFO methods and to better illustrate the utility of hybrids, this Chapter will focus on the results from a water supply problem, notated WS in the remainder of this Chapter, which was described in [72, 71]. Problem WS has been used as a benchmarking problem for optimization methods [51, 32, 29, 50, 43], and was shown to be highly dependent on the formulation of the feasible region Ω [50, 29]. Furthermore, the use if WS in a comparison study of DFO methods in [29] showed that (1) there are multiple, undesirable local minima that can trap local search methods and (2) the constraints on the state variables are highly sensitive to changes in the decision variables.

The goal in the WS problem is to extract a quantity of water from a particular geographic region, an aquifer, while minimizing the capital cost f^c to install a well and the operational cost f^o to operate a well. Thus, the optimization problem is to minimize the total cost of the well-field $f^T = f^c + f^o$ subject to bound constraints on the decision variables, the amount of water extracted, and the physical properties of the aquifer. The decision variables for this problem are the pumping rates $\{Q_k\}_{k=1}^n$, the well locations $\{(\hat{x}_k,\hat{y}_k)\}_{k=1}^{n}$, and the number of wells n in the final design. A negative pumping rate, $Q_k < 0$ for some k, means that a well is extracting and a positive pumping rate, $Q_k > 0$ for some k, means that a well is injecting. The objective function and constraints of WS rely on the solution to a nonlinear partial differential equation to obtain values of the hydraulic head, h, which determines the direction of flow. Thus, h would be considered a state variable. In this example, for each well $k = 1 \dots n$ in a candidate set of wells, the hydraulic head h_k must be obtained via simulation.

The objective function, based on the one proposed in [72, 71] is given by

$$f^T = \underbrace{\sum_{k=1}^{n} D_k + \sum_{k,Q_k<0.0} c_1|1.5Q_k|^{b_1}(z_{gs} - h^{min})^{b_2}}_{f^c} \tag{7.2}$$

$$+ \underbrace{\int_0^{t_f} \left(\sum_{k,Q_k<0.0} c_2Q_k(h_k - z_{gs}) + \sum_{k,Q_k>0.0} c_3Q_k \right) dt,}_{f^o}$$

where c_j and b_j are cost coefficients and exponents given in [71]. In the first f^c term, D_k is the cost to drill and install well k. The second term of f^c includes the cost to install a pump for each extraction well, and this cost is based on the extraction rate and $h^{min} = 10$ m, the minimum allowable hydraulic head and the ground surface elevation z_{gs}. The calculation of f^o is for $t_f = 5$ years. The first part of the integral includes the cost to lift the water to the surface which depends on the hydraulic head h_k in well k. The second part accounts for any injection wells, which are assumed to operate under gravity. Details pertaining to the aquifer and groundwater flow model are fully described in [72] and are not included here as they fall outside of the scope of the application of optimization methods to solve the WS problem.

Note that although the well locations $\{(\hat{x}_k,\hat{y}_k)\}_{k=1}^{n}$ do not explicitly appear in Equation (7.2), they enter through the state variable h as output from a simulation tool. For this work, the U.S. Geological Survey code MODFLOW [92] was used to calculate the head values. MODFLOW is a widely used and well supported block-centered finite difference code that simulates saturated groundwater flow. Since the well locations must lie on the finite difference grid, real-valued locations must be rounded to grid locations for the simulation. This results in small steps and low amplitude noise in the optimization landscapes.

The constraints for the WS application are given as limitations on the pumping rates,

$$-0.0064 \text{ m}^3/\text{s} \le Q_k \le 0.0064 \text{ m}^3/\text{s}, k = 1,...,n, \tag{7.3}$$

and impact on the aquifer in terms of the hydraulic head,

$$10 \text{ m} \le h_k \le 30 \text{ m}, k = 1,...,n. \tag{7.4}$$

The constraints given in Equations (7.3) and (7.4) are enforced at each well. The total amount of water to supply is defined by the constraint

$$\sum_{k=1}^{n} Q_k \leq -0.032 \text{ m}^3/\text{s}. \tag{7.5}$$

While the pumping rates and locations are real-valued, there are options for how to define the variable which indicates the appropriate number of wells. One approach is to start with a large number of candidate wells, N_w, and run multiple optimization scenarios where at the end of each one, wells with sufficiently low pumping rates are removed before the optimization routine continues. However, for realistic water management problems, simulations are time-consuming so it is more attractive to determine the number of wells as the optimization progresses. One way to do this is to include integer variables $\{z_i\}_{i=1}^n$ where each $z_i \in \{0,1\}$ is a binary indicator for assigning a well as off or on. Since this formulation requires an optimization algorithm that can handle integer variables, alternatives have been developed. In [50], three formulations that implicitly determine the number of wells while avoiding the inclusion of integer variables are compared. Two formulations are based on a multiplicative penalty formulation ([69]) and one is based on removing a well during the course of the optimization if the rate becomes sufficiently low. This third technique is implemented here using an inactive-well threshold given by

$$|Q_k| < 10^{-6} \text{ m}^3/\text{s}, k = 1, \ldots, n. \tag{7.6}$$

Note that the cost to install a well is roughly \$20,000, and the operational cost is about \$1,000 per year. Thus, using as few wells as possible drives the optimization regardless of the formulation. However, the inclusion of Equation (7.6) in the formulation results in a narrow region of decrease for an optimization method to find, but a large decrease in cost. Mathematically, using Equation (7.6) allows for real-valued DFO methods, but adds additional discontinuities in the minimization landscapes.

The implementation of the WS problem considered in this study was taken from `http://www4.ncsu.edu/~ctk/community.html` where the entire package of simulation data files and objective function/constraint subroutines are available for download. The final design solution is known to be five wells all operating at $Q_i = -0.0064$ m^3/s with locations aligned with the north and east boundaries, as shown in Table 1. See [32] for details. To study the DFO methods described in this Chapter, a starting point with six candidate wells was used. In order to find the solution, the optimization methods must determine that one well must be shut off while simultaneously optimizing the rates and locations of the remaining wells. Furthermore, the rates must lie on the boundary of the constraint in Equation(7.3) in order to satisfy the constraint given in Equation (7.5). Thus, the WS problem contains challenging features for simulation-based optimization problems that are not unique to environmental engineering but that can be seen across many scientific and engineering disciplines. To summarize, the challenges of the WS problem include a black box objective function and constraints, linear and nonlinear constraints on the

Table 7.1 Five well solution to WS with pumping rates $Q_i = -0.0064\ \mathrm{m}^3/\mathrm{s}, i = 1,\ldots,5$

Well Number	1	2	3	4	5
$\hat{x}$ [m]	350	788	722	170	800
$\hat{y}$ [m]	724	797	579	800	152

decision and state variables, multiple problem formulations, low amplitude noise, a discontinuous and disconnected feasible region, and multiple local minima.

7.3 Some Traditional Derivative-Free Optimization Methods

In this Section, we highlight some DFO approaches to solving the simulation-based WS problem including two global methods (the genetic algorithm (GA) and DIviding RECTangles (DIRECT)), two local methods (asynchronous parallel pattern search (APPS) and implicit filtering), and a statistical alternative which utilizes a process Gaussian model. Global optimization methods seek the extreme value of a given function in a specified feasible region. A global solution is optimal among all possible solutions. In contrast, local methods identify points which are only optimal within a neighborhood of that point. This Section is in no way an exhaustive list of derivative-free methods, but instead are included to give an overview of the importance of selecting a method appropriate for the application. The derivative free optimization community remains active in algorithm development. Some examples of ongoing development include: Design Explore [7] from the Boeing Company, which incorporates surrogates in the search phase; NOMAD (Nonlinear Optimization for Fixed Variables) [4, 2, 3], specifically designed to solve simulation-based problems, and ORBIT (Optimization by Radial Basis Function Interpolation in Trust Region) [77, 88] which makes use of radial basis functions. We encourage interested readers to refer to the citations for more information and to investigate books such as [14] which give a complete overview of the topic.

7.3.1 *Genetic Algorithms (GAs)*

Genetic algorithms [36, 53, 54] are one of the most widely-used DFO methods and are part of a larger class of evolutionary algorithms called population-based, global search, heuristic methods [36]. Population based GAs are based on biological processes such as survival of the fittest, natural selection, inheritance, mutation, or reproduction. Heuristic methods, such as the GA, are experience based. They contrast to deterministic methods which more systematically search the domain space.

The GA codes design points as "individuals" or "chromosomes", typically as binary strings, in a population. It is this binary representation that makes GAs attractive for integer problems (such as WS) since the on-off representation is immediate. Through the above biological processes, the population evolves through a user specified number of generations towards a smaller fitness value. A simple GA can be defined as follows:

1. Generate a random/seeded initial population of size n_p
2. Evaluate the fitness of individuals in initial population
3. Iterate through the specified number of generations:
   a. Rank fitness of individuals
   b. Perform selection
   c. Perform crossover and mutation
   d. Evaluate fitness of newly-generated individuals
   e. Replace non-elite members of population with new individuals

During the selection phase, better fit individuals are arranged randomly to form a mating pool on which further operations are performed. Crossover attempts to exchange information between two design points to produce a new point that preserves the best features of both 'parent points,' and this is illustrated for a binary string in Figure 7.1. Mutation is used to prevent the algorithm from terminating prematurely to a suboptimal point and is used as a means to explore the design space, and it is illustration for a binary string in Figure 7.2. (Note that both Figure 7.1 and 7.2 were taken from [49].) Termination of the algorithm is based on a prescribed number of generations or when the highest ranked individual's fitness has reached a plateau.

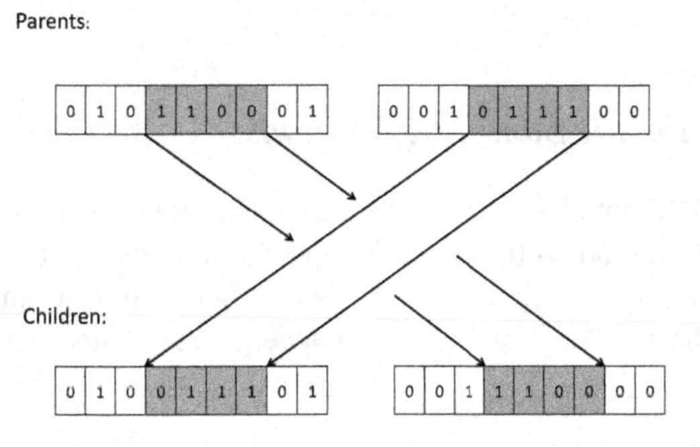

Fig. 7.1 The crossover process for a binary string ([49])

Often, GAs are criticized for their computational complexity and dependence on optimization parameter settings, which are not known a priori [22, 42, 68]. Also, since the GA incorporates a randomness to the search phase, multiple optimization runs may be needed. However, if the user is willing to exhaust a large number of function evaluations, a GA can help provide insight into the design space and locate initial points for fast, local, single search methods. The GA has many alternate forms and has been applied to a wide range of engineering design problems as shown in references such as [60]. Moreover, hybrid GAs have been developed at all levels of the algorithm and with a variety of other global and local search DFO methods. See for example [6, 83, 76] and the references therein.

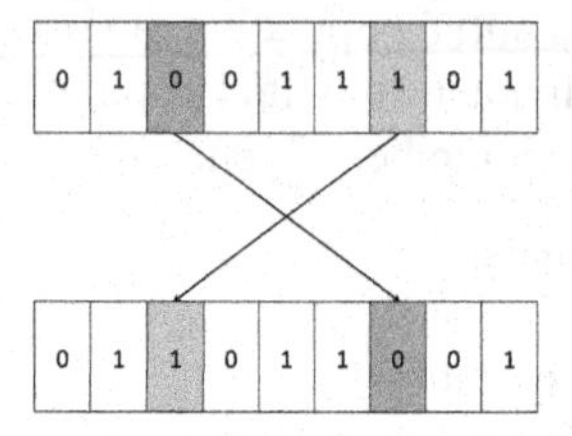

Fig. 7.2 The mutation process for a binary string ([49])

In [32, 29, 50], the NSGA-II implementation [21, 93, 19, 20] of the GA was used on the WS problem for both the mixed-integer formulation and the inactive well-threshold to determine the wells. It was shown that for this problem the GA performed better if (1) the number of wells was determined directly by including the binary on-off switch compared to using the inactive well threshold and (2) if the initial population was seeded with points that had at least five wells operating at -0.0064 m^3/s. If a random initial population was used, the algorithm could not identify the solution after 4,000 function evaluations. If the GA was seeded accordingly, a solution was found within 161 function calls but the function evaluation budget would be exhausted before the algorithm would terminate, which for that work was set to 900.

7.3.2 Deterministic Sampling Methods

Another class of DFO methods is deterministic sampling methods [89, 67, 63, 75]. In general, these methods rely upon a direct search of the decision space and are guided by a pattern or search algorithm. They differ from GAs in that there is no randomness in the method, and rigorous convergence results exist. (See [63] and references therein.)

7.3.2.1 Asynchronous Parallel Pattern Search (APPS)

Asynchronous Parallel Pattern Search (APPS) [55, 62] is a direct search methods which uses a predetermined pattern of points to sample a given function domain. APPS is an example of a generating set search (GSS), a class of algorithms for bound and linearly constrained optimization that obtain conforming search directions from generators of local tangent cones [65, 64]. In the case that only bound constraints are present, GSS is identical to a pattern search. The majority of the computational cost of pattern search methods is the function evaluations, so parallel pattern search (PPS) techniques have been developed to reduce the overall computation time. Specifically, PPS exploits the fact that once the points in the search pattern have been defined, the function values at these points can be computed simultaneously [23, 84]. For example, for a simple two-dimensional function, consider the illustrations in Figure 7.3. First, the points f, g, h, and i in the stencil around point c are evaluated. Then, since f results in the smallest function value, the second

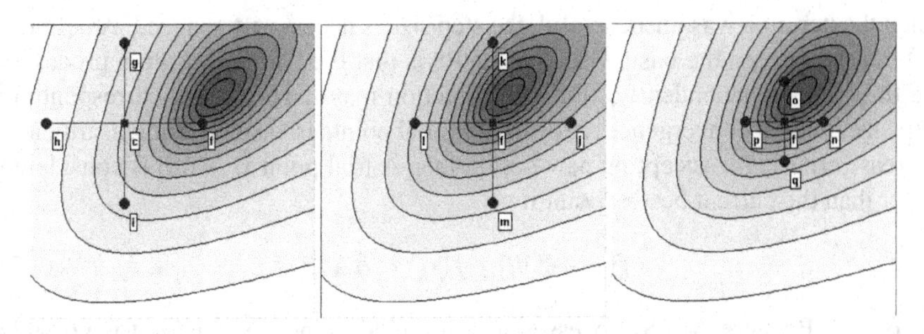

Fig. 7.3 Illustration of the steps of Parallel Pattern Search (PPS) for a simple two-dimensional function. On the left, an initial PPS stencil around starting point c is shown. In the middle, a new stencil is created after successfully finding a new local min (f). On the left, PPS shrinks the stencil after failing to find a new minimum

picture shows a new stencil around point f. Finally, in third picture, since none of the iterates in this new stencil result in a new local minima, the step size of the stencil is reduced.

The APPS algorithm is a modification of PPS that eliminates the synchronization requirements that the function values of all the points in the current search pattern must be completed before the algorithm can progress. It retains the positive features of PPS, but reduces processor latency and requires less total time than PPS to return results [55]. Implementations of APPS have minimal requirements on the number of processors (*i. e.* 2 instead of $n+1$ for PPS) and do not assume that the amount of time required for an objective function evaluation is constant or that the processors are homogeneous.

The implementation of the APPS algorithm is more complicated than a basic GSS in that it requires careful bookkeeping. However, the details are irrelevant to the overall understanding of the method. Instead we present a basic GSS algorithm and direct interested readers to [40] for a detailed description and analysis of the APPS algorithm and corresponding APPSPACK software. The basic GSS algorithm is:

Let x_0 be the starting point, Δ_0 be the initial step size, and $\mathscr{D}$ be the set of positive spanning directions.
While *not converged* Do

1. Generate trial points $Q_k = \{x_i + \tilde{\Delta}_k d_i \mid 1 \le i \le |\mathscr{D}|\}$ where $\tilde{\Delta}_k \in [0, \Delta_k]$ denotes the maximum feasible step along d_i.
2. Evaluate trial points (possibly in parallel)
3. If $\exists\, x_q \in Q_k$ such that $f(x_q) - f(x_k) < \alpha \Delta_k^2$
   Then $x_{k+1} = x_q$ (successful iteration)
   Else $x_{k+1} = x_k$ (unsuccessful iteration) and $\Delta_{k+1} = \Delta_k/2$ (step size reduction)

Note that in a basic GSS, after a successful iteration (one in which a new best point has been found), the step size is either left unchanged or increased. In contrast, when the iteration was unsuccessful, the step size is necessarily reduced. A defining difference between the basic GSS and APPS is that the APPS algorithm processes the directions independently, and each direction may have its own corresponding step size. Global convergence to locally optimal points is ensured using a sufficient decrease criteria for accepting new best points. A trial point $x_k + \Delta d_i$ is considered better than the current best x_k point if

$$f(x_k + \Delta d_i) - f(x_k) < \alpha \Delta^2, \tag{7.7}$$

for $\alpha > 0$. Because APPS processes search direction independently, it is possible that the current best point is updated to a new better point before all the function evaluations associated with a set of trial points Q_k have been completed. These results are referred to as *orphaned points* as they are no longer tied to the current search pattern and attention must be paid to ensure that the sufficient decrease criteria is applied appropriately. The support of these orphan points is a feature of the APPS algorithm which makes it naturally amenable to a hybrid optimization structure. Iterates generated by alternative algorithms can be simply be treated as orphans without the loss of favorable theoretical properties or local convergence theory of APPS. It is important to note that this paradigm is in fact extensible to many other optimization routines and makes the APPS algorithm particularly amenable to hybridization in that it can readily accommodate externally generated points.

The APPS algorithm has been implemented in an open source software package called APPSPACK. It is written in C++ and uses MPI [47, 48] for parallelism. APPSPACK performs function evaluations through system calls to an external executable which can written in any computer language. This simplifies its execution and makes it amenable to customization. Moreover, it should be noted that the most recent version of APPSPACK can handle linear constraints [64, 46], and a software called HOPSPACK builds on the APPSPACK software and includes a GSS solver that can handle nonlinear constraints [45, 74].

In [29], APPS was applied to WS problem using the constraint in Equation (7.6). Like the GA, APPS was sensitive to the initialization of the optimization and required a starting point as described above otherwise the algorithm would converge to a suboptimal six well design. However, given good initial data the algorithm converged to a comparable solution within 200 function evaluations. APPS has also been successfully applied to problems in microfluidics, biology, thermal design, and forging. (See [40] and references therein and the URL `https://software.sandia.gov/appspack/` for user success stories and detailed examples.)

7.3.2.2 Implicit Filtering

Implicit filtering is based on the notion that if derivatives were available and reliable, Newton-like methods would yield fast results. The method evaluates a stencil of points at each iteration used simultaneously to form finite difference gradients and as a pattern for direct search [35]. Then, a secant approach, called a quasi-Newton method, is used to solve the resulting system of nonlinear equations at each iteration [61] to avoid finite difference approximations of the Hessian matrix. In contrast to classical finite-difference based Newton algorithms, implicit filtering begins with a much larger stencil to account for noise. This step-size is reduced as the optimization progresses to improve the accuracy of the gradient approximation and take advantage of the fast convergence of quasi-Newton methods near a local minimum.

To solve the WS problem, a FORTRAN implementation called IFFCO (Implicit Filtering For Constrained Optimization) was used. IFFCO depends on a symmetric rank one quasi-Newton update [11]. The user must supply an objective function and initial iterate and then optimization is terminated based on a function evaluation budget or by exhausting the number of times the finite difference stencil is reduced. We denote the finite difference gradient with increment size p by $\nabla_p f$ and the model Hessian matrix as H. For each p, the projected quasi-Newton iteration proceeds until the center of the finite difference stencil yields the smallest function value or $||\nabla_p f|| \leq \tau p$, which means the gradient has been reduced as much as possible on the current scale. After this, the difference increment is halved (unless the user has specified a particular sequence of increments) and the optimization proceeds until the function evaluation budget is met.

The general unconstrained algorithm can be outlined as follows:

While *not converged*

Do *until* $\nabla_p f < \tau p$
1. Compute $\nabla_p f$
2. Find the least integer λ such that sufficient decrease holds
3. $x = x - \lambda H^{-1} \nabla_p f(x)$
4. Update H via a quasi-Newton method

Reduce p

This can be illustrated on an a small perturbation of a quadratic function as illustrated in Figure 7.4. Given an initial iterate $x_0 = -1.25$ and $p = 0.25$, the resulting centered finite difference stencil is shown on the left. The center of the stencil, $f(x_0)$ is denoted with an "*" and $f(x_0 \pm p)$ is denoted with "o". Since the center of the stencil has the lowest function value, the algorithm would proceed and take a decent step. Then, suppose the next iterate is as in the center picture of Figure 7.4. Then, the lowest function value occurs on the stencil at $f(x_1)$ and thus stencil failure has occurred. In this case, p, is reduced by half. Then, the stencil would be as in the right picture, and stencil failure would not occur, so the algorithm would proceed.

IFFCO was used to solve the WS problem in [32, 29] using the inactive-well threshold and also in [50] a multiplicative penalty term to determine the number of wells. The behavior for both implementations was similar to that of APPS in that a good initial iterate was needed to identify the five well solution. With good initial data, IFFCO identified the solution within 200 function evaluations.

IFFCO and the implicit filtering algorithm in general have been successfully applied to a variety of other challenging simulation-based optimization applications including mechanical engineering [12], polymer processing [31], and physiological modeling [30]. There are several convergence theorems for implicit filtering [13],

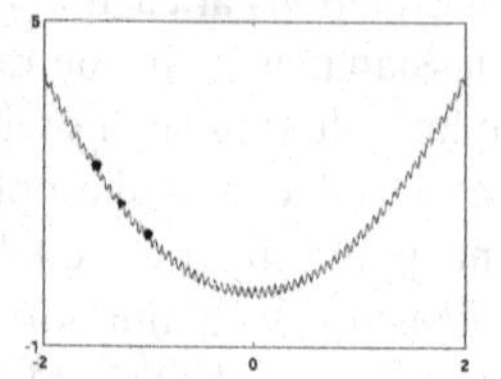

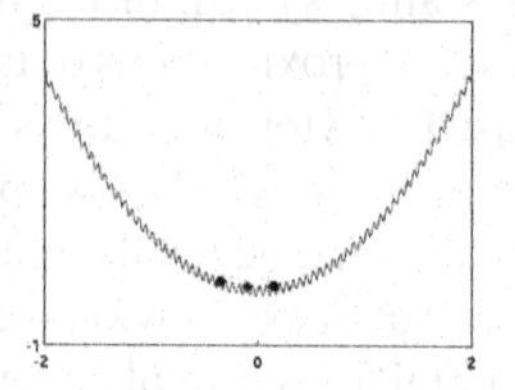

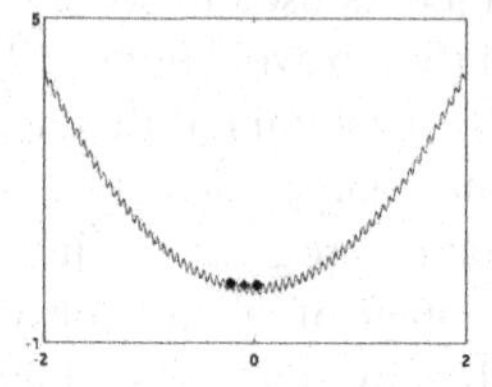

Fig. 7.4 The illustration on the right shows the first implicit filtering stencil for a small perturbation on a quadratic function. The center picture shows stencil failure and the picture on the right illustrates a new stencil with a reduced step size

which was particularly designed for the optimization of noisy functions with bound constraints [35].

Linear and nonlinear constraints may be incorporated into the objective function via a penalty or barrier approach. The default in IFFCO is to handle constraint violations using an extreme barrier approach and simply assign a large function value to any infeasible point. The performance of IFFCO on nonsmooth, nonconvex, noisy problems and even those with disconnected feasible regions are strong but the dependence on the initial starting point is well documented [8, 61]. Also, note that IFFCO includes a projection operator to handle bound constraints.

7.3.2.3 DIRECT

DIRECT, an acronym for DIviding RECTangles, was designed for global optimization of bound constrained problems as an extension of Shuberts Lipschitz optimization method [58]. Since its introduction in the early 1990's, a significant number of papers have been written analyzing, describing, and developing new variations of this highly effective algorithm. Some of these include [57, 34, 16, 5, 86, 80, 28, 10].

DIRECT is essentially a partitioning algorithm that sequentially refines the region defined by bound constraints at each iteration by selecting a hyper-rectangle to trisect [27, 33, 58]. To balance the global and local search, at each iteration a set S of potentially optimal rectangles is identified based on the function value at the center of the rectangle and the size of the rectangle. The basic algorithm is as follows:

1. Normalize the bound constraints to form a unit hypercube search space with center c_1
2. Find $f(c_1)$, set $f_{min} = f(c_1), i = 0$
3. Evaluate $f(c_i + \frac{1}{3}e_i), 1 \leq i \leq n$ where e_i is the ith unit vector
4. While *not converged* Do

   a. Identify the set S of all potentially optimal rectangles
   b. For all $j \in S$, identify the longest sides of rectangle j, evaluate f at centers, and trisect j into smaller rectangles
   c. Update $f_{min}, i = i + 1$

Note that DIRECT requires that both the upper and lower bounds be finite. The algorithm begins by mapping the rectangular feasible region onto the unit hypercube; that is DIRECT optimizes the transformed problem

$$\begin{aligned} \min_{\tilde{x} \in \mathbb{R}^n} \quad & \tilde{f}(\tilde{x}) = f(S\tilde{x} + \ell) \\ \text{subject to} \quad & 0 \leq \tilde{x} \leq e, \end{aligned} \tag{7.8}$$

where $\tilde{x} = S^{-1}(x - \ell)$ with $S = \text{diag}(u_1 - \ell_1, \ldots, u_n - \ell_n)$. Figure 7.5 illustrates three iterations of DIRECT for a two dimensional example. At each iteration, a candidate, potentially optimal, hyper-rectangle is selected and refined. Though other stopping criteria exist, this process typically continues until a user defined budget of function evaluations has been expended.

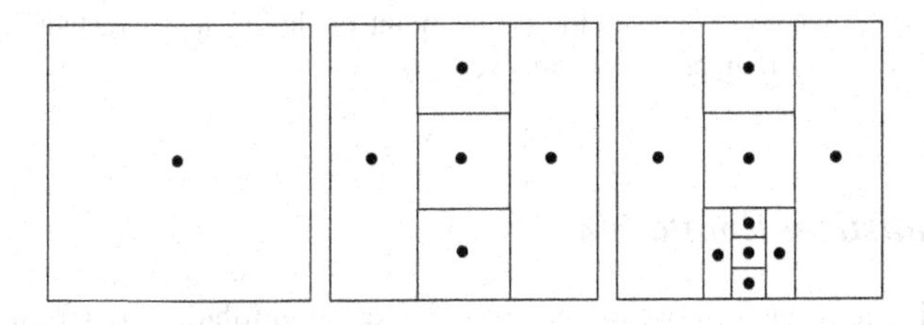

Fig. 7.5 For a two-dimensional problem, DIRECT iteratively subdivides the optimal hyper-rectangle into thirds

The criteria for being a potentially optimal hyper-rectangle given a constant $\varepsilon > 0$ is as follows [58]: Suppose there are K enumerated hyper-rectangles subdividing the unit hypercube from Equation (7.8) with centers c_i, $1 \leq i \leq K$. Let γ_i denote the corresponding distance from the center c_i to its vertices. A hyper-rectangle ℓ is considered *potentially optimal* if there exists $\alpha_K > 0$ such that

$$\tilde{f}(c_\ell) - \alpha_K \gamma_\ell \leq \tilde{f}(c_i) - \alpha_K \gamma_i, \; 1 \leq i \leq K \tag{7.9}$$

$$\tilde{f}(c_\ell) - \alpha_K \gamma_\ell \leq \tilde{f}_{\min} - \varepsilon |f_{\min}|. \tag{7.10}$$

The set of potentially optimal hyper-rectangles forms a convex hull for the set point $\{\tilde{f}(c_i), \gamma_i\}$. Figure 7.6 illustrates this. Notice that the user defined parameter ε controls whether or not the algorithm performs more of a global or local search.

Although DIRECT has been shown to be highly effective for relatively small problems and has proven global convergence, it does suffer at higher dimensions [16, 87, 80, 28] and requires an exhaustive number of function evaluations. In [29], DIRECT was unable to identify a five well solution to the WS problem when starting with an initial six well configuration and using the constraint in Equation (7.6). These results are not surprising given that the five well solution has all of the pumping rates lying on the bound constraint. The sampling strategy of DIRECT does not make it a good candidate for this problem.

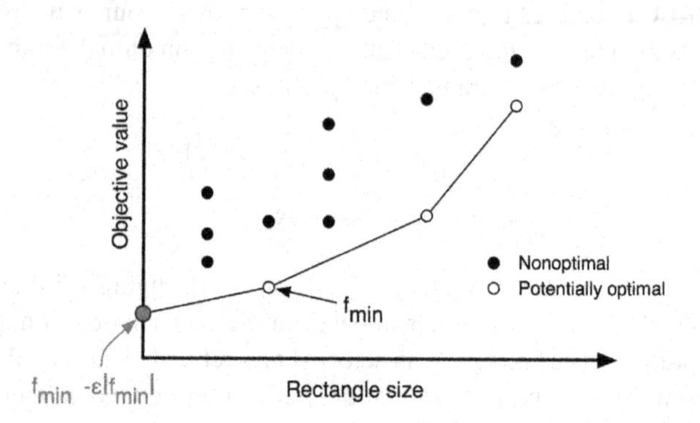

Fig. 7.6 Potentially optimal hyper-rectangles can be found by forming the convex hull of the set $\{f(c_i), \gamma_i\}$, where c_i denotes the center point of the *ith* hyper-rectangle and γ_i the corresponding distance to hyper-rectangle's vertices

7.3.3 Statistical Emulation

An alternative approach to optimization is statistical emulation, wherein the previous runs of the computer code are used to train a statistical model, and the model is used to draw inferences about the location of the optimum. The idea of using a stochastic process to approximate an unknown function dates back as far as Poincaré in the 19th century [24]. In particular, a Gaussian Process (GP) is typically used for the emulation of computer simulators [78, 79]. The output of the simulator is treated as a random variable $Y(x)$ that depends on the input vector x such that the response varies smoothly. This smoothness is given by the covariance structure of the GP. The mean and covariance functions determine the characteristics of the process, as any finite set of locations has a joint multivariate Gaussian distribution [18, 81]. A Bayesian approach allows full estimation of uncertainty, which is useful when trying to determine the probability that an unsampled location will be an improvement over the current known optimum.

Specifically, the uncertainty about future computer evaluations can be quantified by finding the predictive distribution for new input locations conditional on the points that have already been evaluated. Since this is now a full probabilistic model for code output at new locations, any statistic depending upon this output is easily obtained. The expected improvement at a point x, $\mathbb{E}[\min(f_{min} - f(x), 0)]$, is a useful criterion for choosing new locations for evaluation. The paper by [59] illustrates the use of this statistic in optimization. Since the improvement is a random variable, this criterion balances rewarding points where the output is highly uncertain, as well as where the function is generally predicted to be better than the present best point. A number of candidate locations are generated from an optimal space filling design. Then, a GP model is fit to the existing output, and the expected improvement is calculated at each candidate location. The points with highest expected improvement are selected as candidates for the new best point.

Standard GP models have several drawbacks, including strong assumptions of stationarity and poor computational scaling. To reduce these problems, treed Gaussian process (TGP) models partition the input space using a recursive tree structure; and independent GP models are fit within each partition [37, 38]. Such models are a natural extension of standard GP models, and combine partitioning ideas with Bayesian methods to produce smooth fitted functions [9]. The partitions can be fit simultaneously with the parameters of the embedded GP models using reversible jump Markov chain Monte Carlo [41].

Note that the statistical emulation via TGP has the disadvantage of computational expense. As additional points are evaluated, the computational work load of creating the GP model increases significantly. This coupled with some convergence issues when TGP approaches the solution indicate that TGP alone is not an effective method for solving the WS problem. However, TGP is an excellent method for inclusion in a hybrid because these disadvantages can be overcome.

7.4 Some DFO Hybrids

In order to both take advantage of the benefits of more than one optimization approach and to try to overcome method-specific shortcomings, two or more optimization methods may be combined, forming a hybrid. Hybrid optimization is a popular approach in the combinatorial optimization community, where metaheuristics (such as GAs), are combined to improve robustness and blend the distinct strengths of different approaches [83]. More recently, metaheuristics have been combined with deterministic methods (such as pattern search) to form hybrids that simultaneously perform global and local searches [73, 91, 90, 86, 26].

The use of hybrids in the combinatorial optimization community has grown to include a categorization scheme for hybrids [76] which includes four main characteristics: 1) class of algorithms used to form hybrids, 2) level of hybridization, 3) order of execution, and 4) control strategy. Choosing algorithms to hybridize is a significant challenge in forming hybrids. Methods that have complementary advantages and are well suited to the problem of interest should be selected. Hybridization levels include loosely or tightly coupled. In general, loosely coupled approaches retain the individual identities of the methods being hybridized. In contrast, tightly coupled hybrids exhibit a strong relationship between the individual pieces and may share components or functions. Loosely coupled hybrids are advantageous from both a software development and theoretical perspective. They do not require the re-implementation of existing methods and also keep theoretical convergence properties of the individual methods intact. The order of execution of hybrid algorithms can either be sequential or parallel. Sequentially hybrid methods string together a set of algorithms head to tail, using the results of a completed run of one algorithm to seed the next. From this perspective it is often unclear whether or not the previously executed algorithms should be viewed simply as a preprocessing step, or if ensuing algorithm runs should be viewed as post-processing. On the other hand, parallel hybrids execute the individual methods simultaneously and can thus be made to be collaborative and share information dynamically to improve performance [1, 15]. The control strategy of hybrid algorithms can be either integrative or collaborative. In a purely integrative approach, one individual algorithm is subordinate to or an embedded component of another. Collaborative methods give equal importance and control to both algorithms as algorithms merely exchange information instead of being an integral part of one another.

Note that the effectiveness of a hybrid approach to optimization may be compromised if the methods combined are not suited to one other or to the application of interest. In this section, we discuss hybrid optimization in the context of the water resources management problem WS. Four hybrid methods are described here, only two of which are successful for the WS problem. We include the two unsuccessful hybrids to demonstrate the efficacy of tailoring hybrids to address the characteristics of the problem being solved. To test this design, the hybrids were applied to the WS problem without a starting point.

7.4.1 APPS-TGP

Some optimization methods have introduced an *oracle* to predict additional points at which a decrease in the objective function might be observed. Analytically, an oracle is free to choose points by any finite process. (See [63] and references therein.) The addition of an oracle is particularly amenable to a pattern search methods like APPS. The iterate(s) suggested by the oracle are merely additions to the pattern. Furthermore, the asynchronous nature of the APPSPACK implementation makes it adept at handling the evaluation of the additional points. The idea of an oracle is used as a basis for creating a hybrid optimization scheme which combines APPS and the statistical emulator TGP.

In the APPS-TGP hybrid, the TGP statistical model serves as the an oracle. The hopes in utilizing the TGP oracle include added robustness and the introduction of some global properties to APPSPACK. When the oracle is called, the TGP algorithm is applied to the set of evaluated iterates in order to choose additional candidate points. In other words, APPSPACK is still optimizing as normal, but throughout the optimization process, the iterate pairs $(x^i, f(x^i))$ are collected. Then, the TGP model is fit to the existing output, and the expected improvement is calculated at each candidate location. The points with highest expected improvement are passed back to the APPS algorithm to determine if it is a new best point. If not, the point is merely discarded and the APPS algorithm continues without any changes. However, if a TGP point is a new best point, the APPSPACK search pattern continues from that new location. The general flow of this algorithm is illustrated in Figure 7.7. Note that both APPS and TGP generate points and both methods are informed of the function values associated with these iterates.

This hybrid technique is loosely coupled as APPS and TGP run independently of each other. Since the iterates suggested by the TGP algorithm are used in addition

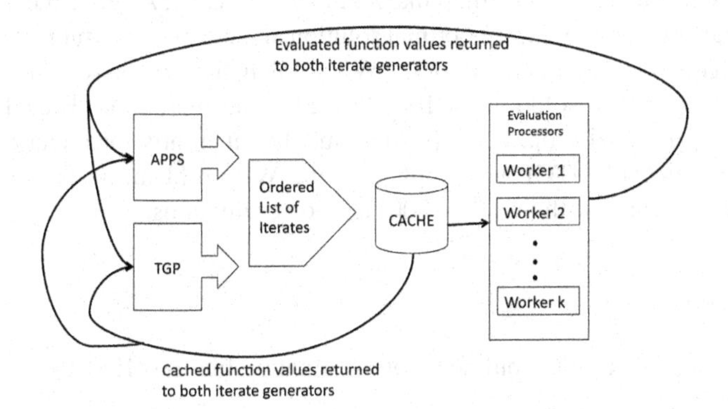

Fig. 7.7 The flow of the APPS-TGP hybrid. Both APPS and TGP generate iterates. The iterates are merged into one list. Then, the function value of each iterate is either obtained from cache or evaluated. Finally, the results are shared with both methods

to the iterates suggested by APPSPACK, there is no adverse affect on the local convergence properties of APPS. As noted earlier, pattern search methods have strong local convergence properties [25, 66, 85]. However, their weakness is that they are local methods. In contrast, TGP performs a global search of the feasible region, but does not have strong local convergence properties. Hence, using the hybridization scheme, TGP lends a globalization to the pattern search and the pattern search further refines TGP iterates by local search. This benefit is clearly illustrated on a model calibration problem from electrical engineering in [82] and on a groundwater remediation problem in [39].

APPS-TGP is also collaborative since APPS and TGP are basically run independently of one another. From the perspective of TGP, a growing cache of function evaluations is being cultivated, and the sole task of TGP is to build a model and select a new set of promising points to be evaluated. The TGP algorithm is not dependent on where this cache of points comes from. Thus in this approach, we may easily incorporate other optimization strategies where each strategy is simply viewed as an external point generating mechanism leveraged by TGP. From the perspective of APPS, points suggested by TGP are interpreted in an identical fashion to other trial points and are ignored unless deemed better than the current best point held by APPS. Thus neither algorithm is aware that a concurrent algorithm is running in parallel. However, the hybridization is integrative in the sense that points submitted by TGP are given a higher priority in the queue of iterates to be evaluated. In the parallel execution of APPS and TGP, TGP is given one processor (because it is computationally prohibitive) while APPS directs the use of the remaining processors to perform point evaluations. Communication between TGP and APPSPACK occurs intermittently through out the optimization process, whenever TGP completes and is ready to look at a new cache of points.

The APPS-TGP method was specifically designed to address the disadvantages associated with using a local method like APPS. Specifically, APPS usually requires a "good" starting point to find an optimal solution and not become trapped in a local minimum. This was demonstrated for the WS problem in [29] where APPS failed to find the optimal 5-well solution for 100 alternative starting points. The addition of TGP provides a global scope to help overcome the inherently local characteristics of APPS. When APPS-TGP was applied to the WS problem, an optimal 5-well solution was obtained with less than 500 function evaluations.

7.4.2 EAGLS

To address mixed-variable, nonlinear optimization problems (MINLPs) of the form

$$\begin{array}{ll} \text{minimize} x \in \mathbb{R}^{nr}, z \in \mathbb{Z}^{nb} & f(z,x) \\ \text{subject to} & c(z,x) \leq 0 \\ & z_\ell \leq z \leq z_u \\ & x_\ell \leq x \leq x_u. \end{array} \tag{7.11}$$

where $c(x) : \mathbb{R}^n \rightarrow \mathbb{R}^m$, consider a hybrid of a GA and a direct search. The APPS-GA hybrid, commonly referred to as EAGLS (Evolutionary Algorithm Guiding Local Search), uses the GA's handling of integer and real variables for global search, and APPS's handling of real variables in parallel for local search [43].

As previously discussed, a GA carries forward a population of points that are iteratively mutated, merged, selected, or dismissed. However, individuals in the population are not given the opportunity to make intergeneration improvements. This is not reflective of he real world, where an organism is not constant throughout its life span, but instead can grow, improve, or become stronger. Improvements within a generation are allowed in EAGLS. The GA still governs point survival, mutation, and merging as an outer iteration, but, during an inner iteration, individual points are improved via APPS applied to the real variables, with the integer variables held fixed. For simplicity, consider the synchronous EAGLS algorithm:

1. Evaluate initial population
2. While *not converged* Do
    a. Perform selection, mutation, crossover
    b. Evaluate new points
    c. Choose points for local search
    d. Make multiple calls to APPS for real-valued subproblems

Of course, to allow the entire population to grow as such, would be computationally prohibitive. Thus, EAGLS employs a ranking algorithm that takes in to account individual proximity to other, better points. The goal of this step is to select promising individuals representing distinct subdomains. Note that the flow of the asynchronous EAGLS algorithm is slightly different than that of APPS-TGP. In this case, NSGA-II generates iterates and multiple instances of APPS also generate iterates. Returned

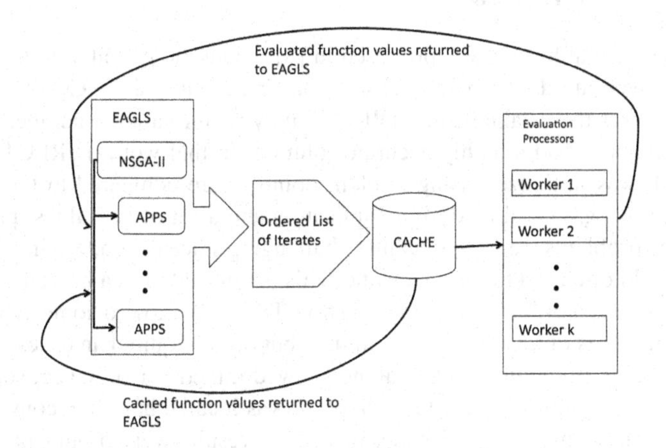

Fig. 7.8 The flow of EAGLS. The NSGA-II algorithm generate iterates. Then, some iterates are selected for refinement by multiple instances of APPS. The iterates are merged into one list, and the function value of each iterate is either obtained from cache or evaluated. Finally, the results are returned so that the APPS instances and the GA can proceed

function values are distributed to the appropriate instance of APPS or the GA. This is illustrated in Figure 7.8.

Note that it is the combinatorial nature of integer variables that makes the solution of MINLPs difficult. If the integer variables are relaxable (i.e. the objective function is defined for rational variables), more sophisticated schemes such as branch and bound may be preferred options. However, for simulation-based optimization problems, the integer variables often represent a descriptive category (i.e. a color or a building material) and may lack the natural ordering required for relaxation. That is, there is may be no well-defined mathematical definition for what is meant by "nearby." In the WS problem, the number of wells is not a relaxable variable because, for example, one-half a well cannot be installed. The other results for the WS problem given in this chapter consider the strictly real-valued WS formulation. However, since EAGLS was designed to handle MINLPs, it was applied to the MINLP formulation of WS. Moreover, EAGLS combines a global and local search in order to take advantage of the global properties and overcome the computational expense of the GA.

To illustrate the global properties of EAGLS, the problem was solved without an initial point. In [43], EAGLS was able to locate a five well solution using only random points in the initial GA population. Moreover, this was done in after about 65 function evaluations. This is an improvement both for the GA and for the local search method APPS. The function evaluation budget was 3000 and roughly 1000 of those were spent on points that did not satisfy the linear constraint in Equation (7.5) which means the simulator was never called.

7.4.3 DIRECT-IFFCO

A simple sequential hybrid was proposed in [8] where the global search strengths of DIRECT were used to generate feasible starting points for IFFCO. This hybrid further addresses the weakness that DIRECT may require a large number of function evaluations to find a highly accurate solution. In that work, DIRECT and IFFCO, which was initialized using random points, were compared to the sequential pairing for a gas pipeline design application and a set of global test problems. The pipeline problems were significantly challenging since the underlying objective function would often fail to return a value. This is referred to as a hidden constraint in simulation-based or black-box optimization. DIRECT showed some evidence of robustness in terms of locating global minima but often required an excessive number of function evaluations. IFFCO alone showed mixed performance; sometimes refining the best solution once a feasible point was located but often converging to a suboptimal local minimum. For the hybrid, the results were promising. Even if the function value at the end of the DIRECT iteration was high, IFFCO was able to avoid entrapment in a local minima using the results. In fact, using DIRECT as a generator of starting points for local searches has been actively studied over the years and applied to a variety of applications. For example, in [17] DIRECT was paired with a sequential quadratic programming method for the local search and outperformed a variety of other global methods applied to an aircraft design problem and in [70] a gradient-based local search was shown to accelerate convergence to the global minimum for a flight control problem.

A different idea was used in [29] where DIRECT was used in conjunction with IFFCO to find starting points for the WS problem. In this case, DIRECT was used to minimize an aggregate of constraint violation and thereby identify sets of feasible starting points and then IFFCO was used to minimize the true objective function. This approach was not successful in that the points identified by DIRECT were so close to multiple local minima that IFFCO was unable to improve the objective function value. In particular, IFFCO would only converge if initial points contained five wells operating on the bound constraint for their pumping rates, and DIRECT did not identify any points of this sort. The advantages obtained by combining DIRECT and IFFCO do not address the characteristics of the WS problem that make it difficult to solve. However, it should be noted that the idea of using DIRECT and IFFCO together in this sort of bi-objective approach certainly warrants further investigation despite the performance on the WS problem.

7.4.4 DIRECT-TGP

Another attempt to improve the local search of DIRECT involves TGP with a gradient-based method on the surrogate model, which is cheap to minimize [52]. Hybridization in this case is performed at the iteration level in that the center of the current rectangle is used as a starting point for a local search on the surrogate. Essentially, the procedure for dividing hyper-rectangles in Step 4(b) in Section 2.2.3

above is replaced with the following steps once the number of function evaluations is larger than $2n+1$, which allows for the initial hypercube sampling:

1. Build TGP surrogate using all known function evaluations
2. Start local search on the surrogate, constrained to the rectangle, using the center of the rectangle as the initial point
3. Evaluate f the local optimum, x_{loc}
4. Return $f(x_{loc})$ instead of $f(c_i)$

The algorithm, although relatively new, has been tested on a suite of bound constrained and nonlinearly constrained problems and a cardiovascular modeling problem proposed in [30].

These promising preliminary results indicate this new hybrid can improve the local search capabilities of DIRECT. This is achieved without compromising the computational efficiency and with practically no additional algorithmic parameters to fine-tune. It should also be noted that other hybrids that attempt to improve the local search of DIRECT have been proposed. For example, [44] proposes a DIRECT-GSS hybrid. The resulting algorithm does show some promising results in terms of reducing the computational workload required to solve the optimization problem, but it has only been investigated on test problems from the literature. Further tests are needed to determine its applicability to engineering applications. Given the performance of the DIRECT-IFFCO approach above, any local search hybrid with DIRECT would likely not perform well on the WS problem.

7.5 Summary and Conclusion

The purpose of this Chapter is to introduce a number of derivative-free optimization techniques available to address simulation-based problems in engineering. In this Chapter, we have shown their effectiveness for a problem from hydrological engineering. The purpose of this exercise was not to compare methods, but instead to show the wide variety of options available. In addition, we showed that some techniques do not address the characteristics of some problems. In Table 1, we summarize these results.

Finally, we note that another utility of hybrid optimization methods could be to inform the decision making processes. Currently, optimization algorithms accept guesses (or iterates) based solely on some notion of (sufficient) increase or decrease of the objective function. In order to serve as decision makers, next-generation algorithms must also consider rankings and probability metrics. For example, computational costs can be assessed so that iterates are only evaluated if they meet a set computational budget. This is particularly important for the expensive objective functions of simulation-based optimization problems. Moreover, hybridization optimization algorithms can incorporate tools that will allow the user to dismiss subsets of the domain that exhibit large variances or that exceed critical thresholds. In addition, hybrid algorithmic frameworks are a step towards finding methods capable of generating a "robust" set of optimal solution options instead of a single optimum.

Table 7.2 Summary of the Performance of some Derivative-Free Methods for the WS Problem

Method	Found 5-Well Solution	Number of Fn Evals	Starting Pt Required
APPS	Y	176	Y
IFFCO	Y	< 200	Y
DIRECT	N	–	Y
GA	Y	161	Y
APPS-TGP	Y	492	N
EAGLS	Y	65	N

The current state of the art is to accept an iterate as an optimum based on the inability to find better guess within a decreasing search region. This may lead to solutions to design problems that are undesirable due to a lack of robustness to small design perturbations. Instead, algorithms that allow designers to choose a solution based on additional criteria can be created in the hybrid framework. For example, a regional optimum could be used to generate a set of multiple solutions from which the designer can choose.

Acknowledgments. The authors would like to thank the American Institute of Mathematics (AIM) for their support of the EAGLS research. We thank Josh Griffin for his assistance with Figures 7.5 and 7.6 included in Section 3.2.3 to illustrate DIRECT and Tammy Kolda for her assistance with Figure 7.3 included in Section 3.2.1 to illustrate APPS. We also thank Josh Griffin, Matt Parno, and Thomas Hemker for their contributions to the hybrid methods research. Sandia National Laboratories is a multiprogram laboratory operated by Sandia Corporation, a Lockheed Martin Company, for the United States Department of Energy's National Nuclear Security Administration under Contract DE-AC04-94AL85000.

References

[1] Alba, E.: Parallel Metaheuristics. John Wiley & Sons, Chichester (2005)
[2] Audet, C., Booker, A., et al.: A surrogate-model-based method for constrained optimization. In: AIAA/USAF/NASA/ISSMO Symposium on Multidisciplinary Analysis and Optimization (2000)
[3] Audet, C., Couture, G., Dennis Jr, J.E.: Nonlinear optimization with mixed variables and derivatives, NOMAD (2002)
[4] Audet, C., Dennis Jr., J.E.: Mesh adaptive direct search algorithms for constrained optimization. Technical report, Ecole Polytechnique de Montreal, Departement de Mathematiques et de Genie Industriel, Montreal (Quebec), H3C 3A7 Canada (2004)
[5] Bartholomew-Biggs, M.C., Parkhurst, S.C., Wilson, S.P.: Global optimization – stochastic or deterministic? In: Albrecht, A.A., Steinhöfel, K. (eds.) SAGA 2003. LNCS, vol. 2827, pp. 125–137. Springer, Heidelberg (2003)

[6] Blum, C., Blesa Aquilera, M.J., Roli, A., Sampels, M.: Hybrid Metaheuristics. SCI. Springer, Heidelberg (2008)

[7] Booker, A.J., Meckesheimer, M.: Reliability based design optimization using design explorer. Opt. Eng. 5, 170–205 (2004)

[8] Carter, R., Gablonsky, J.M., Patrick, A., Kelley, C.T., Eslinger, O.J.: Algorithms for noisy problems in gas transmission pipeline optimization. Opt. Eng., 139–157 (2001)

[9] Chipman, H.A., George, E.I., McCulloch, R.E.: Bayesian treed models. Machine Learning 48, 303–324 (2002)

[10] Chiter, L.: Direct algorithm: A new definition of potentially optimal hyperrectangles. Appl. Math. Comput. 179(2), 742–749 (2006)

[11] Choi, T.D., Eslinger, O.J., Gilmore, P., Patrick, A., Kelley, C.T., Gablonsky, J.M.: IFFCO: Implicit Filtering for Constrained Optimization, Version 2. Technical Report CRSC-TR99-23, North Carolina State Univeristy (July 1999)

[12] Choi, T.D., Eslinger, O.J., Kelley, C.T., David, J.W., Etheridge, M.: Optimization of automotive valve train components with implict filtering. Optim. Engrg. 1, 9–28 (2000)

[13] Choi, T.D., Kelley, C.T.: Superlinear convergence and implicit filtering. SIAM J. Opt. 10, 1149–1162 (2000)

[14] Conn, A., Scheinberg, K., Vincente, L.N.: Introduction to Derivative-Free Optimization. SIAM, Philadelphia (2009)

[15] Cotta, E.-G., Talbi, E.A.: Parallel Hybrid Metaheuristics. In: Parallel Metaheuristics, pp. 347–370. John Wiley & Sons, Inc, Chichester (2005)

[16] Cox, S.E., Hart, W.E., Haftka, R., Watson, L.: DIRECT algorithm with box penetration for improved local convergence. In: 9th AIAA/ISSMO Symposium on Multidisciplinary Analysis and Optimization (2002)

[17] Cox, S.L., Haftka, R.T., Baker, C.A., Grossman, B., Mason, W.H., Watson, L.T.: A comparison of global optimization methods for the design of a high-speed civil transport. Journal of Global Optimization 21, 415–433 (2001)

[18] Cressie, N.A.C.: Statistics for Spatial Data, revised edition. John Wiley & Sons, Chichester (1993)

[19] Deb, K.: An efficient constraint handling method for genetic algorithms. Comp. Methods Appl. Mech. Eng. 186(2-4), 311–338 (2000)

[20] Deb, K., Goel, T.: Controlled elitist non-dominated sorting genetic algorithms for better convergence. In: Zitzler, E., Deb, K., Thiele, L., Coello Coello, C.A., Corne, D.W. (eds.) EMO 2001. LNCS, vol. 1993, p. 67. Springer, Heidelberg (2001)

[21] Deb, K., Pratap, A., Agarwal, S., Meyarivan, T.: A fast and elitist multiobjective genetic algorithm: NSGA-II. IEEE Trans. Evolutionary Comp. 6(2), 182–197 (2002)

[22] Ting, C.-K.: An analysis of the effectiveness of multi-parent crossover. In: Yao, X., Burke, E.K., Lozano, J.A., Smith, J., Merelo-Guervós, J.J., Bullinaria, J.A., Rowe, J.E., Tiňo, P., Kabán, A., Schwefel, H.-P. (eds.) PPSN 2004. LNCS, vol. 3242, pp. 131–140. Springer, Heidelberg (2004)

[23] Dennis Jr., J.E., Torczon, V.: Direct search methods on parallel machines. SIAM J. Opt. 1, 448–474 (1991)

[24] Diaconis, P.: Bayesian numerical analysis. In: Gupta, S.S., Berger, J.O. (eds.) Statistical Decision Theory and Related Topics IV, Springer, Heidelberg (1988)

[25] Dolan, E.D., Lewis, R.M., Torczon, V.: On the local convergence properties of parallel pattern search. Technical Report 2000-36, NASA Langley Research Center, Inst. Comput. Appl. Sci. Engrg., Hampton, VA (2000)

[26] Fan, S.-K.S., Zahara, E.: A hybrid simplex search and particle swarm optimization for unconstrained optimization. Eur. J. Oper. Res. 181(2), 527–548 (2007)

[27] Finkel, D.E.: Global Optimization with the DIRECT Algorithm. PhD thesis, North Carolina State Univ., Raleigh, NC (2005)
[28] Finkel, D.E., Kelley, C.T.: Additive scaling and the DIRECT algorithm. J. Global Optim. 36(4), 597–608 (2006)
[29] Fowler, K.R., et al.: A comparison of derivative-free optimization methods for water supply and hydraulic capture community problems. Adv. Water Resourc. 31(5), 743–757 (2008)
[30] Fowler, K.R., Gray, G.A., Olufsen, M.S.: Modeling heart rate regulation part ii: Parameter identification and analysis. J. Cardiovascular Eng. 8(2) (2008)
[31] Fowler, K.R., Jenkins, E.W., LaLonde, S.L.: Understanding the effects of polymer extrusion filter layering configurations using simulation-based optimization. Optim. Engrg. 11, 339–354 (2009)
[32] Fowler, K.R., Kelley, C.T., Miller, C.T., Kees, C.E., Darwin, R.W., Reese, J.P., Farthing, M.W., Reed, M.S.C.: Solution of a well-field design problem with implicit filtering. Opt. Eng. 5, 207–234 (2004)
[33] Gablonsky, J.M.: DIRECT Version 2.0 User Guide. Technical Report CRSCTR01- 08, Center for Research in Scientific Computation, NC State (2001)
[34] Gablonsky, J.M., Kelley, C.T.: A locally-biased form of the DIRECT algorithm. J. Global Optim. 21(1), 27–37 (2001)
[35] Gilmore, P., Kelley, C.T.: An implicit filtering algorithm for optimization of functions with many local minima. SIAM J. Opt. 5, 269–285 (1995)
[36] Goldberg, D.E.: Genetic Algorithms in Search, Optimization, and Machine Learning. Addison-Wesley, Reading (1989)
[37] Gramacy, R.B., Lee, H.K.H.: Bayesian treed Gaussian process models. Technical report, Dept. of Appl. Math & Statist., Univ. of California, Santa Cruz (2006)
[38] Gramacy, R.B., Lee, H.K.H.: Bayesian treed Gaussian process models with an application to computer modeling. J. Amer. Statist. Assoc. 103, 1119–1130 (2008)
[39] Gray, G.A., Fowler, K., Griffin, J.D.: Hybrid optimization schemes for simulation based problems. Procedia Comp. Sci. 1(1), 1343–1351 (2010)
[40] Gray, G.A., Kolda, T.G.: Algorithm 856: APPSPACK 4.0: Asynchronous parallel pattern search for derivative-free optimization. ACM Trans. Math. Software 32(3), 485–507 (2006)
[41] Green, P.J.: Reversible jump markov chain monte carlo computation and bayesian model determination. Biometrika 82, 711–732 (1995)
[42] Grefenstette, J.J.: Optimization of control parameters for genetic algorithms. IEEE Trans. Sys. Man Cybernetics, SMC 16(1), 122–128 (1986)
[43] Griffin, J.D., Fowler, K.R., Gray, G.A., Hemker, T., Parno, M.D.: Derivative-free optimization via evolutinary algorithms guiding local search (EAGLS) for MINLP. Pac. J. Opt (2010) (to appear)
[44] Griffin, J.D., Kolda, T.G.: Asynchronous parallel hybrid optimization combining DIRECT and GSS. Optim. Meth. Software 25(5), 797–817 (2010)
[45] Griffin, J.D., Kolda, T.G.: Nonlinearly-constrained optimization using heuristic penalty methods and asynchronous parallel generating set search. Appl. Math. Res. express 2010(1), 36–62 (2010)
[46] Griffin, J.D., Kolda, T.G., Lewis, R.M.: Asynchronous parallel generating set search for linearly-constrained optimization. SIAM J. Sci. Comp. 30(4), 1892–1924 (2008)
[47] Gropp, W., Lusk, E., Doss, N., Skjellum, A.: A high-performance, portable implementation of the MPI message passing interface standard. Parallel Comput. 22, 789–828 (1996)

[48] Gropp, W.D., Lusk, E.: User's guide for mpich, a portable implementation of MPI. Technical Report ANL-96/6, Mathematics and Computer Science Division, Argonne National Lab (1996)
[49] Hackett, P.: A comparison of selection methods based on the performance of a genetic program applied to the cart-pole problem (1995) ; A Bachelor's thesis for Griffith University, Gold Coast Campus, Queensland
[50] Hemker, T., Fowler, K.R., Farthing, M.W., von Stryk, O.: A mixed-integer simulation-based optimization approach with surrogate functions in water resources management. Opt. Eng. 9(4), 341–360 (2008)
[51] Hemker, T., Fowler, K.R., von Stryk, O.: Derivative-free optimization methods for handling fixed costs in optimal groundwater remediation design. In: Proc. of the CMWR XVI - Computational Methods in Water Resources, June 19-22 (2006)
[52] Hemker, T., Werner, C.: Direct using local search on surrogates. Submitted to Pac. J. Opt. (2010)
[53] Holland, J.H.: Adaption in Natural and Artificial Systems. Univ. of Michigan Press, Ann Arbor (1975)
[54] Holland, J.H.: Genetic algorithms and the optimal allocation of trials. SIAM J. Comput. 2 (1975)
[55] Hough, P.D., Kolda, T.G., Torczon, V.: Asynchronous parallel pattern search for nonlinear optimization. SIAM J. Sci. Comput. 23, 134–156 (2001)
[56] Hough, P.D., Meza, J.C.: A class of trust-region methods for parallel optimization. SIAM J. Opt. 13(1), 264–282 (2002)
[57] Jones, D.R.: The direct global optimization algorithm. In: Encyclopedia of Optimization, vol. 1, pp. 431–440. Kluwer Academic, Boston (2001)
[58] Jones, D.R., Perttunen, C.D., Stuckman, B.E.: Lipschitzian optimization without the lipschitz constant. J. Opt. Theory Apps. 79(1), 157–181 (1993)
[59] Jones, D.R., Schonlau, M., Welch, W.J.: Efficient global optimization of expensive blackbox functions. J. Global Optim. 13, 455–492 (1998)
[60] Karr, C., Freeman, L.M.: Industrial Applications of Genetic Algorithms. International Series on Computational Intelligence. CRC Press, Boca Raton (1998)
[61] Kelley, C.: Iterative methods for optimization. SIAM, Philadelphia (1999)
[62] Kolda, T.G.: Revisiting asynchronous parallel pattern search. Technical Report SAND2004-8055, Sandia National Labs, Livermore, CA 94551 (February 2004)
[63] Kolda, T.G., Lewis, R.M., Torczon, V.: Optimization by direct search: New perspectives on some classical and modern methods. SIAM Rev. 45(3), 385–482 (2003)
[64] Kolda, T.G., Lewis, R.M., Torczon, V.: Stationarity results for generating set search for linearly constrained optimization. SIAM J. Optim. 17(4), 943–968 (2006)
[65] Lewis, R.M., Shepherd, A., Torczon, V.: Implementing generating set search methods for linearly constrained minimization. Technical Report WMCS- 2005-01, Department of Computer Science, College of William & Mary, Williamsburg, VA (July 2006) (revised)
[66] Lewis, R.M., Torczon, V.: Rank ordering and positive basis in pattern search algorithms. Technical Report 96-71, NASA Langley Research Center, Inst. Comput. Appl. Sci. Engrg., Hampton, VA (1996)
[67] Lewis, R.M., Torczon, V., Trosset, M.W.: Direct search methods: Then and now. J. Comp. Appl. Math. 124(1-2), 191–207 (2000)
[68] Lobo, F.G., Lima, C.F., Michalewicz, Z. (eds.): Parameter settings in evolutionary algorithms. Springer, Heidelberg (2007)

[69] McKinney, D.C., Lin, M.D.: Approximate mixed integer nonlinear programming methods for optimal aquifer remdiation design. Water Resour. Res. 31, 731–740 (1995)
[70] Menon, P.P., Bates, D.G., Postlethwaite, I.: A deterministic hybrid optimization algorithm for nonlinear flight control systems analysis. In: Proceedings of the 2006 American Control Conference, Minneapolis, MN, pp. 333–338. IEEE Computer Society Press, Los Alamitos (2006)
[71] Meyer, A.S., Kelley, C.T., Miller, C.T.: Electronic supplement to "optimal design for problems involving flow and transport in saturated porous media". Adv. Water Resources 12, 1233–1256 (2002)
[72] Meyer, A.S., Kelley, C.T., Miller, C.T.: Optimal design for problems involving flow and transport in saturated porous media. Adv. Water Resources 12, 1233–1256 (2002)
[73] Payne, J.L., Eppstein, M.J.: A hybrid genetic algorithm with pattern search for finding heavy atoms in protein crystals. In: GECCO 2005: Proceedings of the 2005 conference on Genetic and evolutionary computation, pp. 374–384. ACM Press, New York (2005)
[74] Plantenga, T.D.: HOPSPACK 2.0 User Manual (v 2.0.1). Technical Report SAND2009-6265, Sandia National Labs, Livermore, CA (2009)
[75] Powell, M.J.D.: Direct search algorithms for optimization calculations. Acta Numer. 7, 287–336 (1998)
[76] Raidl, G.R.: A unified view on hybrid metaheuristics. In: Almeida, F., Blesa Aguilera, M.J., Blum, C., Moreno Vega, J.M., Pérez Pérez, M., Roli, A., Sampels, M. (eds.) HM 2006. LNCS, vol. 4030, pp. 1–12. Springer, Heidelberg (2006)
[77] Regis, R.G., Shoemaker, C.A.: Constrained global optimization of expensive black box functions using radial basis functions. J. Global Opt. 31 (2005)
[78] Sacks, J., Welch, W.J., Mitchell, T.J., Wynn, H.P.: Design and analysis of computer experiments. Statist. Sci. 4, 409–435 (1989)
[79] Santner, T.J., Williams, B.J., Notz, W.I.: The Design and Analysis of Computer Experiments. Springer, New York (2003)
[80] Siah, E.S., Sasena, M., Volakis, J.L., Papalambros, P.Y., Wiese, R.W.: Fast parameter optimization of large-scale electromagnetic objects using DIRECT with Kriging metamodeling. IEEE T. Microw. Theory 52(1), 276–285 (2004)
[81] Stein, M.L.: Interpolation of Spatial Data. Springer, New York (1999)
[82] Taddy, M., Lee, H.K.H., Gray, G.A., Griffin, J.D.: Bayesian guided pattern search for robust local optimization. Technometrics 51(4), 389–401 (2009)
[83] Talbi, E.G.: A taxonomy of hybrid metaheurtistics. J. Heuristics 8, 541–564 (2004)
[84] Torczon, V.: PDS: Direct search methods for unconstrained optimization on either sequential or parallel machines. Technical Report TR92-09, Rice Univ., Dept. Comput. Appl. Math., Houston, TX (1992)
[85] Torczon, V.: On the convergence of pattern search algorithms. SIAM J. Opt. 7, 1–25 (1997)
[86] Wachowiak, K.P., Peters, T.M.: Combining global and local parallel optimization for medical image registration. In: Fitzpatrick, J.M., Reinhardt, J.M. (eds.) Medical Imaging 2005: Image Processing, vol. 5747, pp. 1189–1200. SPIE, San Jose (2005)
[87] Wachowiak, M.P., Peters, T.M.: Parallel optimization approaches for medical image registration. In: Barillot, C., Haynor, D.R., Hellier, P. (eds.) MICCAI 2004. LNCS, vol. 3216, pp. 781–788. Springer, Heidelberg (2004)
[88] Wild, S., Regis, R.G., Shoemaker, C.A.: ORBIT: optimization by radial basis function interpolation in trust region. SIAM J. Sci. Comput. 30(6), 3197–3219 (2008)

[89] Wright, M.H.: Direct search methods: Once scorned, now respectable. In: Griffiths, D.F., Watson, G.A. (eds.) Numerical Analysis 1995 (Proceedings of the 1995 Dundee Biennial Conference in Numerical Analysis). Pitman Research Notes in Mathematics, vol. 344, pp. 191–208. CRC Press, Boca Raton (1996)
[90] Yehui, P., Zhenhai, L.: A derivative-free algorithm for unconstrained optimization. Appl. Math. - J. Chinese Univ. 20(4), 491–498 (2007)
[91] Zhang, T., Choi, K.K., et al.: A hybrid surrogate and pattern search optimization method and application to microelectronics. Struc. Multidisiciplinary Opt. 32, 327–345 (2006)
[92] Zheng, C., Hill, M.C., Hsieh, P.A.: MODFLOW2000, The U.S.G.S Survey Modular Ground-Water Model User Guide to the LMT6 Package, the Linkage With MT3DMS for Multispecies Mass Transport Modeling. USGS, user's guide edition (2001)
[93] Zitzler, E., Deb, K., Thiele, L.: Comparison of multiobjective evolutionary algorithms: Empirical results. Evolutionary Comp. J. 8(2), 173–195 (2000)

Chapter 8
Simulation-Driven Design in Microwave Engineering: Methods

Slawomir Koziel and Stanislav Ogurtsov

Abstract. Today, electromagnetic (EM) simulation is inherent in analysis and design of microwave components. Available simulation packages allow engineers to obtain accurate responses of microwave structures. In the same time the task of microwave component design can be formulated and solved as an optimization problem where the objective function is supplied by an EM solver. Unfortunately, accurate simulations may be computationally expensive; therefore, optimization approaches with the EM solver directly employed in the optimization loop may be very time consuming or even impractical. On the other hand, computationally efficient microwave designs can be realized using surrogate-based optimization. In this chapter, simulation-driven design methods for microwave engineering are described where optimization of the original model is replaced by iterative re-optimization of its surrogate, a computationally cheap low-fidelity model which, in the same time, should have reliable prediction capabilities. These optimization methods include space mapping, simulation-based tuning, variable-fidelity optimization, and various response correction techniques.

Keywords: computer-aided design (CAD), microwave design, simulation-driven optimization, electromagnetic (EM) simulation, surrogate-based optimization, space mapping, surrogate model, high-fidelity model, low-fidelity model.

8.1 Introduction

Computer-aided full-wave electromagnetic analysis has been used in microwave engineering for a few decades. Initially, its main application area was design verification. Electromagnetic (EM) simulations can be highly accurate but, at the

---

Slawomir Koziel · Stanislav Ogurtsov
Engineering Optimization & Modeling Center School of Science and Engineering, Reykjavik University, Menntavegur 1, 101 Reykjavik, Iceland
email: koziel@ru.is, stanislav@ru.is

S. Koziel & X.-S. Yang (Eds.): Comput. Optimization, Methods and Algorithms, SCI 356, pp. 153–178.
springerlink.com

same time, they are computationally expensive. Automated EM-simulation-driven optimization was not possible until 1980s when faster CPUs as well as robust algorithms became available [1]. During the 1980s, commercial EM simulation software packages, e.g., those developed by Ansoft Corporation, Hewlett-Packard, and Sonnet Software, started appearing on the market. Formal EM-based optimization of microwave structures has been reported since 1994 [2]-[4].

In many situations, theoretical models of the microwave structures can only be used to yield the initial designs that need to be further tuned to meet performance requirements. Today, EM-simulation-driven optimization and design closure become increasingly important due to complexity of microwave structures and increasing demands for accuracy. Also, EM-based design is a must for a growing number of microwave devices such as ultrawideband (UWB) antennas [5] and substrate-integrated circuits [6]. For circuits like these, no design-ready theoretical models are available, so that design improvement can be only obtained through geometry adjustments based on repetitive simulations.

In this chapter, some major challenges of EM-simulation-driven microwave design are discussed, and traditional approaches that have been used over the years are reviewed. Certain microwave-engineering-specific approaches that aim at reducing the computational cost of the design process (in particular, system decomposition and co-simulation) are mentioned. We also characterize optimization techniques available in commercial EM simulation packages.

The main focus of this chapter is on surrogate-based approaches that allow computationally efficient optimization. Fundamentals of surrogate-based microwave design as well as popular strategies for building surrogate models are discussed. Special emphasis is put on surrogates exploiting physics-based low-fidelity models.

8.2 Direct Approaches

Microwave design task can be formulated as a nonlinear minimization problem

$$x^* \in \arg\min_{x \in X_f} U\left(\boldsymbol{R}_f(x)\right) \tag{8.1}$$

where $\boldsymbol{R}_f \in R^m$ denotes the response vector of the device of interest, e.g., the modulus of the transmission coefficient $|S_{21}|$ evaluated at m different frequencies. U is a given scalar merit function, e.g., a minimax function with upper and lower specifications [7]. Vector $\boldsymbol{x}^*$ is the optimal design to be determined. Normally, $\boldsymbol{R}_f$ is obtained through computationally expensive electromagnetic simulation. It is referred to as the high-fidelity or fine model.

The conventional way of handling the design problem (8.1) is to employ the EM simulator directly within the optimization loop as illustrated in Fig. 8.1. This direct approach faces some fundamental difficulties. The most important one is the computational cost. EM simulation of a microwave device at a single design can take—depending on the system complexity—as long as several minutes, several hours or even a few days. On the other hand, the typical (e.g., gradient-based) optimization algorithms may require dozens or hundreds of EM simulations, which makes the optimization impractical. Another difficulty is that the responses obtained through EM simulation typically have poor analytical properties. In particular, they contain a lot of numerical noise: discretization topology of the simulated structure may change abruptly even for small changes of the design variables which is caused by adaptive meshing techniques utilized by most modern EM solvers. This, in turn, results in the discontinuity of the response function $\boldsymbol{R}_f(\boldsymbol{x})$. Additional problem for direct EM-based optimization is that the sensitivity information may not be available or expensive to compute. Only recently, computationally cheap adjoint sensitivities [8] started to become available in some major commercial EM simulation packages, although for frequency-domain solvers only [9], [10].

It has to be emphasized that probably the most common approach to simulation-driven design used by engineers and designers these days is repetitive parameters sweeps. Typically, a number of EM analyses are carried out while varying a single design variable. Such a process is then repeated for other variables. The information obtained through such parameter sweeps is combined with engineering experience in order to yield a refined design that satisfies the prescribed specifications. This process is quite tedious and time consuming and, of course, requires a substantial designer interaction.

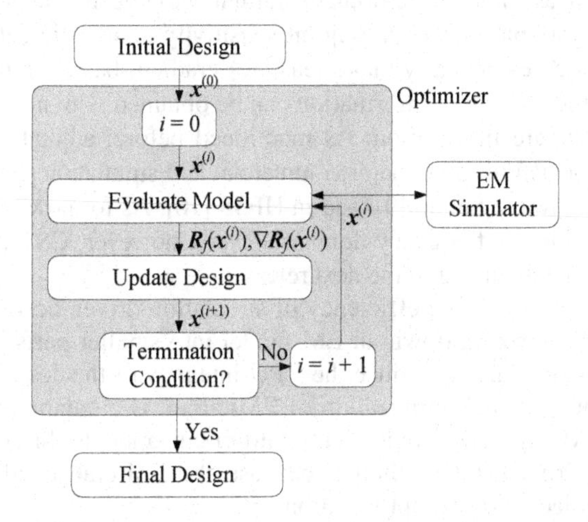

Fig. 8.1 Conventional simulation-driven design optimization: the EM solver is directly employed in the optimization loop. Each modification of the design requires additional simulation of the structure under consideration. Typical (e.g., gradient-based) optimization algorithms may require tens or hundreds of computationally expensive iterations.

In terms of automated EM-based design, conventional techniques are still in use including gradient-based methods (e.g., quasi-Newton techniques [11]), as well as derivative free approaches such as Nelder-Mead algorithm [12]. In some areas, particularly for antenna design, population-based search techniques are used such as genetic algorithms [13], [14] or particle swarm optimizers [15], [16]. These algorithms are mostly exploited to handle issues such as multiple local optima for antenna-related design problems, although they suffer from substantial computational overhead. Probably the best picture of the state of the art in the automated EM-simulation-based design optimization is given by the methods that are available in major commercial software packages such as CST Microwave Studio [9], Sonnet Software [17], HFSS [10], or FEKO [18]. All these packages offer traditional techniques including gradient-based algorithm, simplex search, or genetic algorithms. Practical use of these methods is quite limited.

One of possible ways of alleviating the difficulties of EM-simulation-based design optimization is the use of adjoint sensitivities. The adjoint sensitivity approach dates back to the 1960s work of Director and Rohrer [8]. Bandler *et al.* [19] also addressed adjoint circuit sensitivities, e.g., in the context of microwave design. Interest in EM-based adjoint calculations was revived after the work [20] was published. Since 2000, a number of interesting publications addressed the application of the so-called adjoint variable method (AVM) to different numerical EM solvers. These include the time-domain transmission-line modeling (TLM) method [21], the finite-difference time-domain (FDTD) method [22], the finite-element method (FEM) [23], the method of moments (MoM) [24], the frequency domain TLM [25], and the mode-matching method (MM) [26]. These approaches can be classified as either time-domain adjoint variable methods or frequency-domain adjoint variable methods. Adjoint sensitivity is an efficient way to speed up (and, in most cases, actually make feasible) gradient-based optimization using EM solvers, as the derivative information can be obtained with no extra EM simulation of the structure in question. As mentioned before, adjoint sensitivities are currently implemented in some major commercial EM simulation packages, particularly in CST Microwave Studio [9], and in HFSS [10]. As for now, adjoint sensitivity is only available for frequency-domain solvers; however, CST plans to implement it in time-domain in one of the next releases.

Another way of improving efficiency of simulation-driven design is circuit decomposition, i.e., breaking down an EM model into smaller parts and combining them in a circuit simulator to reduce the CPU-intensity of the design process [27]-[29]. Co-simulation or co-optimization of EM/circuit is a common industry solution to blend EM-simulated components into circuit models. In general through, this is only a partial solution though because the EM-embedded co-simulation model is still subject to direct optimization.

8.3 Surrogate-Based Design Optimization

It appears that computationally efficient simulation-driven design can be performed using surrogate models. Microwave design through surrogate-based optimization (SBO) [7], [30], [31] is the main focus of this chapter. Surrogate-based methods are treated in some detail in Chapter 3. Here, only some background information is presented. The primary reason for using SBO approach in microwave engineering is to speed up the design process by shifting the optimization burden to an inexpensive yet reasonably accurate surrogate model of the device.

The generic SBO framework described here that the direct optimization of the computationally expensive EM-simulated high-fidelity model $\boldsymbol{R}_f$ is replaced by an iterative procedure [7], [32]

$$\boldsymbol{x}^{(i+1)} = \arg\min_{\boldsymbol{x}} U\left(\boldsymbol{R}_s^{(i)}(\boldsymbol{x})\right) \tag{8.2}$$

that generates a sequence of points (designs) $\boldsymbol{x}^{(i)} \in X_f$, $i = 0, 1, \ldots$, being approximate solutions to the original design problem (1). Each $\boldsymbol{x}^{(i+1)}$ is the optimal design of the surrogate model $\boldsymbol{R}_s^{(i)} : X_s^{(i)} \to R^m$, $X_s^{(i)} \subseteq R^n$, $i = 0, 1, \ldots$. $\boldsymbol{R}_s^{(i)}$ is assumed to be a computationally cheap and sufficiently reliable representation of the fine model $\boldsymbol{R}_f$, particularly in the neighborhood of the current design $\boldsymbol{x}^{(i)}$. Under these assumptions, the algorithm (8.2) is likely to produce a sequence of designs that quickly approach $\boldsymbol{x}_f^*$.

Typically, $\boldsymbol{R}_f$ is only evaluated once per iteration (at every new design $\boldsymbol{x}^{(i+1)}$) for verification purposes and to obtain the data necessary to update the surrogate model. Since the surrogate model is computationally cheap, its optimization cost (cf. (2)) can usually be neglected and the total optimization cost is determined by the evaluation of $\boldsymbol{R}_f$. The key point here is that the number of evaluations of $\boldsymbol{R}_f$ for a well performing surrogate-based algorithm is substantially smaller than for any direct optimization method (e.g., gradient-based one) [9]. Figure 8.2 shows the block diagram of the SBO optimization process.

If the surrogate model satisfies zero- and first-order consistency conditions with the fine model, i.e., $\boldsymbol{R}_s^{(i)}(\boldsymbol{x}^{(i)}) = \boldsymbol{R}_f(\boldsymbol{x}^{(i)})$ and $(\partial\boldsymbol{R}_s^{(i)}/\partial\boldsymbol{x})(\boldsymbol{x}^{(i)}) = (\partial\boldsymbol{R}_f/\partial\boldsymbol{x})(\boldsymbol{x}^{(i)})$ (verification of the latter requires $\boldsymbol{R}_f$ sensitivity data), and the algorithm (2) is enhanced by the trust region method [33], then it is provably convergent to a local fine model optimum [34]. Convergence can also be guaranteed if the algorithm (2) is enhanced by properly selected local search methods [35]. Space mapping [7], [30], [36], [37], is an example of a surrogate-based methodology that does not normally rely on the aforementioned enhancements; however, it requires the surrogate model to be constructed from the physically-based coarse model [7]. This usually gives remarkably good performance in the sense of the space mapping algorithm being able to quickly locate a satisfactory design. Unfortunately space mapping suffers from convergence problems [38] and it is sensitive to the quality of the coarse model and the type of transformations used to create the surrogate [39].

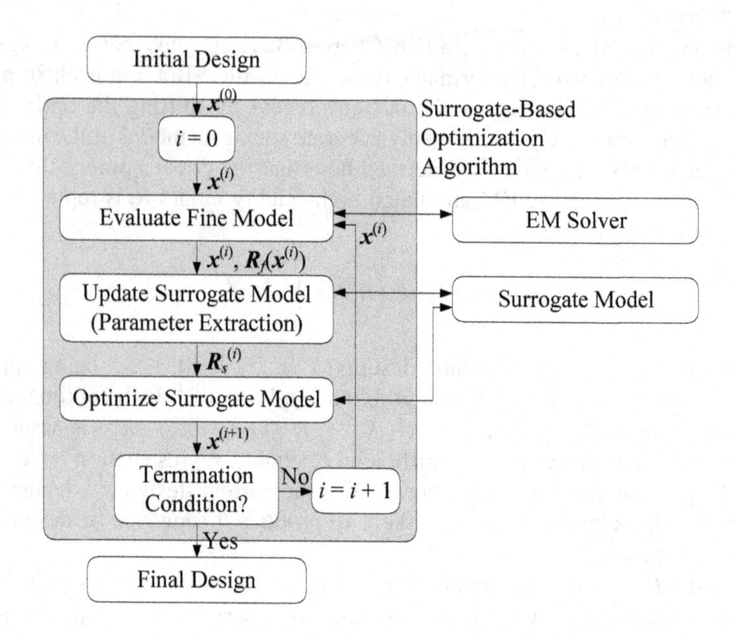

Fig. 8.2 Surrogate-based simulation-driven design optimization: the optimization burden is shifted to the computationally cheap surrogate model which is updated and re-optimized at each iteration of the main optimization loop. High-fidelity EM simulation is only performed once per iteration to verify the design produced by the surrogate model and to update the surrogate itself. The number of iterations for a well-performing SBO algorithm is substantially smaller than for conventional techniques.

8.4 Surrogate Models for Microwave Engineering

There are a number of ways to create surrogate models of microwave and radio-frequency (RF) devices and structures. They can be classified into two groups: functional and physical surrogates. Functional models are constructed from sampled high-fidelity model data using suitable function approximation techniques. Physical surrogates exploit fast but limited-accuracy models that are physically related to the original structure under consideration.

Functional surrogate models can be created using various function approximation techniques including low-order polynomials [40], radial basis functions [40], kriging [31], fuzzy systems [41], support-vector regression [42], [43], and neural networks [44]-[46], the last one probably being the most popular and successful approach in this group. Approximation models are very fast, unfortunately, to achieve good modeling accuracy, a large amount of training data obtained through massive EM simulations is necessary. Moreover, the number of data pairs necessary to ensure sufficient accuracy grows exponentially with the number of the design variables. Practical models based on function approximation techniques may need hundreds or even thousands of EM simulations in order to ensure reasonable accuracy. This is justified in the case of library models created for multiple usage but not so much in the case of ad hoc surrogates created for specific tasks such as parametric optimization, yield-driven design, and/or statistical analysis at a given (e.g., optimal) design.

Physical surrogates are based on underlying physically-based low-fidelity models of the structure of interest (denoted here as $\boldsymbol{R}_c$). Physically-based models describe the same physical phenomena as the high-fidelity model, however, in a simplified manner. In microwave engineering, the high-fidelity model describes behavior of the system in terms of the distributions of the electric and magnetic fields within (and, sometimes in its surrounding) that are calculated by solving the corresponding set of Maxwell equations [47]. Furthermore, the system performance is expressed through certain characteristics related to its input/output ports (such as so-called S-parameters [47]). All of these are obtained as a result of high-resolution electromagnetic simulation where the structure under consideration is finely discretized. In this context, the physically-based low-fidelity model of the microwave device can be obtained through:

- Analytical description of the structure using theory-based or semi-empirical formulas,
- Different level of physical description of the system. The typical example in microwave engineering is equivalent circuit [7], where the device of interest is represented using lumped components (inductors, capacitors, microstrip line models, etc.) with the operation of the circuit described directly by impedances, voltages and currents; electromagnetic fields are not directly considered,
- Low-fidelity electromagnetic simulation. This approach allows us to use the same EM solver to evaluate both the high- and low-fidelity models, however, the latter is using much coarser simulation mesh which results in degraded accuracy but much shorter simulation time.

The three groups of models have different characteristics. While analytical and equivalent-circuit models are computationally cheap, they may lack accuracy and they are typically not available for structures such as antennas and substrate-integrated circuits. On the other hand, coarsely-discretized EM models are available for any device. They are typically accurate, however, relatively expensive. The cost is a major bottleneck in adopting coarsely-discretized EM models to surrogate-based optimization in microwave engineering. One workaround is to build a function-approximation model using coarse-discretization EM-simulation data (using, e.g., kriging [31]). This, however, requires dense sampling of the design space, and should only be done locally to avoid excessive CPU cost. Table 8.1 summarizes the characteristics of the low-fidelity models available in microwave engineering. A common feature of physically-based low-fidelity models is that the amount of high-fidelity model data necessary to build a reliable surrogate model is much smaller than in case of functional surrogates [48].

Table 8.1 Physically-based low-fidelity models in microwave engineering

Model Type	CPU Cost	Accuracy	Availability
Analytical	Very cheap	Low	Rather limited
Equivalent circuit	Cheap	Decent	Limited (mostly filters)
Coarsely-discretized EM simulation	Expensive	Good to very good	Generic: available for all structures

Consider an example microstrip bandpass filter [48] shown in Fig. 8.3(a). The high-fidelity filter model is simulated using EM solver FEKO [18]. The low-fidelity model is an equivalent circuit implemented in Agilent ADS [49] (Fig. 8.3(b)). Figure 8.4(a) shows the responses (here, the modulus of transmission coefficient, $|S_{21}|$, versus frequency) of both models at certain reference design $\boldsymbol{x}^{(0)}$. While having similar shape, the responses are severely misaligned. Figure 8.4(b) shows the responses of the high-fidelity model and the surrogate constructed using the low-fidelity model and space mapping [48]. The surrogate is build using a single training point – high-fidelity model data at $\boldsymbol{x}^{(0)}$ – and exhibits very good matching with the high-fidelity model at $\boldsymbol{x}^{(0)}$. Figure 8.4(c) shows the high-fidelity and surrogate model response at a different design: the good alignment between the models is still maintained. This comes from the fact that the physically-based low-fidelity model has similar properties to the high-fidelity one and local model alignment usually results in relatively good global matching.

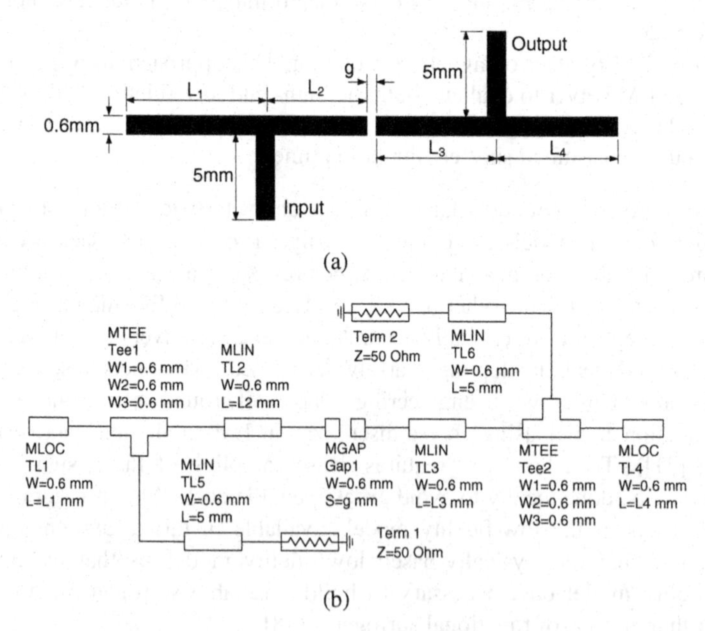

Fig. 8.3 Microstrip bandpass filter [48]: (a) geometry, (b) low-fidelity circuit model.

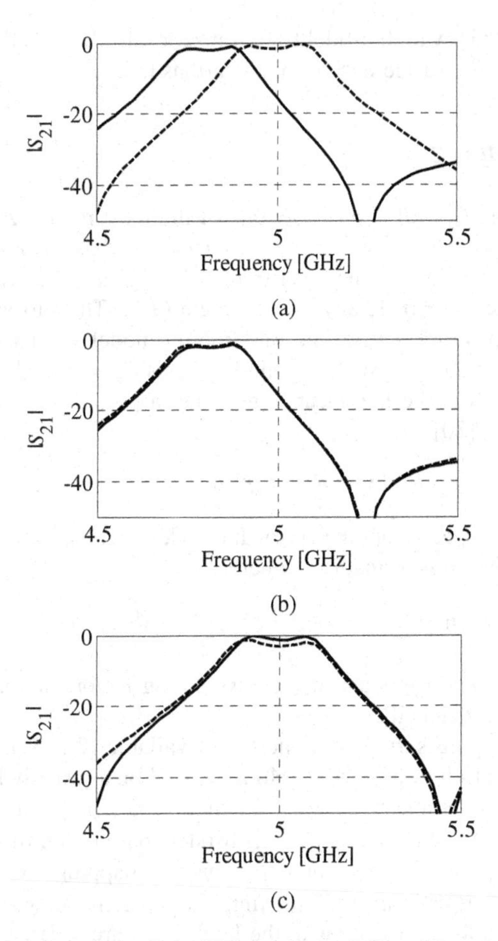

Fig. 8.4 Microstrip bandpass filter [48]: (a) high- (—) and low-fidelity (- - -) model response at the reference design $\boldsymbol{x}^{(0)}$; (b) responses of the high-fidelity model (—) and surrogate model constructed from the low-fidelity model using space mapping (- - -) at $\boldsymbol{x}^{(0)}$; (c) responses of the high-fidelity model (—) and the surrogate (- - -) at another design $\boldsymbol{x}$. The surrogate model was constructed using a single high-fidelity model response (at $\boldsymbol{x}^{(0)}$) but a good matching between the models is preserved even away from the reference design, which is due to the fact that the low-fidelity model is physically based.

8.5 Microwave Simulation-Driven Design Exploiting Physically-Based Surrogates

In this section several techniques for computationally efficient simulation-driven design of microwave structures are presented. The focus is on approaches that exploit the SBO framework (8.2) and the surrogate model constructed using an underlying physically-based low-fidelity model. Discussion covers the following methods: space mapping [7], [30], simulation-based tuning [50], shape-preserving response prediction [51], variable-fidelity optimization [52], as well as optimization through adaptively adjusted design specifications [53].

8.5.1 Space Mapping

Space mapping (SM) [7], [30] is probably one of the most recognized SBO techniques using physically-based low-fidelity (or coarse) models in microwave engineering. Space mapping exploits the algorithm (8.2) to generate a sequence of approximate solutions $\boldsymbol{x}^{(i)}$, $i = 0, 1, 2, \ldots$, to problem (8.1). The surrogate model at iteration i, $\boldsymbol{R}_s^{(i)}$, is constructed from the low-fidelity model so that the misalignment between $\boldsymbol{R}_s^{(i)}$ and the fine model is minimized using so-called parameter extraction process, which is the nonlinear minimization problem by itself [7]. The surrogate is defined as [30]

$$\boldsymbol{R}_s^{(i)}(\boldsymbol{x}) = \boldsymbol{R}_{s.g}(\boldsymbol{x}, \boldsymbol{p}^{(i)}) \tag{8.3}$$

where $\boldsymbol{R}_{s.g}$ is a generic space mapping surrogate model, i.e., the low-fidelity model composed with suitable transformations, whereas

$$\boldsymbol{p}^{(i)} = \arg\min_{\boldsymbol{p}} \sum\nolimits_{k=0}^{i} w_{i.k} \,||\, \boldsymbol{R}_f(\boldsymbol{x}^{(k)}) - \boldsymbol{R}_{s.g}(\boldsymbol{x}^{(k)}, \boldsymbol{p}) \,|| \tag{8.4}$$

is a vector of model parameters and $w_{i.k}$ are weighting factors; a common choice of $w_{i.k}$ is $w_{i.k} = 1$ for all i and all k.

Various space mapping surrogate models are available [7], [30]. They can be roughly categorized into four groups: (i) Models based on a (usually linear) distortion of coarse model parameter space, e.g., input space mapping of the form $\boldsymbol{R}_{s.g}(\boldsymbol{x}, \boldsymbol{p}) = \boldsymbol{R}_{s.g}(\boldsymbol{x}, \boldsymbol{B}, \boldsymbol{c}) = \boldsymbol{R}_c(\boldsymbol{B}\cdot\boldsymbol{x} + \boldsymbol{c})$ [7]; (ii) Models based on a distortion of the coarse model response, e.g., output space mapping of the form $\boldsymbol{R}_{s.g}(\boldsymbol{x}, \boldsymbol{p}) = \boldsymbol{R}_{s.g}(\boldsymbol{x}, \boldsymbol{d}) = \boldsymbol{R}_c(\boldsymbol{x}) + \boldsymbol{d}$ [30]; (iii) Implicit space mapping, where the parameters used to align the surrogate with the fine model are separate from the design variables, i.e., $\boldsymbol{R}_{s.g}(\boldsymbol{x}, \boldsymbol{p}) = \boldsymbol{R}_{s.g}(\boldsymbol{x}, \boldsymbol{x}_p) = \boldsymbol{R}_{c.i}(\boldsymbol{x}, \boldsymbol{x}_p)$, with $\boldsymbol{R}_{c.i}$ being the coarse model dependent on both the design variables $\boldsymbol{x}$ and so-called preassigned parameters $\boldsymbol{x}_p$ (e.g., dielectric constant, substrate height) that are normally fixed in the fine model but can be freely altered in the coarse model [30]; (iv) Custom models exploiting parameters characteristic to a given design problem; the most characteristic example is the so-called frequency space mapping $\boldsymbol{R}_{s.g}(\boldsymbol{x}, \boldsymbol{p}) = \boldsymbol{R}_{s.g}(\boldsymbol{x}, \boldsymbol{F}) = \boldsymbol{R}_{c.f}(\boldsymbol{x}, \boldsymbol{F})$ [7], where $\boldsymbol{R}_{c.f}$ is a frequency-mapped coarse model, i.e., the coarse model evaluated at frequencies different from the original frequency sweep for the fine model, according to the mapping $\omega \rightarrow f_1 + f_2\omega$, with $\boldsymbol{F} = [f_1\ f_2]^T$.

Space mapping usually comprises combined transformations. At instance, a surrogate model employing input, output, and frequency SM transformations would be $\boldsymbol{R}_{s.g}(\boldsymbol{x}, \boldsymbol{p}) = \boldsymbol{R}_{s.g}(\boldsymbol{x}, \boldsymbol{c}, \boldsymbol{d}, \boldsymbol{F}) = \boldsymbol{R}_{c.f}(\boldsymbol{x} + \boldsymbol{c}, \boldsymbol{F}) + \boldsymbol{d}$. The rationale for this is that a properly chosen mapping may significantly improve the performance of the space mapping algorithm, however, the optimal selection of the mapping type for a given design problem is not trivial [38]. Work has been done to ease the selection process for a given design problem [39], [48]. However, regardless of the mapping choice, coarse model accuracy is what principally affects the performance of the space mapping design process. One can quantify the quality of the surrogate model through rigorous convergence conditions [38]. These conditions, although useful for developing more efficient space mapping algorithms and automatic surrogate model selection techniques, cannot usually be verified because of the limited amount of data available from the fine model. In practice, the most important criterion for assessing the quality or accuracy of the coarse model is still visual inspection of the fine and coarse model responses at certain points and/or examining absolute error measures such as $||\boldsymbol{R}_f(\boldsymbol{x}) - \boldsymbol{R}_c(\boldsymbol{x})||$.

The coarse model is the most important factor that affects the performance of the space mapping algorithm. The first stems from accuracy. Coarse model accuracy (more generally, the accuracy of the space mapping surrogate [38]) is the main factor that determines the efficiency of the algorithm in terms of finding a satisfactory design. The more accurate the coarse model, the smaller the number of fine model evaluations necessary to complete the optimization process. If the coarse model is insufficiently accurate, the space mapping algorithm may need more fine model evaluations or may even fail to find a good quality design.

The second important characteristic is the evaluation cost. It is essential that the coarse model is computationally much cheaper than the fine model because both parameter extraction (8.4) and surrogate optimization (8.2) require large numbers of coarse model evaluations. Ideally, the evaluation cost of the coarse model should be negligible when compared to the evaluation cost of the fine model, in which case the total computational cost of the space mapping optimization process is merely determined by the necessary number of fine model evaluations. If the evaluation time of the coarse model is too high, say, larger than 1% of the fine model evaluation time, the computational cost of surrogate model optimization and, especially, parameter extraction, start playing important roles in the total cost of space mapping optimization and may even determine it. Therefore, practical applicability of space mapping is limited to situations where the coarse model is computationally much cheaper than the fine model. Majority of SM models reported in the literature (e.g., [7], [30], [36]) concern microstrip filters, transformers or junctions where fast and reliable equivalent circuit coarse models are easily available.

8.5.2 Simulation-Based Tuning and Tuning Space Mapping

Tuning is ubiquitous in engineering practice. It is usually associated with the process of manipulating free or tunable parameters of a device or system after that device or system has been manufactured. The traditional purpose of permitting tunable elements is (8.1) to facilitate user-flexibility in achieving a desired response or behavior from a manufactured outcome during its operation, or (8.2) to correct inevitable postproduction manufacturing defects, small due perhaps to tolerances, or large due perhaps to faults in the manufacturing process [54]. Tuning of an engineering design can be seen, in essence, as a user- or robot-directed optimization process.

Tuning space mapping (TSM) [50] combines the concept of tuning, widely used in microwave engineering [55], [56], and space mapping. It is an iterative optimization procedure that assumes the existence of two surrogate models: both are less accurate but computationally much cheaper than the fine model. The first model is a so-called tuning model $\boldsymbol{R}_t$ that contains relevant fine model data (typically a fine model response) at the current iteration point and tuning parameters (typically implemented through circuit elements inserted into tuning ports). The tunable parameters are adjusted so that the model $\boldsymbol{R}_t$ satisfies the design specifications. The second model, $\boldsymbol{R}_c$ is used for calibration purposes: it allows us to translate the change of the tuning parameters into relevant changes of the actual design variables; $\boldsymbol{R}_c$ is dependent on three sets of variables: design parameters, tuning parameters (which are actually the same parameters as the ones used in $\boldsymbol{R}_t$), and SM parameters that are adjusted using the usual parameter extraction process [7] in order to have the model $\boldsymbol{R}_c$ meet certain matching conditions. Typically, the model $\boldsymbol{R}_c$ is a standard SM surrogate (i.e., a coarse model composed with suitable transformations) enhanced by the same or corresponding tuning elements as the model $\boldsymbol{R}_t$. The conceptual illustrations of the fine model, the tuning model and the calibration model are shown in Fig. 8.5.

The iteration of the TSM algorithm consists of two steps: optimization of the tuning model and a calibration procedure. First, the current tuning model $\boldsymbol{R}_t^{(i)}$ is built using fine model data at point $\boldsymbol{x}^{(i)}$. In general, because the fine model with inserted tuning ports is not identical to the original structure, the tuning model response may not agree with the response of the fine model at $\boldsymbol{x}^{(i)}$ even if the values of the tuning parameters $\boldsymbol{x}_t$ are zero, so that these values must be adjusted to, say, $\boldsymbol{x}_{t.0}^{(i)}$, in order to obtain alignment [50]:

$$\boldsymbol{x}_{t.0}^{(i)} = \arg\min_{\boldsymbol{x}_t} \left\| \boldsymbol{R}_f(\boldsymbol{x}^{(i)}) - \boldsymbol{R}_t^{(i)}(\boldsymbol{x}_t) \right\| \tag{8.5}$$

In the next step, one optimizes $\boldsymbol{R}_t^{(i)}$ to have it meet the design specifications. Optimal values of the tuning parameters $\boldsymbol{x}_{t.1}^{(i)}$ are obtained as follows:

$$\boldsymbol{x}_{t.1}^{(i)} = \arg\min_{\boldsymbol{x}_t} U\left(\boldsymbol{R}_t^{(i)}(\boldsymbol{x}_t)\right) \tag{8.6}$$

Having $\boldsymbol{x}_{t.1}^{(i)}$, the calibration procedure is performed to determine changes in the design variables that yield the same change in the calibration model response as that caused by $\boldsymbol{x}_{t.1}^{(i)} - \boldsymbol{x}_{t.0}^{(i)}$ [50]. First one adjusts the SM parameters $\boldsymbol{p}^{(i)}$ of the calibration model to obtain a match with the fine model response at $\boldsymbol{x}^{(i)}$

$$\boldsymbol{p}^{(i)} = \arg\min_{\boldsymbol{p}} \left\| \boldsymbol{R}_f(\boldsymbol{x}^{(i)}) - \boldsymbol{R}_c(\boldsymbol{x}^{(i)}, \boldsymbol{p}, \boldsymbol{x}_{t.0}^{(i)}) \right\|. \tag{8.7}$$

The calibration model is then optimized with respect to the design variables in order to obtain the next iteration point $\boldsymbol{x}^{(i+1)}$

$$\boldsymbol{x}^{(i+1)} = \arg\min_{\boldsymbol{x}} \left\| \boldsymbol{R}_t^{(i)}(\boldsymbol{x}_{t.1}^{(i)}) - \boldsymbol{R}_c(\boldsymbol{x}, \boldsymbol{p}^{(i)}, \boldsymbol{x}_{t.0}^{(i)}) \right\|. \tag{8.8}$$

Note that $\boldsymbol{x}_{t.0}^{(i)}$ is used in (8.7), which corresponds to the state of the tuning model after performing the alignment procedure (8.5), and $\boldsymbol{x}_{t.1}^{(i)}$ in (8.8), which corresponds to the optimized tuning model (cf. (6)). Thus, (8.7) and (8.8) allow finding the change of design variable values $\boldsymbol{x}^{(i+1)} - \boldsymbol{x}^{(i)}$ necessary to compensate the effect of changing the tuning parameters from $\boldsymbol{x}_{t.0}^{(i)}$ to $\boldsymbol{x}_{t.1}^{(i)}$.

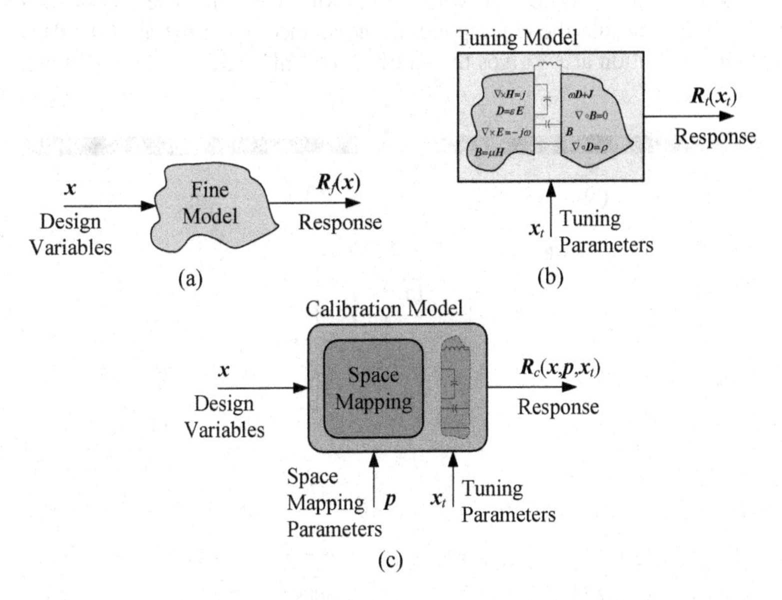

Fig. 8.5 Conceptual illustrations of the fine model, the tuning model and the calibration model: (a) the fine model is typically based on full-wave simulation, (b) the tuning model exploits the fine model "image" (e.g., in the form of *S*-parameters corresponding to the current design imported to the tuning model using suitable data components) and a number of circuit-theory-based tuning elements, (c) the calibration model is usually a circuit equivalent dependent on the same design variables as the fine model, the same tuning parameters as the tuning model and, additionally, a set of space mapping parameters used to align the calibration model with both the fine and the tuning model during the calibration process.

It should be noted that the calibration procedure described here represents the most generic approach. In some cases, there is a formula that establishes an analytical relation between the design variables and the tuning parameters so that the updated design can be found simply by applying that formula [50]. In particular, the calibration formula may be just a linear function so that $\boldsymbol{x}^{(i+1)} = \boldsymbol{x}^{(i)} + \boldsymbol{s}^{(i)} * (\boldsymbol{x}_{t.1}^{(i)} - \boldsymbol{x}_{t.0}^{(i)})$, where $\boldsymbol{s}^{(i)}$ is a real vector and $*$ denotes a Hadamard product (i.e., component-wise multiplication) [50]. If the analytical calibration is possible, there is no need to use the calibration model. Other approaches to the calibration process can be found in the literature [50], [57]. In some cases (e.g., [57]), the tuning parameters may be in identity relation with the design variables, which simplified the implementation of the algorithm.

The operation of the tuning space mapping algorithm can be clarified using a simple example of a microstrip transmission line [50]. The fine model is implemented in Sonnet ***em*** [17] (Fig. 8.6(a)), and the fine model response is taken as the inductance of the line as a function of the line's length. The original length of the line is chosen to be $\boldsymbol{x}^{(0)}$ = 400 mil with a width of 0.635 mm. The goal is to find a length of line such that the corresponding inductance is 6.5 nH at 300 MHz. The Sonnet ***em*** simulation at $\boldsymbol{x}^{(0)}$ gives the value of 4.38 nH, i.e., $\boldsymbol{R}_f(\boldsymbol{x}^{(0)})$ = 4.38 nH.

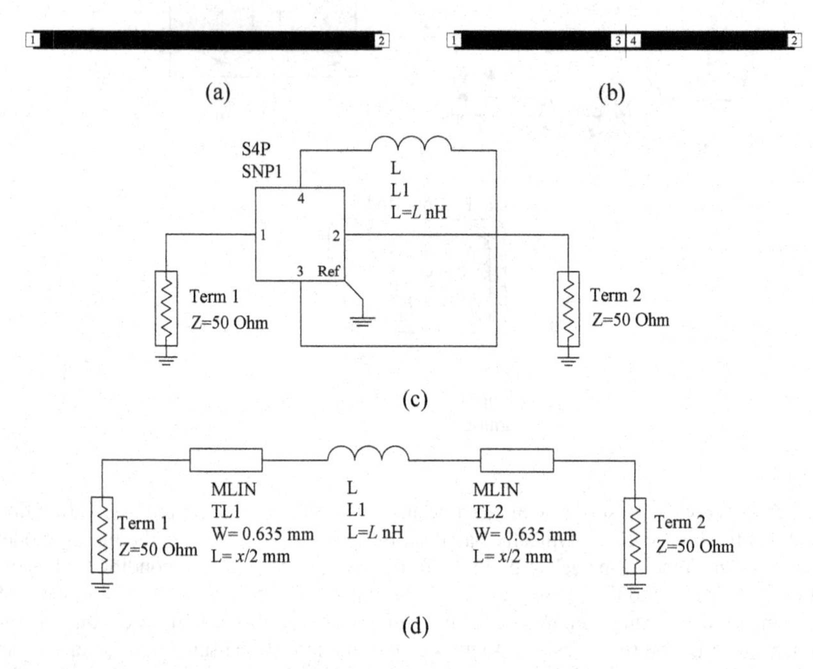

Fig. 8.6 TSM optimization of the microstrip line [50]: (a) original structure of the microstrip line in Sonnet, (b) the microstrip line after being divided and with inserted the co-calibrated ports, (c) tuning model, (d) calibration model.

The tuning model $\boldsymbol{R}_t$ is developed by dividing the structure in Fig. 8.6(a) into two separate parts and adding the two tuning ports as shown in Fig. 8.6(b). A small inductor is then inserted between these ports as a tuning element. The tuning model is implemented in Agilent ADS [47] and shown in Fig. 8.6(c). The model contains the fine model data at the initial design in the form of the S4P element as well as the tuning element (inductor). Because of Sonnet's co-calibrated ports technology [56], there is a perfect agreement between the fine and tuning model responses when the value of the tuning inductance is zero, so that $\boldsymbol{x}_{t.0}^{(0)}$ is zero in this case.

Next, the tuning model should be optimized to meet the target inductance of 6.5 nH. The optimized value of the tuning inductance is $\boldsymbol{x}_{t.1}^{(0)}$ = 2.07 nH.

The calibration model is shown in Fig. 8.6(d). Here, the dielectric constant of the microstrip element is used as a space mapping parameter $\boldsymbol{p}$. Original value of this parameter, 9.8, is adjusted using (8.7) to 23.7 so that the response of the calibration model is 4.38 nH at 400 mil, i.e., it agrees with the fine model response at $\boldsymbol{x}^{(0)}$. Now, the new value of the microstrip length is obtained using (8.8). In particular, one optimizes $\boldsymbol{x}$ with the tuning inductance set to $\boldsymbol{x}_{t.0}{}^{(0)} = 0$ nH to match the total inductance of the calibration model to the optimized tuning model response, 6.5 nH. The result is $\boldsymbol{x}^{(1)} = 585.8$ mil; the fine model response at $\boldsymbol{x}^{(1)}$ obtained by Sonnet ***em*** simulation is 6.48 nH. This result can be further improved by performing a second iteration of the TSM, which gives the length of the microstrip line equal to $\boldsymbol{x}^{(2)} = 588$ mil and its corresponding inductance of 6.5 nH.

Simulation-based tuning and tuning space mapping can be extremely efficient as demonstrated in Chapter 12. In particular, a satisfactory design can be obtained after just one or two iterations. However, the tuning methodology has limited applications. It it well suited for structures such as microstrip filters but it can hardly be applied for radiating structures (antennas). Also, tuning of cross-sectional parameters (e.g., microstrip width) is not straightforward [50]. On the other hand, the tuning procedure is invasive in the sense that the structure may need to be cut. The fine model simulator must allow such cuts and allow tuning elements to be inserted. This can be done using, e.g., Sonnet ***em*** [17]. Also, EM simulation of a structure containing a large number of tuning ports is computationally far more expensive than the simulation of the original structure (without the ports). Depending on the number of design variables, the number of tuning ports may be as large as 30, 50 or more [50], which may increase the simulation time by one order of magnitude or more. Nevertheless, recent results presented in [58] indicate possibility of speeding up the tuning process by using so-called reduced structures.

8.5.3 Shape-Preserving Response Prediction

Shape-preserving response prediction (SPRP) [51] is a response correction technique that takes advantage of the similarity between responses of the high- and low-fidelity models in a very straightforward way. SPRP assumes that the change of the high-fidelity model response due to the adjustment of the design variables can be predicted using the actual changes of the low-fidelity model response. Therefore, it is critically important that the low-fidelity model is physically based, which ensures that the effect of the design parameter variations on the model response is similar for both models. In microwave engineering this property is likely to hold, particularly if the low-fidelity model is the coarsely-discretization structure evaluated using the same EM solver as the one used to simulate the high-fidelity model.

The change of the low-fidelidy model response is described by the translation vectors corresponding to a certain (finite) number of characteristic points of the model's response. These translation vectors are subsequently used to predict the change of the high-fidelity model response with the actual response of $\boldsymbol{R}_f$ at the current iteration point, $\boldsymbol{R}_f(\boldsymbol{x}^{(i)})$, treated as a reference.

Figure 8.7(a) shows the example low-fidelity model response, $|S_{21}|$ in the frequency range 8 GHz to 18 GHz, at the design $\boldsymbol{x}^{(i)}$, as well as the low-fidelity model response at some other design $\boldsymbol{x}$. The responses come from the double folded stub bandstop filter example considered in [51]. Circles denote characteristic points of $\boldsymbol{R}_c(\boldsymbol{x}^{(i)})$, selected here to represent $|S_{21}| = -3$ dB, $|S_{21}| = -20$ dB, and the local $|S_{21}|$ maximum (at about 13 GHz). Squares denote corresponding characteristic points for $\boldsymbol{R}_c(\boldsymbol{x})$, while line segments represent the translation vectors ("shift") of the characteristic points of $\boldsymbol{R}_c$ when changing the design variables from $\boldsymbol{x}^{(i)}$ to $\boldsymbol{x}$. Since the low-fidelity model is physically based, the high-fidelity model response at the given design, here, $\boldsymbol{x}$, can be predicted using the same translation vectors applied to the corresponding characteristic points of the high-fidelity model response at $\boldsymbol{x}^{(i)}$, $\boldsymbol{R}_f(\boldsymbol{x}^{(i)})$. This is illustrated in Fig. 8.7(b).

Rigorous formulation of SPRP uses the following notation concerning the responses: $\boldsymbol{R}_f(\boldsymbol{x}) = [R_f(\boldsymbol{x},\omega_1) \ldots R_f(\boldsymbol{x},\omega_m)]^T$ and $\boldsymbol{R}_c(\boldsymbol{x}) = [R_c(\boldsymbol{x},\omega_1) \ldots R_c(\boldsymbol{x},\omega_m)]^T$, where ω_j, $j = 1, \ldots, m$, is the frequency sweep. Let $\boldsymbol{p}_j^f = [\omega_j^f \; r_j^f]^T$, $\boldsymbol{p}_j^{c0} = [\omega_j^{c0} \; r_j^{c0}]^T$, and $\boldsymbol{p}_j^c = [\omega_j^c \; r_j^c]^T$, $j = 1, \ldots, K$, denote the sets of characteristic points of $\boldsymbol{R}_f(\boldsymbol{x}^{(i)})$, $\boldsymbol{R}_c(\boldsymbol{x}^{(i)})$ and $\boldsymbol{R}_c(\boldsymbol{x})$, respectively. Here, ω and r denote the frequency and magnitude components of the respective point. The translation vectors of the low-fidelity model response are defined as $\boldsymbol{t}_j = [\omega_j^t \; r_j^t]^T, j = 1,\ldots, K$, where $\omega_j^t = \omega_j^c - \omega_j^{c0}$ and $r_j^t = r_j^c - r_j^{c0}$.

The shape-preserving response prediction surrogate model is defined as follows

$$\boldsymbol{R}_s^{(i)}(\boldsymbol{x}) = [R_s^{(i)}(\boldsymbol{x},\omega_1) \; \ldots \; R_s^{(i)}(\boldsymbol{x},\omega_m)]^T \tag{8.9}$$

where

$$R_s^{(i)}(\boldsymbol{x},\omega_j) = R_{f.i}(\boldsymbol{x}^{(i)}, F(\omega_j, \{-\omega_k^t\}_{k=1}^K)) + R(\omega_j, \{r_k^t\}_{k=1}^K) \tag{8.10}$$

for $j = 1, \ldots, m$. $R_{f.i}(\boldsymbol{x},\omega_1)$ is an interpolation of $\{R_f(\boldsymbol{x},\omega_1), \ldots, R_f(\boldsymbol{x},\omega_m)\}$ onto the frequency interval $[\omega_1,\omega_m]$.

The scaling function F interpolates the data pairs $\{\omega_1,\omega_1\}$, $\{\omega_1^f,\omega_1^f-\omega_1^t\}$, …, $\{\omega_K^f,\omega_K^f-\omega_K^t\}$, $\{\omega_m,\omega_m\}$, onto the frequency interval $[\omega_1,\omega_m]$. The function R does a similar interpolation for data pairs $\{\omega_1,r_1\}$, $\{\omega_1^f,r_1^f-r_1^t\}$, …, $\{\omega_K^f,r_K^f-r_K^t\}$, $\{\omega_m,r_m\}$; here $r_1 = R_c(\boldsymbol{x},\omega_1) - R_c(\boldsymbol{x}^r,\omega_1)$ and $r_m = R_c(\boldsymbol{x},\omega_m) - R_c(\boldsymbol{x}^r,\omega_m)$. In other words, the function F translates the frequency components of the characteristic points of $\boldsymbol{R}_f(\boldsymbol{x}^{(i)})$ to the frequencies at which they should be located according to the translation vectors $\boldsymbol{t}_j$, while the function R adds the necessary magnitude component.

It should be emphasized that shape-preserving response prediction a physically-based low-fidelity model is critical for the method's performance. On the other hand, SPRP can be characterized as a non-parametric, nonlinear and design-variable dependent response correction, and it is therefore distinct from any known space mapping approaches. Another important feature that differentiates SPRP from space mapping and other approaches (e.g., tuning) is implementation simplicity. Unlike space mapping, SPRP does not use any extractable parameters (which are normally found by solving a separate nonlinear minimization problem), the problem of the surrogate model selection [38], [39] (i.e., the choice of the transformation and its parameters) does not exist, and the interaction between the models is very simple (only through the translation vectors (8.3), (8.4)). Unlike tuning methodologies, SPRP does not require any modification of the optimized structure (such as "cutting" and insertion of the tuning components [50]).

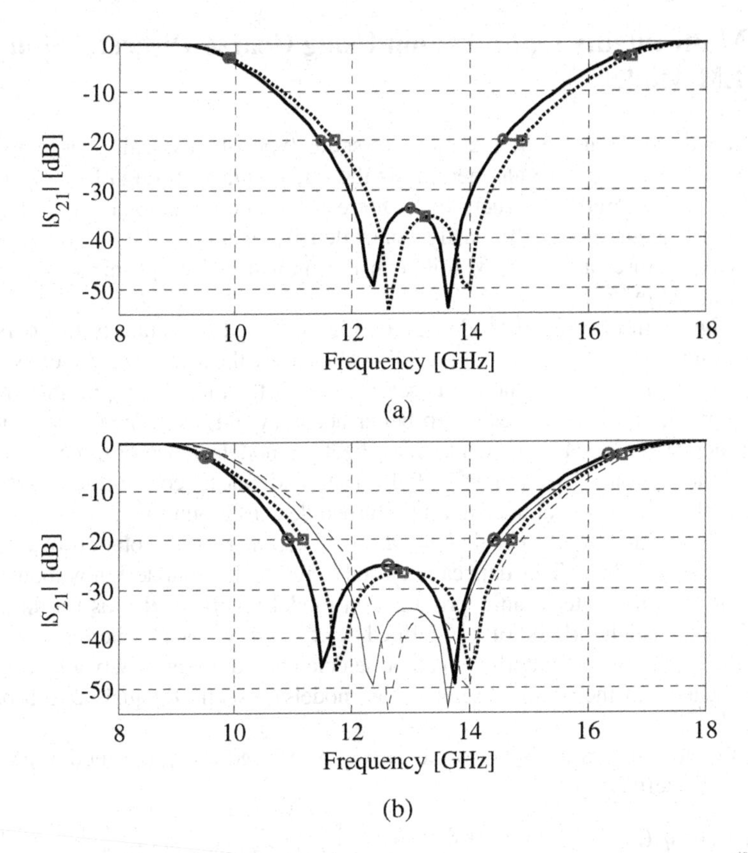

Fig. 8.7 SPRP concept: (a) Example low-fidelity model response at the design $\boldsymbol{x}^{(i)}$, $\boldsymbol{R}_c(\boldsymbol{x}^{(i)})$ (solid line), the low-fidelity model response at $\boldsymbol{x}$, $\boldsymbol{R}_c(\boldsymbol{x})$ (dotted line), characteristic points of $\boldsymbol{R}_c(\boldsymbol{x}^{(i)})$ (circles) and $\boldsymbol{R}_c(\boldsymbol{x})$ (squares), and the translation vectors (short lines); (b) High-fidelity model response at $\boldsymbol{x}^{(i)}$, $\boldsymbol{R}_f(\boldsymbol{x}^{(i)})$ (solid line) and the predicted high-fidelity model response at $\boldsymbol{x}$ (dotted line) obtained using SPRP based on characteristic points of Fig. 8.1(a); characteristic points of $\boldsymbol{R}_f(\boldsymbol{x}^{(i)})$ (circles) and the translation vectors (short lines) were used to find the characteristic points (squares) of the predicted high-fidelity model response; low-fidelity model responses $\boldsymbol{R}_c(\boldsymbol{x}^{(i)})$ and $\boldsymbol{R}_c(\boldsymbol{x})$ are plotted using thin solid and dotted line, respectively [51].

If one-to-one correspondence between the characteristic points of the high- and low-fidelity model is not satisfied despite use of the coarse-mesh EM-based low-fidelity model, the sets of corresponding characteristic points can be generated based not on distinctive features of the responses (e.g., characteristic response levels or local minima/maxima) but by introducing additional points that are equally spaced in frequency and inserted between well defined points [51]. These additional points not only ensure that the shape-preserving response prediction model (8.3), (8.4) is well defined but also allows us to capture the response shape of the models even though the number of distinctive features (e.g., local maxima and minima) is different for high- and low-fidelity models.

8.5.4 Multi-fidelity Optimization Using Coarse-Discretization EM Models

As mentioned in Section 8.4, the most versatile type of physically-based low-fidelity model in microwave engineering is the one obtained through EM simulation of coarsely-discretized structure of interest. The computational cost of the model and its accuracy can be easily controlled by changing the discretization density. This feature has been exploited in the multi-fidelity optimization algorithm introduced in [52].

The design optimization methodology of [52] is based on a family of coarse-discretization models $\{\boldsymbol{R}_{c.j}\}$, $j = 1,\ldots, K$, all evaluated by the same EM solver as the one used for the high-fidelity model. Discretization of the model $\boldsymbol{R}_{c.j+1}$ is finer than that of the model $\boldsymbol{R}_{c.j}$, which results in better accuracy but also longer evaluation time. In practice, the number of coarse-discretization models is two or three.

Having the optimized design $\boldsymbol{x}^{(K)}$ of the last (and finest) coarse-discretization model $\boldsymbol{R}_{c.K}$, the model is evaluated at all perturbed designs around $\boldsymbol{x}^{(K)}$, i.e., at $\boldsymbol{x}_k^{(K)} = [x_1^{(K)} \ldots x_k^{(K)} + \text{sign}(k)\cdot d_k \ldots x_n^{(K)}]^T$, $k = -n, -n+1, \ldots, n-1, n$. A notation of $\boldsymbol{R}^{(k)} = \boldsymbol{R}_{c.K}(\boldsymbol{x}_k^{(K)})$ is adopted here. This data can be used to refine the final design without directly optimizing $\boldsymbol{R}_f$. Instead, an approximation model involving $\boldsymbol{R}^{(k)}$ is set up and optimized in the neighborhood of $\boldsymbol{x}^{(K)}$ defined as $[\boldsymbol{x}^{(K)} - \boldsymbol{d}, \boldsymbol{x}^{(K)} + \boldsymbol{d}]$, where $\boldsymbol{d} = [d_1\ d_2 \ldots d_n]^T$. The size of the neighborhood can be selected based on sensitivity analysis of $\boldsymbol{R}_{c.1}$ (the cheapest of the coarse-discretization models); usually $\boldsymbol{d}$ equals 2 to 5 percent of $\boldsymbol{x}^{(K)}$.

Here, the approximation is performed using a reduced quadratic model $\boldsymbol{q}(\boldsymbol{x}) = [q_1\ q_2 \ldots q_m]^T$, defined as

$$q_j(\boldsymbol{x}) = q_j([x_1 \ldots x_n]^T) = \lambda_{j.0} + \lambda_{j.1}x_1 + \ldots + \lambda_{j.n}x_n + \lambda_{j.n+1}x_1^2 + \ldots + \lambda_{j.2n}x_n^2 \tag{8.11}$$

Coefficients $\lambda_{j.r}$, $j = 1, \ldots, m$, $r = 0, 1, \ldots, 2n$, can be uniquely obtained by solving the linear regression problems

$$\begin{bmatrix} 1 & x_{-n.1}^{(K)} & \cdots & x_{-n.n}^{(K)} & (x_{-n.1}^{(K)})^2 & \cdots & (x_{-n.n}^{(K)})^2 \\ \vdots & \vdots & & \vdots & \vdots & & \vdots \\ 1 & x_{0.1}^{(K)} & \cdots & x_{0.n}^{(K)} & (x_{0.1}^{(K)})^2 & \cdots & (x_{-n.n}^{(K)})^2 \\ \vdots & \vdots & & \vdots & \vdots & & \vdots \\ 1 & x_{n.1}^{(K)} & \cdots & x_{n.n}^{(K)} & (x_{n.1}^{(K)})^2 & \cdots & (x_{-n.n}^{(K)})^2 \end{bmatrix} \cdot \begin{bmatrix} \lambda_{j.0} \\ \lambda_{j.1} \\ \vdots \\ \lambda_{j.2n} \end{bmatrix} = \begin{bmatrix} R_j^{(-n)} \\ \vdots \\ R_j^{(0)} \\ \vdots \\ R_j^{(n)} \end{bmatrix} \tag{8.12}$$

where $x_{k.j}^{(K)}$ is a jth component of the vector $\boldsymbol{x}_k^{(K)}$, and $R_j^{(k)}$ is a jth component of the vector $\boldsymbol{R}^{(k)}$, i.e.,

In order to account for unavoidable misalignment between $\boldsymbol{R}_{c.K}$ and $\boldsymbol{R}_f$, instead of optimizing the quadratic model $\boldsymbol{q}$, it is recommended to optimize a corrected model $\boldsymbol{q}(\boldsymbol{x}) + [\boldsymbol{R}_f(\boldsymbol{x}^{(K)}) - \boldsymbol{R}_{c.K}(\boldsymbol{x}^{(K)})]$ that ensures a zero-order consistency [34] between $\boldsymbol{R}_{c.K}$ and $\boldsymbol{R}_f$. The refined design can be then found as

$$\boldsymbol{x}^* = \arg \min_{\boldsymbol{x}^{(K)} - \boldsymbol{d} \le \boldsymbol{x} \le \boldsymbol{x}^{(K)} + \boldsymbol{d}} U(\boldsymbol{q}(\boldsymbol{x}) + [\boldsymbol{R}_f(\boldsymbol{x}^{(K)}) - \boldsymbol{R}_{c.K}(\boldsymbol{x}^{(K)})]) \tag{8.13}$$

This kind of correction is also known as output space mapping [30]. If necessary, the step (8.4) can be performed a few times starting from a refined design, i.e., $\boldsymbol{x}^* = \text{argmin}\{\boldsymbol{x}^{(K)} - \boldsymbol{d} \le \boldsymbol{x} \le \boldsymbol{x}^{(K)} + \boldsymbol{d} : U(\boldsymbol{q}(\boldsymbol{x}) + [\boldsymbol{R}_f(\boldsymbol{x}^*) - \boldsymbol{R}_{c.K}(\boldsymbol{x}^*)])\}$ (each iteration requires only one evaluation of $\boldsymbol{R}_f$).

The design optimization procedure can be summarized as follows (input arguments are: initial design $\boldsymbol{x}^{(0)}$ and the number of coarse-discretization models K):

1. Set $j = 1$;
2. Optimize coarse-discretization model $\boldsymbol{R}_{c.j}$ to obtain a new design $\boldsymbol{x}^{(j)}$ using $\boldsymbol{x}^{(j-1)}$ as a starting point;
3. Set $j = j + 1$; if $j < K$ go to 2;
4. Obtain a refined design $\boldsymbol{x}^*$ as in (8.13);
5. END;

Note that the original model $\boldsymbol{R}_f$ is only evaluated at the final stage (step 4) of the optimization process. Operation of the algorithm in illustrated in Fig. 8.8. Coarse-discretization models can be optimized using any available algorithm.

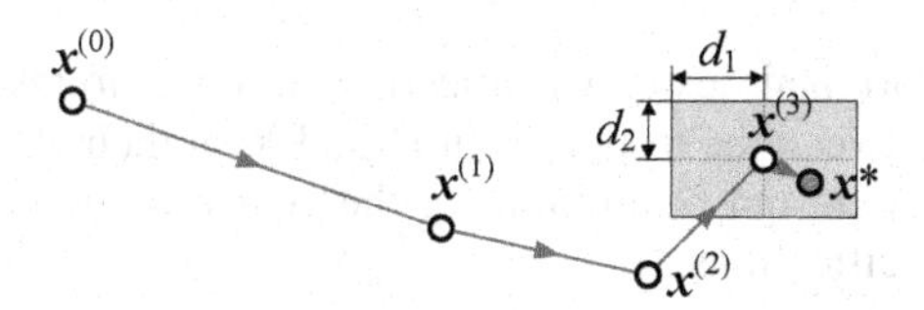

Fig. 8.8 Operation of the multi-fidelity design optimization procedure for $K = 3$ (three coarse-discretization models). The design $\boldsymbol{x}^{(j)}$ is obtained as the optimal solution of the model $\boldsymbol{R}_{c.j}$, $j = 1, 2, 3$. A reduced second-order approximation model $\boldsymbol{q}$ is set up in the neighborhood of $\boldsymbol{x}^{(3)}$ (gray area) and the final design $\boldsymbol{x}^*$ is obtained by optimizing a reduced $\boldsymbol{q}$ as in (8.4).

Typically, the major difference between the responses of $\boldsymbol{R}_f$ and coarse-discretization models $\boldsymbol{R}_{c.j}$ is that they are shifted in frequency. This difference can be easily absorbed by frequency-shifting the design specifications while optimizing a model $\boldsymbol{R}_{c.j}$. More specifically, suppose that the design specifications are described as $\{\omega_{k.L}, \omega_{k.H}; s_k\}$, $k = 1, \ldots, n_s$, (e.g., specifications $|S_{21}| \ge -3$ dB for 3 GHz $\le \omega \le 4$ GHz, $|S_{21}| \le -20$ dB for 1 GHz $\le \omega \le 2$ GHz and $|S_{21}| \le -20$ dB for 5 GHz $\le \omega \le 7$ GHz would be described as $\{3, 4; -3\}$, $\{1, 2; -20\}$, and $\{5, 7; -20\}$). If the average frequency shift between responses of $\boldsymbol{R}_{c.j}$ and $\boldsymbol{R}_{c.j+1}$ is $\Delta\omega$, this difference can be absorbed by modifying the design specifications to $\{\omega_{k.L} - \Delta\omega, \omega_{k.H} - \Delta\omega, s_k\}$, $k = 1, \ldots, n_s$.

As mentioned above, the number K of coarse-discretization models is typically two or three. The first coarse-discretization model $\boldsymbol{R}_{c.1}$ should be set up so that its evaluation time is at least 30 to 100 times shorter than the evaluation time of the fine model. The reason is that the initial design may be quite poor so that the expected number of evaluations of $\boldsymbol{R}_{c.1}$ is usually large. By keeping $\boldsymbol{R}_{c.1}$ fast, one can control the computational overhead related to its optimization. Accuracy of $\boldsymbol{R}_{c.1}$ is not critical because its optimal design is only supposed to give a rough estimate of the fine model optimum. The second (and, possibly third) coarse-discretization model should be more accurate but still at least about 10 times faster than the fine model. This can be achieved by proper manipulation of the solver mesh density.

8.5.5 Optimization Using Adaptively Adjusted Design Specifications

The techniques described in Section 8.5.1 to 8.5.4 aimed at correcting the low-fidelity model so that it becomes, at least locally, an accurate representation of the high-fidelity model. An alternative way of exploiting low-fidelity models in simulation-driven design of microwave structures is to modify the design specifications in such a way that the updated specifications reflect the discrepancy between the models. This approach is extremely simple to implement because no changes of the low-fidelity model are necessary.

The adaptively adjusted design specifications optimization procedure introduced in [53] consists of the following two simple steps that can be iterated if necessary:

1. Modify the original design specifications in order to take into account the difference between the responses of $\boldsymbol{R}_f$ and $\boldsymbol{R}_c$ at their characteristic points.
2. Obtain a new design by optimizing the coarse model with respect to the modified specifications.

Characteristic points of the responses should correspond to the design specification levels. They should also include local maxima/minima of the respective responses at which the specifications may not be satisfied. Figure 8.9(a) shows fine and coarse model response at the optimal design of $\boldsymbol{R}_c$, corresponding to the bandstop filter example considered in [53]; design specifications are indicated using horizontal lines. Figure 8.9(b) shows characteristic points of $\boldsymbol{R}_f$ and $\boldsymbol{R}_c$ for the bandstop filter example. The points correspond to –3 dB and –30 dB levels as well to the local maxima of the responses. As one can observe in Fig. 8.9(b) the selection of points is rather straightforward.

In the first step of the optimization procedure, the design specifications are modified (or mapped) so that the level of satisfying/violating the modified specifications by the coarse model response corresponds to the satisfaction/violation levels of the original specifications by the fine model response.

More specifically, for each edge of the specification line, the edge frequency is shifted by the difference of the frequencies of the corresponding characteristic points, e.g., the left edge of the specification line of –30 dB is moved to the right by about 0.7 GHz, which is equal to the length of the line connecting the corresponding characteristic points in Fig. 8.9(b). Similarly, the specification levels are shifted by the difference between the local maxima/minima values for the respective points, e.g., the –30 dB level is shifted down by about 8.5 dB because of the difference of the local maxima of the corresponding characteristic points of $\boldsymbol{R}_f$ and $\boldsymbol{R}_c$. Modified design specifications are shown in Fig. 8.9(c).

The coarse model is subsequently optimized with respect to the modified specifications and the new design obtained this way is treated as an approximated solution to the original design problem (i.e., optimization of the fine model with respect to the original specifications). Steps 1 and 2 (listed above) can be repeated if necessary. Substantial design improvement is typically observed after the first iteration, however, additional iterations may bring further enhancement [53].

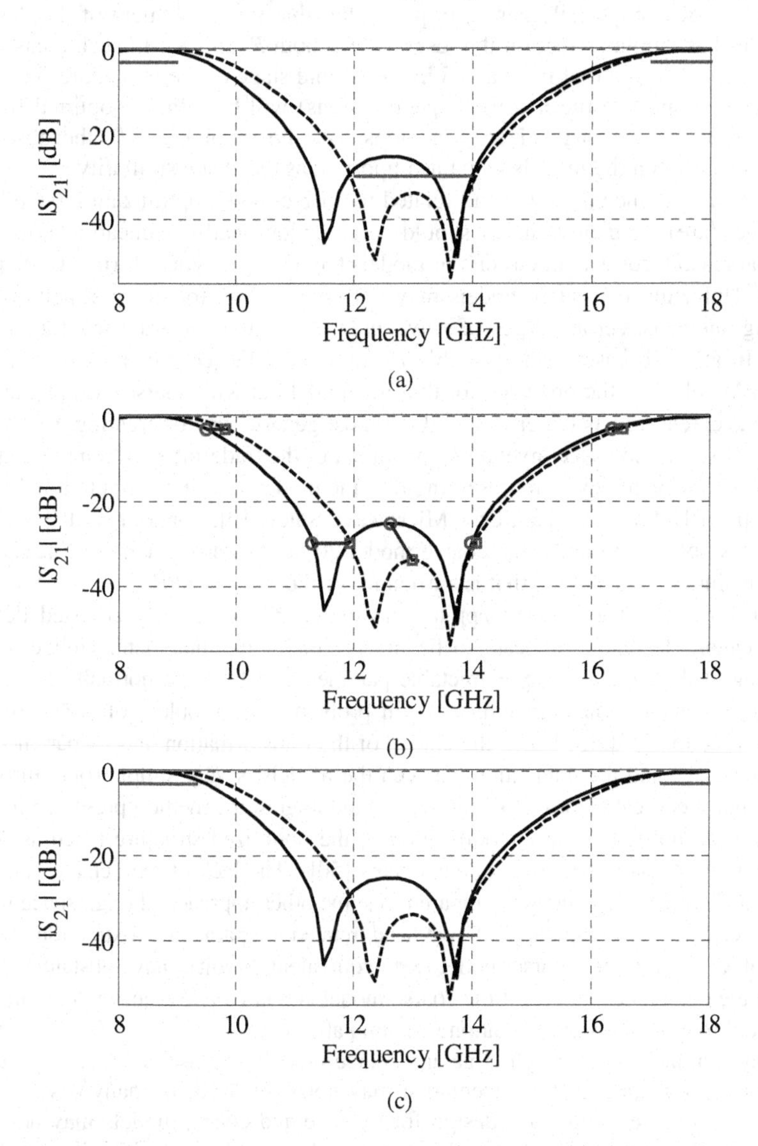

Fig. 8.9 Bandstop filter example (responses of $\boldsymbol{R}_f$ and $\boldsymbol{R}_c$ are marked with solid and dashed line, respectively): (a) fine and coarse model responses at the initial design (optimum of $\boldsymbol{R}_c$) as well as the original design specifications, (b) characteristic points of the responses corresponding to the specification levels (here, –3 dB and –30 dB) and to the local response maxima, (c) fine and coarse model responses at the initial design and the modified design specifications.

In the first step of the optimization procedure, the design specifications are modified (or mapped) so that the level of satisfying/violating the modified specifications by the coarse model response corresponds to the satisfaction/violation levels of the original specifications by the fine model response. It is assumed that the coarse model is physically-based, in particular, that the adjustment of the design variables has similar effect on the response for both $\boldsymbol{R}_f$ and $\boldsymbol{R}_c$. In such a case the coarse model design that is obtained in the second stage of the procedure (i.e., optimal with respect to the modified specifications) will be (almost) optimal for $\boldsymbol{R}_f$ with respect to the original specifications. As shown in Fig. 8.9, the absolute matching between the models is not as important as the shape similarity.

In order to reduce the overhead related to coarse model optimization (step 2 of the procedure) the coarse model should be computationally as cheap as possible. For that reason, equivalent circuits or models based on analytical formulas are preferred. Unfortunately, such models may not be available for many structures including antennas, certain types of waveguide filters and substrate integrated circuits. In all such cases, it is possible to implement the coarse model using the same EM solver as the one used for the fine model but with coarser discretization. To some extent, this is the easiest and the most generic way of creating the coarse model. Also, it allows a convenient adjustment of the trade-off between the quality of $\boldsymbol{R}_c$ (i.e., the accuracy in representing the fine model) and its computational cost. For popular EM solvers (e.g., CST Microwave Studio [9], Sonnet ***em*** [17], FEKO [18]) it is possible to make the coarse model 20 to 100 faster than the fine model while maintaining accuracy that is sufficient for the method SPRP.

When compared to space mapping and tuning, the adaptively adjusted design specifications technique appears to be much simpler to implement. Unlike space mapping, it does not use any extractable parameters (which are normally found by solving a separate nonlinear minimization problem), the problem of the surrogate model selection [38], [39] (i.e., the choice of the transformation and its parameters) does not exist, and the interaction between the models is very simple (only through the design specifications). Unlike tuning methodologies, the method presented in this section does not require any modification of the optimized structure (such as "cutting" and insertion of the tuning components [50]). The lack of extractable parameters is its additional advantage compared to some other approached (e.g., space mapping) because the computational overhead related to parameter extraction, while negligible for very fast coarse model (e.g., equivalent circuit), may substantially increase the overall design cost if the coarse model is relatively expensive (e.g., implemented through coarse-discretization EM simulation).

If the similarity between the fine and coarse model response is not sufficient the adaptive design specifications technique may not work well. In many cases, however, using different reference design for the fine and coarse models may help. In particular, $\boldsymbol{R}_c$ can be optimized with respect to the modified specifications starting not from $\boldsymbol{x}^{(0)}$ (the optimal solution of $\boldsymbol{R}_c$ with respect to the original specifications), but from another design, say $\boldsymbol{x}_c^{(0)}$, at which the response of $\boldsymbol{R}_c$ is as similar to the response of $\boldsymbol{R}_f$ at $\boldsymbol{x}^{(0)}$ as possible. Such a design can be obtained as follows [7]:

$$x_c^{(0)} = \arg\min_z \| R_f(x^{(0)}) - R_c(z) \| \tag{8.14}$$

At iteration i of the optimization process, the optimal design of the coarse model R_c with respect to the modified specifications, $x_c^{(i)}$, has to be translated to the corresponding fine model design, $x^{(i)}$, as follows $x^{(i)} = x_c^{(i)} + (x^{(0)} - x_c^{(0)})$. Note that the preconditioning procedure (8.14) is performed only once for the entire optimization process. The idea of coarse model preconditioning is borrowed from space mapping (more specifically, from the original space mapping concept [7]). In practice, the coarse model can be "corrected" to reduce its misalignment with the fine model using any available degrees of freedom, for example, preassigned parameters as in implicit space mapping [33].

8.6 Summary

Simulation-driven optimization has become an important design tool in contemporary microwave engineering. Its importance is expected to grow in the future due to the rise of the new technologies and the novel classes of devices and systems for which traditional design methods are not applicable. The surrogate-based approach and methods described in this chapter can make the electromagnetic-simulation-based design optimization feasible and cost efficient. In Chapter 12, a number of applications of the techniques presented here are demonstrated in the design of common microwave devices including filters, antennas and interconnect structures.

References

1. Bandler, J.W., Chen, S.H.: Circuit optimization: the state of the art. IEEE Trans. Microwave Theory Tech. 36, 424–443 (1988)
2. Bandler, J.W., Biernacki, R.M., Chen, S.H., Swanson, J.D.G., Ye, S.: Microstrip filter design using direct EM field simulation. IEEE Trans. Microwave Theory Tech. 42, 1353–1359 (1994)
3. Swanson Jr., D.G.: Optimizing a microstrip bandpass filter using electromagnetics. Int. J. Microwave and Millimeter-Wave CAE 5, 344–351 (1995)
4. De Zutter, D., Sercu, J., Dhaene, V., De Geest, J., Demuynck, F.J., Hammadi, S., Paul, C.-W.: Recent trends in the integration of circuit optimization and full-wave electromagnetic analysis. IEEE Trans. Microwave Theory Tech. 52, 245–256 (2004)
5. Schantz, H.: The art and science of ultrawideband antennas. Artech House, Boston (2005)
6. Wu, K.: Substrate Integrated Circuits (SiCs) – A new paradigm for future Ghz and Thz electronic and photonic systems. IEEE Circuits Syst. Soc. Newsletter 3 (2009)
7. Bandler, J.W., Cheng, Q.S., Dakroury, S.A., Mohamed, A.S., Bakr, M.H., Madsen, K., Søndergaard, J.: Space mapping: the state of the art. IEEE Trans. Microwave Theory Tech. 52, 337–361 (2004)
8. Director, S.W., Rohrer, R.A.: The generalized adjoint network and network sensitivities. IEEE Trans. Circuit Theory CT-16, 318–323 (1969)

9. CST Microwave Studio, ver. 20109 CST AG, Bad Nauheimer Str. 19, D-64289 Darmstadt, Germany (2010)
10. HFSS, release 13.0, ANSYS (2010), http://www.ansoft.com/products/hf/hfss/
11. Wrigth, S.J., Nocedal, J.: Numerical Optimization. Springer, Heidelberg (1999)
12. Kolda, T.G., Lewis, R.M., Torczon, V.: Optimization by direct search: new perspectives on some classical and modern methods. SIAM Rev. 45, 385–482 (2003)
13. Lai, M.-I., Jeng, S.-K.: Compact microstrip dual-band bandpass filters design using genetic-algorithm techniques. IEEE Trans. Microwave Theory Tech. 54, 160–168 (2006)
14. Haupt, R.L.: Antenna design with a mixed integer genetic algorithm. IEEE Trans. Antennas Propag. 55, 577–582 (2007)
15. Jin, N., Rahmat-Samii, Y.: Parallel particle swarm optimization and finite- difference time-domain (PSO/FDTD) algorithm for multiband and wide-band patch antenna designs. IEEE Trans. Antennas Propag. 53, 3459–3468 (2005)
16. Jin, N., Rahmat-Samii, Y.: Analysis and particle swarm optimization of correlator antenna arrays for radio astronomy applications. IEEE Trans. Antennas Propag. 56, 1269–1279 (2008)
17. Sonnet em. Ver. 12.54, Sonnet Software. North Syracuse, NY (2009)
18. FEKO User's Manual. Suite 5.5, EM Software & Systems-S.A (Pty) Ltd, 32 Techno Lane, Technopark, Stellenbosch, 7600, South Africa (2009)
19. Bandler, J.W., Seviora, R.E.: Wave sensitivities of networks. IEEE Trans. Microwave Theory Tech. 20, 138–147 (1972)
20. Chung, Y.S., Cheon, C., Park, I.H., Hahn, S.Y.: Optimal design method for microwave device using time domain method and design sensitivity analysis-part II: FDTD case. IEEE Trans. Magn. 37, 3255–3259 (2001)
21. Bakr, M.H., Nikolova, N.K.: An adjoint variable method for time domain TLM with fixed structured grids. IEEE Trans. Microwave Theory Tech. 52, 554–559 (2004)
22. Nikolova, N.K., Tam, H.W., Bakr, M.H.: Sensitivity analysis with the FDTD method on structured grids. IEEE Trans. Microwave Theory Tech. 52, 1207–1216 (2004)
23. Webb, J.P.: Design sensitivity of frequency response in 3-D finite-element analysis of microwave devices. IEEE Trans. Magn. 38, 1109–1112 (2002)
24. Nikolova, N.K., Bandler, J.W., Bakr, M.H.: Adjoint techniques for sensitivity analysis in high-frequency structure CAD. IEEE Trans. Microwave Theory Tech. 52, 403–419 (2004)
25. Ali, S.M., Nikolova, N.K., Bakr, M.H.: Recent advances in sensitivity analysis with frequency-domain full-wave EM solvers. Applied Computational Electromagnetics Society J. 19, 147–154 (2004)
26. El Sabbagh, M.A., Bakr, M.H., Nikolova, N.K.: Sensitivity analysis of the scattering parameters of microwave filters using the adjoint network method. Int. J. RF and Microwave Computer-Aided Eng. 16, 596–606 (2006)
27. Snyder, R.V.: Practical aspects of microwave filter development. IEEE Microwave Magazine 8(2), 42–54 (2007)
28. Shin, S., Kanamaluru, S.: Diplexer design using EM and circuit simulation techniques. IEEE Microwave Magazine 8(2), 77–82 (2007)
29. Bhargava, A.: Designing circuits using an EM/circuit co-simulation technique. RF Design. 76 (January 2005)

30. Koziel, S., Bandler, S.W., Madsen, K.: A space mapping framework for engineering optimization: theory and implementation. IEEE Trans. Microwave Theory Tech. 54, 3721–3730 (2006)
31. Queipo, N.V., Haftka, R.T., Shyy, W., Goel, T., Vaidynathan, R., Tucker, P.K.: Surrogate based analysis and optimization. Progress in Aerospace Sciences 41, 1–28 (2005)
32. Forrester, A.I.J., Keane, A.J.: Recent advances in surrogate-based optimization. Prog. Aerospace Sciences 45, 50–79 (2009)
33. Conn, A.R., Gould, N.I.M., Toint, P.L.: Trust Region Methods. MPS-SIAM Series on Optimization (2000)
34. Alexandrov, N.M., Dennis, J.E., Lewis, R.M., Torczon, V.: A trust region framework for managing use of approximation models in optimization. Struct. Multidisciplinary Optim. 15, 16–23 (1998)
35. Booker, A.J., Dennis Jr., J.E., Frank, P.D., Serafini, D.B., Torczon, V., Trosset, M.W.: A rigorous framework for optimization of expensive functions by surrogates. Structural Optimization 17, 1–13 (1999)
36. Amari, S., LeDrew, C., Menzel, W.: Space-mapping optimization of planar coupled-resonator microwave filters. IEEE Trans. Microwave Theory Tech. 54, 2153–2159 (2006)
37. Crevecoeur, G., Sergeant, P., Dupre, L., Van de Walle, R.: Two-level response and parameter mapping optimization for magnetic shielding. IEEE Trans. Magn. 44, 301–308 (2008)
38. Koziel, S., Bandler, J.W., Madsen, K.: Quality assessment of coarse models and surrogates for space mapping optimization. Optimization Eng. 9, 375–391 (2008)
39. Koziel, S., Bandler, J.W.: Space-mapping optimization with adaptive surrogate model. IEEE Trans. Microwave Theory Tech. 55, 541–547 (2007)
40. Simpson, T.W., Peplinski, J., Koch, P.N., Allen, J.K.: Metamodels for computer-based engineering design: survey and recommendations. Engineering with Computers 17, 129–150 (2001)
41. Miraftab, V., Mansour, R.R.: EM-based microwave circuit design using fuzzy logic techniques. IEE Proc. Microwaves, Antennas & Propagation 153, 495–501 (2006)
42. Yang, Y., Hu, S.M., Chen, R.S.: A combination of FDTD and least-squares support vector machines for analysis of microwave integrated circuits. Microwave Opt. Technol. Lett. 44, 296–299 (2005)
43. Xia, L., Meng, J., Xu, R., Yan, B., Guo, Y.: Modeling of 3-D vertical interconnect using support vector machine regression. IEEE Microwave Wireless Comp. Lett. 16, 639–641 (2006)
44. Burrascano, P., Dionigi, M., Fancelli, C., Mongiardo, M.: A neural network model for CAD and optimization of microwave filters. In: IEEE MTT-S Int. Microwave Symp. Dig., Baltimore, MD, pp. 13–16 (1998)
45. Zhang, L., Xu, J., Yagoub, M.C.E., Ding, R., Zhang, Q.-J.: Efficient analytical formulation and sensitivity analysis of neuro-space mapping for nonlinear microwave device modeling. IEEE Trans. Microwave Theory Tech. 53, 2752–2767 (2005)
46. Kabir, H., et al.: Neural network inverse modeling and applications to microwave filter design. IEEE Trans. Microwave Theory Tech. 56, 867–879 (2008)
47. Pozar, D.M.: Microwave Engineering, 3rd edn. Wiley, Chichester (2004)
48. Koziel, S., Cheng, Q.S., Bandler, J.W.: Implicit space mapping with adaptive selection of preassigned parameters. IET Microwaves, Antennas & Propagation 4, 361–373 (2010)

49. Agilent ADS. Version 2009, Agilent Technologies, 395 Page Mill Road, Palo Alto, CA, 94304 (2009)
50. Koziel, S., Meng, J., Bandler, J.W., Bakr, M.H., Cheng, Q.S.: Accelerated microwave design optimization with tuning space mapping. IEEE Trans. Microwave Theory and Tech. 57, 383–394 (2009)
51. Koziel, S.: Shape-preserving response prediction for microwave design optimization. IEEE Trans. Microwave Theory and Tech. (2010) (to appear)
52. Koziel, S., Ogurtsov, S.: Robust multi-fidelity simulation-driven design optimization of microwave structures. In: IEEE MTT-S Int. Microwave Symp. Dig., Anaheim, CA, pp. 201–204 (2010)
53. Koziel, S.: Efficient optimization of microwave structures through design specifications adaptation. In: IEEE Int. Symp. Antennas Propag., Toronto, Canada (2010)
54. Bandler, J.W., Salama, A.E.: Functional approach to microwave postproduction tuning. IEEE Trans. Microwave Theory Tech. 33, 302–310 (1985)
55. Swanson, D., Macchiarella, G.: Microwave filter design by synthesis and optimization. IEEE Microwave Magazine 8(2), 55–69 (2007)
56. Rautio, J.C.: EM-component-based design of planar circuits. IEEE Microwave Magazine 8(4), 79–90 (2007)
57. Cheng, Q.S., Bandler, J.W., Koziel, S.: Tuning Space Mapping Optimization Exploiting Embedded Surrogate Elements. In: IEEE MTT-S Int. Microwave Symp. Dig., Boston, MA, pp. 1257–1260 (2009)
58. Koziel, S., Bandler, J.W., Cheng, Q.S.: Design optimization of microwave circuits through fast embedded tuning space mapping. In: European Microwave Conference, Paris, September 26-October 1 (2010)

Chapter 9
Variable-Fidelity Aerodynamic Shape Optimization

Leifur Leifsson and Slawomir Koziel

Abstract. Aerodynamic shape optimization (ASO) plays an important role in the design of aircraft, turbomachinery and other fluid machinery. Simulation-driven ASO involves the coupling of computational fluid dynamics (CFD) solvers with numerical optimization methods. Although being relatively mature and widely used, ASO is still being improved and numerous challenges remain. This chapter provides an overview of simulation-driven ASO methods, with an emphasis on surrogate-based optimization (SBO) techniques. In SBO, a computationally cheap surrogate model is used in lieu of an accurate high-fidelity CFD simulation in the optimization process. Here, a particular focus is given to SBO exploiting surrogate models constructed from corrected physics-based low-fidelity models, often referred to as variable- or multi-fidelity optimization.

9.1 Introduction

Aerodynamic and hydrodynamic design optimization is of primary importance in several disciplines [1-3]. In aircraft design, both for conventional transport aircraft and unmanned air vehicles, the aerodynamic wing shape is designed to provide maximum efficiency under a variety of takeoff, cruise, maneuver, loiter, and landing conditions [1, 4-7]. Constraints on aerodynamic noise are also becoming increasingly important [8, 9]. In the design of turbines, such as gas, steam, or wind turbines, the blades are designed to maximize energy output for a given working fluid and operating conditions [2, 10]. The shapes of the propeller blades of ships are optimized to increase efficiency [11]. The fundamental design problem, common to all these disciplines, is to design a streamlined wing (or blade) shape that provides the desired performance for a given set of operating conditions, while at the same time fulfilling one or multiple design constraints [12-20].

Leifur Leifsson · Slawomir Koziel
Engineering Optimization & Modeling Center,
School of Science and Engineering, Reykjavik University,
Menntavegur 1, 101 Reykjavik, Iceland
email: {leifurth,koziel}@ru.is

S. Koziel & X.-S. Yang (Eds.): Comput. Optimization, Methods and Algorithms, SCI 356, pp. 179–210.
springerlink.com

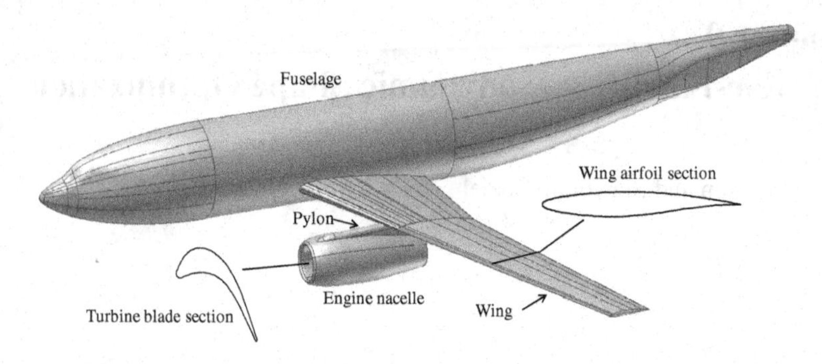

Fig. 9.1 A CAD drawing of a typical transport aircraft with a turbofan jet engine. The aircraft wing and the turbine blades of the turbofan engines are streamlined aerodynamic surfaces defined by airfoil sections

In the early days of engineering design, the designer would have to rely on experience and physical experiments. Nowadays, most engineering design is performed using computational tools, especially in the early phases, i.e., conceptual and preliminary design. This is commonly referred to as simulation-driven (or simulation-based) design. Physical experiments are normally performed at the final design stages only, mostly for validation purposes. The fidelity of the computational methods used in design has been steadily increasing. Over forty years ago, the computational fluid dynamic (CFD) tools were only capable of simulating potential flow past simplified wing configurations [1, 21]. Today's commercial CFD tools, e.g., [22, 23], are capable of simulating three-dimensional viscous flows past full aircraft configurations using the Reynolds-Averaged Navier-Stokes (RANS) equations with the appropriate turbulence models [24].

The use of optimization methods in the design process, either as a design support tool or for automated design, has now become commonplace. In aircraft design, the use of numerical optimization techniques began in the mid 1970's by coupling CFD tools with gradient-based optimization methods [1]. Substantial progress has been made since then, and the exploitation of higher-fidelity methods, coupled with optimization techniques, has led to improved design efficiency [4, 12-16]. An overview of the relevant work is provided in the later sections. In spite of being widespread, simulation-driven aerodynamic design optimization involves numerous challenges:

- High-fidelity CFD simulations are computationally expensive. A CFD simulation involves solving the governing flow equations on a computational mesh. The resulting system of algebraic equations can be very large, with a number of unknowns equal to the product of the number of flow variables and the number of mesh points. For a three-dimensional turbulent RANS flow simulation with one million mesh points, and a two-equation turbulence model, there will be seven flow variables, leading to an algebraic system of seven million equations with seven million unknowns. Depending on the computational resources, this kind of simulation can take many days on a parallel computer [25]. In the corresponding two-dimensional case, simulations on meshes with over one hundred thousand mesh points are not uncommon. A single simulation in this case can take over one hour on a typical desktop computer.

- Design optimization normally requires a large number of simulations. For example, even in the case of a two-dimensional airfoil shape optimization with three design variables, a gradient-based optimization method can require over one hundred function evaluations, and optimization process could take as long as one week [26]. For higher-dimensional problems, the required number function evaluations may be substantially larger.
- A large number of design variables. Typically, an airfoil shape can be described accurately with, say, ten to fifteen design variables [27]. An entire transport wing shape might require a few (say three to seven) airfoils at various spanwise locations, leading to at least thirty design variables, aside from the planform variables (e.g., span, sweep, twist) [4].
- Multiple operating conditions [15] (e.g., a range of Mach numbers) and multiple objectives (e.g., minimum take-off gross weight, minimum drag, minimum noise) [8] may need to considered in the design process.
- Uncertainty in the operating conditions and in the airfoil shape may need to be taken into account, leading to the need of carrying out stochastic analysis, which is always more time consuming than a deterministic one [28].
- The simulation results normally include numerical noise [29]. This can be due to partially converged solutions or due to badly generated computational meshes.
- The objectives, e.g., the drag force, can be numerically sensitive to mesh resolution [30].
- Coupling of the aerodynamics with other disciplines should be considered as well. For example, the coupling of the aerodynamic load with the wing structure by way of structural analysis [31]. This is referred to as multidisciplinary design and optimization (MDO).

The above remarks indicate that high-fidelity CFD simulations are computationally far too expensive to be used in a direct, simulation-based design optimization, especially when using conventional, gradient-based techniques. Any further improvement to the overall efficiency of the design process can only be achieved by developing more efficient optimization methods, i.e., reducing the number of high-fidelity CFD simulations required to yield an optimized design, and/or employing more powerful computing resources.

An important research area in the field of aerodynamic optimization is focused on employing the surrogate-based optimization (SBO) techniques [32, 33]. One of the major objectives is to reduce the number of high-fidelity model evaluations, and thereby making the optimization process more efficient. In SBO, the accurate but computationally expensive high-fidelity CFD simulations are replaced—in the optimization process—by a cheap surrogate model. In this chapter, we provide a review of some representative works on aerodynamic shape optimization relating both to the direct [1, 4, 10, 12-21] and the surrogate-based optimization approaches [29, 31-33]. In particular, the main emphasis of the chapter is on the SBO approach exploiting surrogate models constructed from corrected physics-based low-fidelity models [26, 34-38]. This is often referred to as variable- or multi-fidelity optimization.

The chapter is organized as follows. In Section 9.2, we formulate the aerodynamic shape optimization problem using the example of airfoil design. The basics of the CFD modeling and simulation process are described in Section 9.3. Direct CFD-driven optimization is discussed in Section 9.4, whereas the SBO methodologies are presented in Section 9.5. Section 9.6 concludes the chapter.

9.2 Problem Formulation

Aerodynamic design optimization includes a variety of specific problems ranging from two-dimensional airfoil shape optimization [13] to three-dimensional wing (or blade) design [4], involving one or several objective functions [19, 39], as well as one or multiple operating conditions [15, 27]. In this chapter, in order to highlight the formulation, design challenges and solution methodologies, we concentrate on airfoil shape optimization for one representative operating condition, and with a single objective function.

An airfoil is a streamlined aerodynamic surface such as the one shown in Fig. 9.2. The length of the airfoil is called the chord and is denoted by c. The thickness, denoted by t, varies along the chord line. The curvature, called the camber, described by the mean camber line, varies also along the chord line. The leading-edge (LE) is normally rounded and the trailing-edge (TE) is normally sharp, either closed or open.

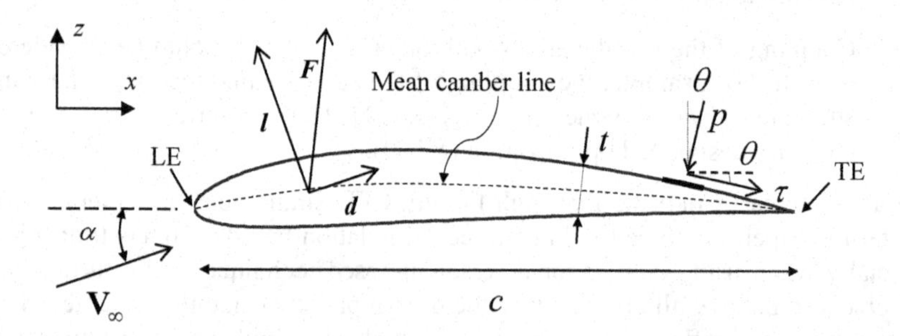

Fig. 9.2 A single-element airfoil section (this solid line) of chord length c and thickness t. $\boldsymbol{V}_\infty$ is at an angle of attack α relative to the x-axis. $\boldsymbol{F}$ is the force acting on the airfoil due to the airflow, where the component perpendicular to $\boldsymbol{V}_\infty$ is called the lift force $\boldsymbol{l}$ and the component parallel to $\boldsymbol{V}_\infty$ is called the drag force $\boldsymbol{d}$. p is the pressure acting normal to a surface element of length ds, and τ is the viscous wall shear stress acting parallel to the surface element. θ is the angle that p and τ make relative to the z- and x-axis, respectively, positive clockwise

The function of the airfoil is to generate a lift force $\boldsymbol{l}$ (a force component perpendicular to the free-stream) at a range of operating conditions (Mach number M_∞, Reynolds number, angle of attack α). Normally, the drag force $\boldsymbol{d}$ (a force component parallel to the free-stream) is to be minimized. These forces are due to the pressure distribution p (acting normal to the surface) and the shear stress distribution τ (acting parallel to the surface) over the surface of the airfoil. A detailed description of their calculation is given in Section 9.3.2.4. The forces are written in non-dimensional form. They are called the lift coefficient and the drag coefficient. The lift coefficient is defined as

$$C_l \equiv \frac{l}{q_\infty S} \tag{9.1}$$

where l is the magnitude of the lift force, $q_\infty \equiv (1/2)\rho_\infty V_\infty^2$ is the dynamic pressure, ρ_∞ is the air density, V_∞ is the free-stream velocity, and S is a reference surface. For a two-dimensional airfoil, the reference area is taken to be the chord length multiplied by a unit depth, i.e., $S = c$. Similarly, the drag coefficient is defined as

$$C_d \equiv \frac{d}{q_\infty S} \tag{9.2}$$

where d is the magnitude of the drag force.

There are two main approaches to airfoil design. One is to design the airfoil section in order to maximize its performance. This is called direct design, and the most common design setups include lift maximization, drag minimization, and lift-to-drag ratio maximization [14]. Another way is to define a priori a specific flow behavior that is to be attained. The airfoil shape is then designed to achieve this flow behavior. This is called inverse design and, typically, a target airfoil surface pressure distribution is prescribed [40].

Table 9.1 Typical problem formulations for two-dimensional airfoil shape optimization. Additionally a constraint on the minimum allowable airfoil cross-sectional area is included.

Case	$f(\mathbf{x})$	$g_1(\mathbf{x})$
Lift maximization	$-C_l(\mathbf{x})$	$C_d(\mathbf{x}) - C_d^{limit} \le 0$
Drag minimization	$C_d(\mathbf{x})$	$C_l^{limit} - C_l(\mathbf{x}) \le 0$
L/D maximization	$-C_l(\mathbf{x})/C_d(\mathbf{x})$	$C_l^{limit} - C_l(\mathbf{x}) \le 0$
Inverse design	$1/2 \int (C_p(\mathbf{x}) - C_p^{\text{target}})^2 ds$	

In general, aerodynamic shape optimization can be formulated as a nonlinear minimization problem, i.e., for a given operating condition, solve

$$\begin{aligned} &\min_{\mathbf{x}} f(\mathbf{x}) \\ &\text{s.t. } g_j(\mathbf{x}) \leq 0 \\ &l \leq \mathbf{x} \leq u \end{aligned} \tag{9.3}$$

where $f(\mathbf{x})$ is the objective function, $\mathbf{x}$ is the design variable vector (parameters describing the airfoil shape), $g_j(\mathbf{x})$ are the design constraints, and l and u are the lower and upper bounds, respectively. The detailed formulation depends on the particular design problem. Typical problem formulations for two-dimensional airfoil optimization are listed in Table 9.1. Additional constraints are often prescribed. For example, to account for the wing structural components inside the airfoil, one sets a constraint on the airfoil cross-sectional area, which can be formally written as $g_2(\mathbf{x}) = A_{min} - A(\mathbf{x}) \leq 0$, where $A(\mathbf{x})$ is the cross-sectional area of the airfoil for the design vector $\mathbf{x}$ and A_{min} is the minimum allowable cross-sectional area. Other constraints can be included depending on the design situation, e.g., a maximum pitching moment or a maximum local allowable pressure coefficient [41].

An aircraft wing and a turbomachinery blade are three-dimensional aerodynamic surfaces. A schematic of a typical wing (or a blade) planform is shown in Fig. 9.3, where—at each spanstation (numbered 1 through 4)—the wing cross-section is defined by an airfoil shape. The number of spanstations can be smaller or larger than four, depending on the design scenario. Between each station, there is a straight-line wrap. Parameters controlling the planform shape include the wing span, the quarter-chord wing sweep angle, the chord lengths and thickness-to-chord ratio at each spanstation, the wing taper ratio, and the twist distribution.

Numerical design optimization of the three-dimensional wing (or blade) is performed in a similar fashion as for the two-dimensional airfoil [4]. In the problem formulation, the section lift and drag coefficients are replaced by the overall lift and drag coefficients. However, the number of design variables is much larger and the fluid flow domain is three-dimensional. These factors increase the computational burden, and the setup of the optimization process becomes even more important [4].

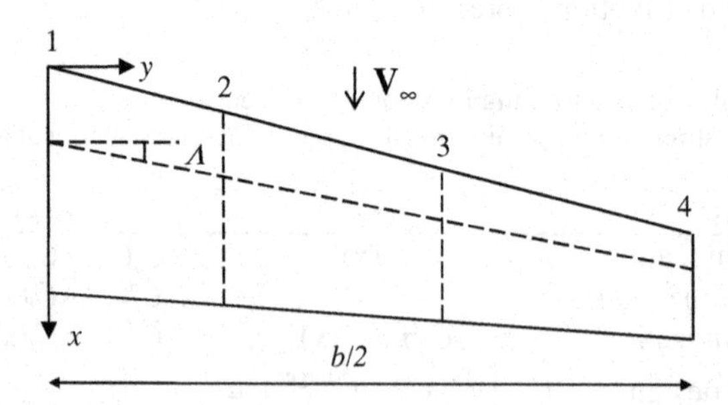

Fig. 9.3 A schematic of a wing planform of semi-span $b/2$ and quarter chord sweep angle Λ. Other planform parameters (not shown) are the taper ratio (ratio of tip chord to root chord) and the twist distribution. $\boldsymbol{V}_\infty$ is the free-stream speed

The problem formulations presented above apply to a single operating condition (Mach number, Reynolds number, angle of attack) and a single objective function. An airfoil optimized for a single operating point may have severe performance degradation for off-design points, or, in some cases, even a small deviation from the design point could result in a dramatic change in the lift and drag coefficients [15, 19, 28].

Robust optimization is employed to improve the general wing/blade performance at the optimal solution, in particular, to make it insensitive to small perturbations of the design variables or the operating conditions. In robust optimization, the objective is to achieve consistent performance improvement over a given range of uncertainty parameters. Example work on this subject include the airfoil optimization for a consistent drag reduction over a Mach number range [28, 39], also called multi-point optimization, and the aerodynamic design of a turbine blade airfoil shape taking into account the performance degradation due to manufacturing uncertainties [42].

In many cases, several (often competing) objectives may have to be considered at the same time. This is referred to as multi-objective design optimization. For example, during the take-off and landing of an aircraft, limits on the external noise are becoming increasingly important, and have to be accounted for in the design process [8, 9]. On the other hand, the search for robust airfoil designs can be treated as multi-objective optimization, i.e., maximizing robustness and performance simultaneously (since these are very likely conflicting objectives) [42].

9.3 Computational Fluid Dynamic Modeling

This section presents a brief introduction to the elements of a CFD analysis. We introduce the governing fluid flow equations and explain the hierarchy of simplified forms of the governing equations which are commonly used in aerodynamic design. The CFD process is then illustrated with an example two-dimensional simulation of the flow past an airfoil at transonic flow conditions.

9.3.1 Governing Equations

The fluid flow past an aerodynamic surface is governed by the Navier-Stokes equations. For a Newtonian fluid, compressible viscous flows in two dimensions, without body forces, mass diffusion, finite-rate chemical reactions, heat conduction, or external heat addition, the Navier-Stokes equations, can be written in Cartesian coordinates as [24]

$$\frac{\partial \mathbf{U}}{\partial t} + \frac{\partial \mathbf{E}}{\partial x} + \frac{\partial \mathbf{F}}{\partial x} = 0 \tag{9.4}$$

where **U**, **E**, and **F** are vectors given by

$$\mathbf{U}=\begin{bmatrix}\rho\\ \rho u\\ \rho v\\ E_t\end{bmatrix}\quad \mathbf{E}=\begin{bmatrix}\rho u\\ \rho u^2+p-\tau_{xx}\\ \rho uv-\tau_{xy}\\ (E_t+p)u-u\tau_{xx}-v\tau_{xy}\end{bmatrix}\quad \mathbf{F}=\begin{bmatrix}\rho v\\ \rho uv-\tau_{xy}\\ \rho v^2+p-\tau_{yy}\\ (E_t+p)v-u\tau_{xy}-v\tau_{yy}\end{bmatrix} \tag{9.5}$$

Here, ρ is the fluid density, u and v are the x and y velocity components, respectively, p is the static pressure, $E_t = \rho\,(e+V^2/2)$ is the total energy per unit volume, e is the internal energy per unit mass, $V^2/2$ is the kinetic energy, and τ is the viscous shear stress tensor given by [24]

$$\tau_{xx}=\frac{2}{3}\mu\left(2\frac{\partial u}{\partial x}-\frac{\partial v}{\partial y}\right)\quad \tau_{yy}=\frac{2}{3}\mu\left(2\frac{\partial v}{\partial y}-\frac{\partial u}{\partial x}\right)\quad \tau_{xy}=\mu\left(\frac{\partial u}{\partial y}+\frac{\partial v}{\partial x}\right) \tag{9.6}$$

where μ is the dynamic viscosity of the fluid.

The first row of Eq. (9.6) corresponds to the continuity equation, the second and third rows are the momentum equations, and the fourth row is the energy equation. These four scalar equations contain five unknowns, namely (ρ, p, e, u, v). An equation of state is needed to close the system of equations. For most problems in gas dynamics, it is possible to assume a perfect gas, which is defined as a gas whose intermolecular forces are negligible. A perfect gas obeys the perfect gas equation of state [24]

$$p=\rho RT \tag{9.7}$$

where R is the gas constant.

The governing equations are a set of coupled, highly nonlinear partial differential equations. The numerical solution of these equations is quite challenging. What complicates things even further is that all flows will become turbulent above a critical value of the Reynolds number $Re = VL/\upsilon$, where V and L are representative values of velocity and length scales and υ is the kinematical viscosity. Turbulent flows are characterized by the appearance of statistical fluctuations of all the variables (ρ, p, e, u, v) around mean values.

By making appropriate assumptions about the fluid flow, the governing equations can be simplified and their numerical solution becomes computationally less expensive. In general, there are two approaches that differ in either neglecting the effects of viscosity or including them into the analysis. The hierarchy of the governing flow equations depending on the assumptions made about the fluid flow situation is shown in Fig. 9.4.

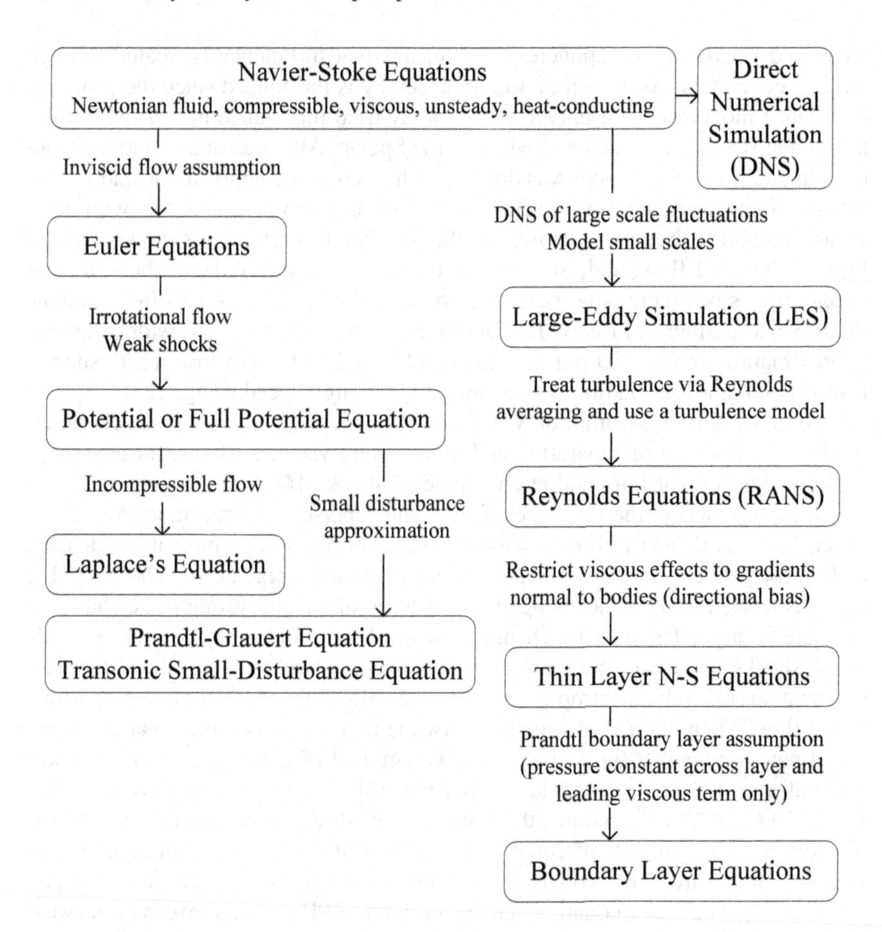

Fig. 9.4 A hierarchy of the governing fluid flow equations with the associated assumptions and approximations

Direct Numerical Simulation (DNS) has as objective to simulate the whole range of the turbulent statistical fluctuations at all relevant physical scales. This is a formidable challenge, which grows with increasing Reynolds number as the total computational effort for DNS simulations is proportional to Re^3 for homogeneous turbulence [25]. Due to limitations of computational capabilities, DNS is not available for typical engineering flows such as those encountered in airfoil design for typical aircraft and turbomachinery, i.e., with Reynolds numbers from 10^5 to 10^7.

Large-Eddy Simulation (LES) is of the same category as DNS, in that it computes directly the turbulent fluctuations in space and time, but only above a certain length scale. Below that scale, the turbulence is modeled by semi-empirical laws. The total computational effort for LES simulations is proportional to $Re^{9/4}$, which is significantly lower than for DNS [25]. However, it is still excessively high for large Reynolds number applications.

The Reynolds equations (also called the Reynolds-averaged Navier-Stokes equations (RANS)) are obtained by time-averaging of a turbulent quantity into their mean and fluctuating components. This means that turbulence is treated through turbulence models. As a result, a loss in accuracy is introduced since the available turbulence models are not universal. A widely used turbulence model for simulation of the flow past airfoils and wings is the Spalart-Allmaras one-equation turbulence model [43]. The model was developed for aerospace applications and is considered to be accurate for attached wall-bounded flows and flows with mild separation and recirculation. However, the RANS approach retains the viscous effects in the fluid flow, and, at the same time, significantly reduces the computational effort since there is no need to resolve all the turbulent scales (as it is done in DNS and partially in LES). This approach is currently the most widely applied approximation in the CFD practice and can be applied to both low-speed, such as take-off and landing conditions of an aircraft, and high-speed design [25].

The inviscid flow assumption will lead to the Euler equations. These equations hold, in the absence of separation and other strong viscous effects, for any shape of the body, thick or thin, and at any angle of attack [44]. Shock waves appear in transonic flow where the flow goes from being supersonic to subsonic. Across the shock, there is almost a discontinuous increase in pressure, temperature, density, and entropy, but a decrease in Mach number (from supersonic to subsonic). The shock is termed weak if the change in pressure is small, and strong if the change in pressure is large. The entropy change is of third order in terms of shock strength. If the shocks are weak, the entropy change across shocks is small, and the flow can be assumed to be isentropic. This, in turn, allows for the assumption of irrotational flow. Then, the Euler equations cascade to a single nonlinear partial differential equation, called the full potential equation (FPE). In the case of a slender body at a small angle of attack, we can make the assumption of a small disturbance. Then, the FPE becomes the transonic small-disturbance equation (TSDE). These three different sets of equations, i.e., the Euler equations, FPE, and TSDE, represent a hierarchy of models for the analysis of inviscid, transonic flow past airfoils [44]. The Euler equations are exact, while FPE is an approximation (weak shocks) to those equations, and TSDE is a further approximation (thin airfoils at small angle of attack). These approaches can be applied effectively for high-speed design, such as the cruise design of transport aircraft wings [13, 14] and the design of turbomachinery blades [2].

There are numerous airfoil and wing models that are not typical CFD models, but they are nevertheless widely used in aerodynamic design. Examples of such methods include thin airfoil theory, lifting line theory (unswept wings), vortex lattice methods (wings), and panel methods (airfoils and wings). These methods are out of the scope of this chapter, but the interested reader is directed to [45] and [46] for the details. In the following section, we describe the elements of a typical CFD simulation of the RANS or Euler equations.

9.3.2 Numerical Modeling

In general, a single CFD simulation is composed of four steps, as shown in Fig. 9.5: the geometry generation, meshing of the solution domain, numerical solution of the governing fluid flow equations, and post-processing of the flow results, which involves, in the case of numerical optimization, calculating the objectives and constraints. We discuss each step of the CFD process and illustrate it by giving an example two-dimensional simulation of the flow past the NACA 2412 airfoil at transonic flow conditions.

9.3.2.1 Geometry

Several methods are available for describing the airfoil shape numerically, each with its own benefits and drawbacks. In general, these methods are based on two different approaches, either the airfoil shape itself is parameterized, or, given an initial airfoil shape, the shape deformation is parameterized.

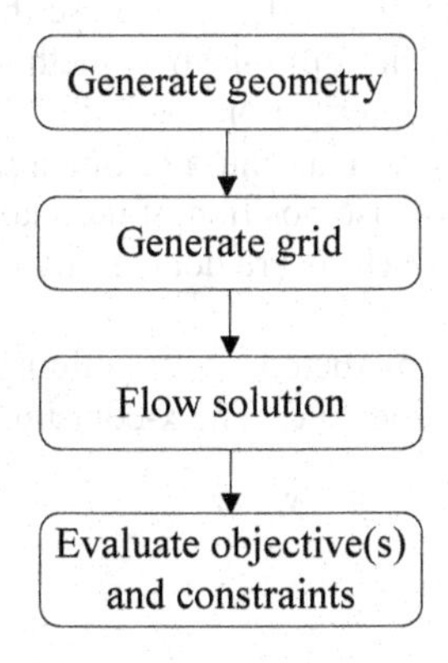

Fig. 9.5 Elements of a single CFD simulation in numerical airfoil shape optimization

The shape deformation approach is usually performed in two steps. First, the surface of the airfoil is deformed by adding values computed from certain functions to the upper and lower sides of the surfaces. Several different types of functions can be considered, such as the Hicks-Henne bump functions [1], or the transformed cosine functions [47]. After deforming the airfoil surface, the computational grid needs to be regenerated. Either the whole grid is regenerated based on the airfoil shape deformation, or the grid is deformed locally, accounting for the airfoil shape deformation. The latter is computationally more efficient. An example grid deformation method is the volume spline method [47]. In some cases, the first step described here above is skipped, and the grid points on the airfoil surface are used directly for the shape deformation [14].

Numerous airfoil shape parameterization methods have been developed. The earliest development of parameterized airfoil sections was performed by the National Advisory Committee for Aeronautics (NACA) in the 1930's [48]. Their development was derived from wind tunnel experiments, and, therefore, the shapes generated by this method are limited to those investigations. However, only three parameters are required to describe their shape. Nowadays, the most widely used airfoil shape parameterization methods are the Non-Uniform Rational B-Spline

(NURBS) [27], and the Bézier curves [49] (a special case of NURBS). These methods use a set of control points to define the airfoil shape and are general enough so that (nearly) any airfoil shape can be generated. In numerical optimization, these control points are used as design variables and they provide sufficient control of the shape so that local changes on the upper and lower surfaces can be made separately. The number of control points varies depending on how accurately the shape is to be controlled. NURBS requires as few as thirteen control points to represent a large family of airfoils [27]. Other parameterization methods include the PARSEC method [50], which uses 11 specific airfoil geometry parameters (such as leading edge radius, and upper and lower crest location including curvature), and the Bezier-PARSEC method [51], which combines the Bezier and PARSEC methods.

In this chapter, for the sake of simplicity, we use the NACA airfoil shapes [48] to illustrate some variable-fidelity optimization methods. In particular, we use the NACA four-digit airfoil parameterization method, where the airfoil shape is defined by three parameters m (the maximum ordinate of the mean camberline as a fraction of chord), p (the chordwise position of the maximum ordinate) and t/c (the thickness-to-chord ratio). The airfoils are denoted by NACA *mpxx*, where *xx* is the thickness-to-chord ratio, t/c.

The NACA airfoils are constructed by combining a thickness function $y_t(x)$ with a mean camber line function $y_c(x)$. The x-coordinates are [48]

$$x_{u,l} = x \mp y_t \sin\theta \tag{9.8}$$

and the y-coordinates are

$$y_{u,l} = y_c \pm y_t \cos\theta \tag{9.9}$$

where u and l refer to the upper and lower surfaces, respectively, $y_t(x)$ is the thickness function, $y_c(x)$ is the mean camber line function, and

$$\theta = \tan^{-1}\left(\frac{dy_c}{dx}\right) \tag{9.10}$$

is the mean camber line slope. The NACA four-digit thickness distribution is given by

$$y_t = t(a_0 x^{1/2} - a_1 x - a_2 x^2 + a_3 x^3 - a_4 x^4) \tag{9.11}$$

where $a_0 = 1.4845$, $a_1 = 0.6300$, $a_2 = 1.7580$, $a_3 = 1.4215$, $a_4 = 0.5075$, and t is the maximum thickness. The mean camber line is given by

$$y_c = \begin{cases} \dfrac{m}{p^2}(2px - x^2), x < p \\ \dfrac{m}{(1-p)^2}(1 - 2p + 2px - x^2), x \geq p \end{cases} \tag{9.12}$$

Three example NACA four-digit airfoils are shown in Fig. 9.6.

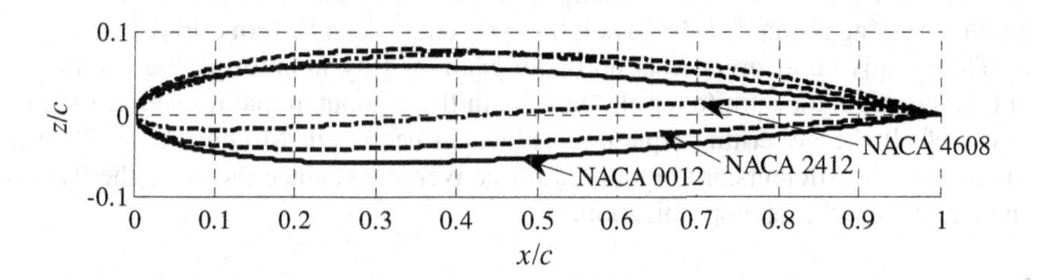

Fig. 9.6 Shown are three different NACA four-digit airfoil sections. NACA 0012 ($m = 0$, $p = 0$, $t/c = 0.12$) is shown by a solid line (-). NACA 2412 ($m = 0.02$, $p = 0.4$, $t/c = 0.12$) is shown by a dash line (--). NACA 4608 ($m = 0.04$, $p = 0.6$, $t/c = 0.08$) is shown by a dash-dot line (-·-)

9.3.2.2 Computational Grid

The governing equations are solved on a computational grid. The grid needs to resolve the entire solution domain, as well as the detailed airfoil geometry. Furthermore, the grid needs to be sufficiently fine to capture the flow physics accurately. For example, a fine grid resolution is necessary near the airfoil surface, especially near the LE where the flow gradients are large. Also, if viscous effects are included, then the grid needs to be fine near the entire airfoil surface (and any other wall surface in the solution domain). The grid can be much coarser several chord lengths away from the airfoil and in the farfield. For a detailed discussion on grid generation the reader is referred to [24] and [25].

For illustration purposes, a typical grid for an airfoil used in aircraft design, generated using the computer program ICEM CFD [52], is shown in Fig. 9.7. This is a structured curvilinear body-fitted grid of C-topology (a topology that can be associated to the letter C, i.e., at the inlet the grid surrounds the leading-edge of the airfoil, but is open at the other end). The size of the computational region is made large enough so that it will not affect the flow solution. In this case, there are 24 chord lengths in front of the airfoil, 50 chord lengths behind it, and 25 chord lengths above and below it. The airfoil leading-edge (LE) is located at the origin.

9.3.2.3 Flow Solution

Most commercially available CFD flow solvers are based on the Finite Volume Method (FVM). According to FVM, the solution domain is subdivided into a finite number of small control volumes (cells) by a grid. The grid defines the boundaries of the control volumes, while the computational node lies at the center of the control volume. Integral conservation of mass, momentum, and energy are satisfied exactly over each control volume. The result is a set of linear algebraic equations, one for each control volume. The set of equations are then solved iteratively, or simultaneously. Iterative solution is usually performed with relaxation to suppress numerical oscillations in the flow solution that result from numerical errors. The iterative process is repeated until the change in the flow variables in two subsequent iterations becomes smaller than the prescribed convergence threshold. Further reading on the FVM and solution procedures can be found in [24, 25].

The iterative convergence is normally examined by monitoring the overall residual, which is the sum (over all the cells in the computational domain) of the L^2 norm of all the governing equations solved in each cell. Moreover, the lift and drag forces coefficients are monitored for convergence, since these are the figures interest in airfoil shape optimization.

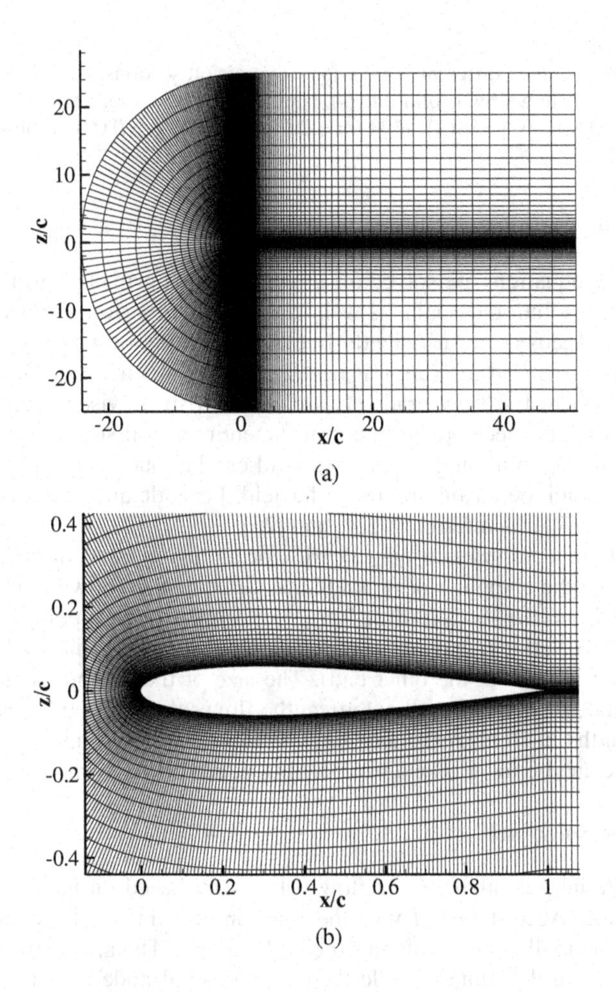

Fig. 9.7 (a) An example computational grid for the NACA 0012 airfoil with a C-topology, (b) a view of the computational grid close to the airfoil

As an illustration, we consider the FVM-based computer code FLUENT [22] for the fluid flow simulations. Compressible inviscid flow past the NACA 2412 airfoil at Mach number $M_\infty = 0.75$ and an angle of attack $\alpha = 1$ degree is simulated using the Euler equations and a similar grid as shown in Fig. 9.7. The convergence of the residuals for mass, momentum, and energy is shown in Fig. 9.8(a) and the convergence of the lift and drag coefficients is shown in Fig. 9.8(b). The limit on the residuals to indicate convergence was set to 10^{-6}. The solver needed 216 iterations to reach full convergence of the flow solution. However, only about 50 iterations or so are necessary to reach convergence of the lift and drag coefficient. The Mach number contour plot of the flow field around the airfoil is shown in Fig. 9.9(a) and the pressure distribution on the airfoil surface is shown in Fig. 9.9(b). On the upper surface there is a strong shock with associated decrease in flow speed and an increase in pressure.

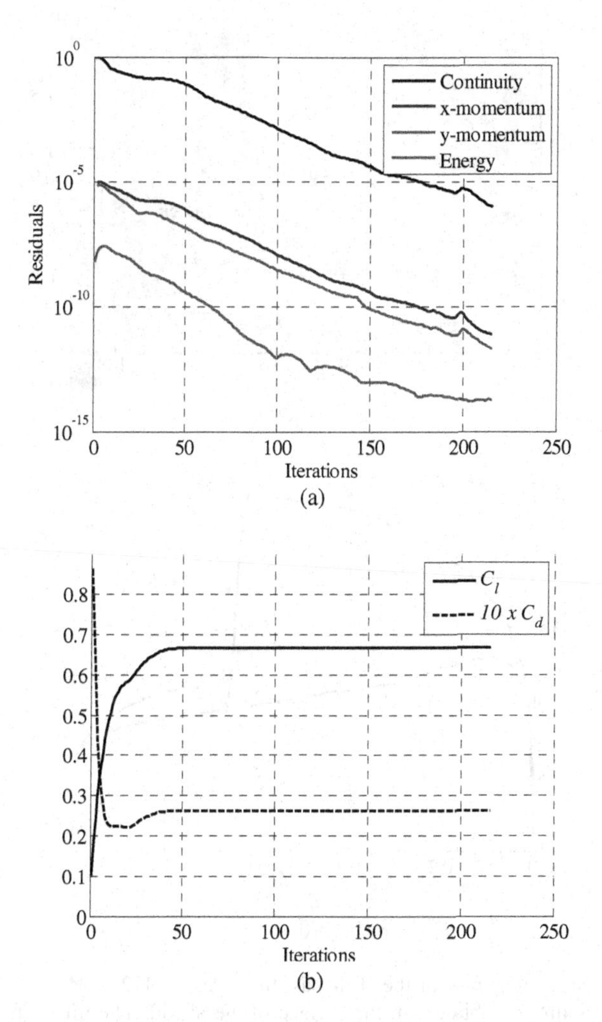

Fig. 9.8 (a) Convergence history of the simulation of the flow past the NACA 2412 at M_∞ = 0.75 and α = 1 deg., (b) convergence of the lift and drag coefficients. The converged values of the lift coefficient is $C_l = 0.67$ and the drag coefficient is $C_d = 0.0261$

9.3.2.4 Aerodynamic Forces

The aerodynamic forces are calculated by integrating the pressure (p) and the viscous wall shear stress (τ), as defined in Figure 9.2, over the surface of the airfoil. The pressure coefficient is defined as

$$C_p \equiv \frac{p - p_\infty}{q_\infty} \tag{9.13}$$

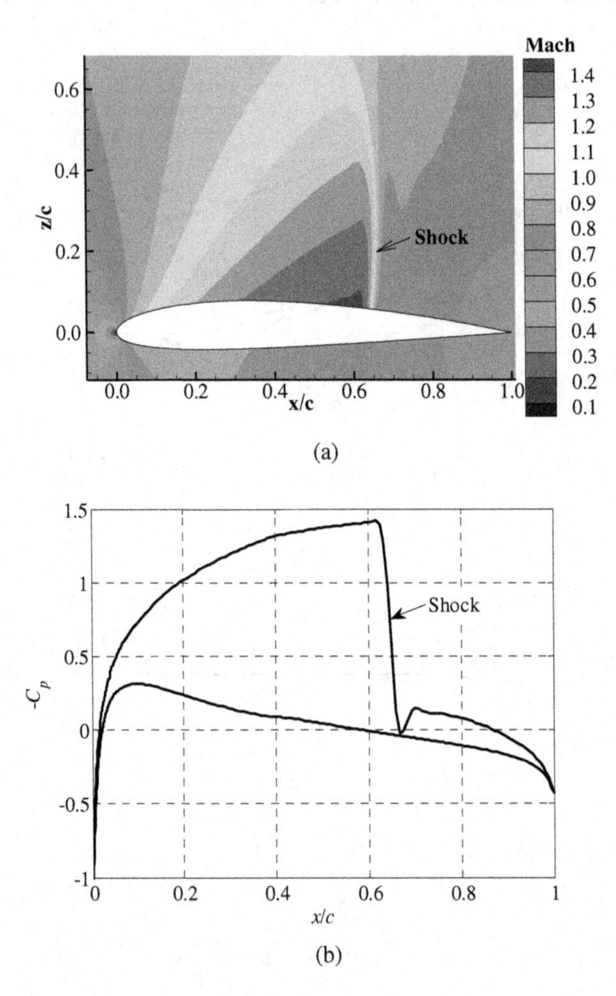

Fig. 9.9 (a) Mach contour plot of the flow past the NACA 2412 at $M_\infty = 0.75$ and $\alpha = 1$ deg., (b) the pressure distribution on the surface of the airfoil. The lift coefficient is $C_l = 0.67$ and drag coefficient $C_d = 0.0261$

where p_∞ is the free-stream pressure. Similarly, the shear stress coefficient is defined as

$$C_f \equiv \frac{\tau}{q_\infty} \tag{9.14}$$

The normal force coefficient (parallel to the z-axis) acting on the airfoil is [46]

$$C_n = -\int (C_{p_u} \cos\theta + C_{f_u} \sin\theta) ds_u + \int (C_{p_l} \cos\theta + C_{f_l} \sin\theta) ds_l \tag{9.15}$$

where ds is the length of a surface element, θ is the angle (positive clockwise) that p and τ make relative to the z- and x-axis, respectively. The subscripts u and l refer to the upper and lower airfoil surfaces, respectively. The horizontal force coefficient (parallel to the x-axis) acting on the airfoil is [46]

$$C_a = \int (-C_{p_u} \sin\theta + C_{f_u} \cos\theta) ds_u + \int (C_{p_l} \sin\theta + C_{f_l} \cos\theta) ds_l \tag{9.16}$$

The lift force coefficient is calculated as

$$C_l = C_n \cos\alpha - C_a \sin\alpha \tag{9.17}$$

where α is the airfoil angle of attack, and the drag force coefficient is calculated as

$$C_d = C_n \sin\alpha + C_a \cos\alpha \tag{9.18}$$

9.4 Direct Optimization

The direct optimization is understood here as employing the high-fidelity simulation model directly in the optimization loop. The flow of the direct optimization process is shown in Fig. 9.10 and can be described as follows. First, an initial design $\mathbf{x}^{(0)}$ is generated and the high-fidelity CFD simulation model is evaluated at that design, yielding values of the objective function and the constraints. Then, the optimization algorithm finds a new airfoil design, $\mathbf{x}$, and the high-fidelity CFD simulation model is evaluated at that design and the objective and constraints are recalculated. Based on the improvement or deterioration in the objective function and the values of the constraints (fulfilled, critical, or violated), either the optimizer finds another design to evaluate, or it uses the current design for the ith design iteration to yield design $x^{(i)}$. The high-fidelity model can be evaluated several times during one design iteration. Now, this loop is repeated until a termination condition is met and an optimized design has been reached. The termination condition could be, for example, based on the change in airfoil shape between two adjacent design iterations $\mathbf{x}^{(i)}$ and $\mathbf{x}^{(i+1)}$.

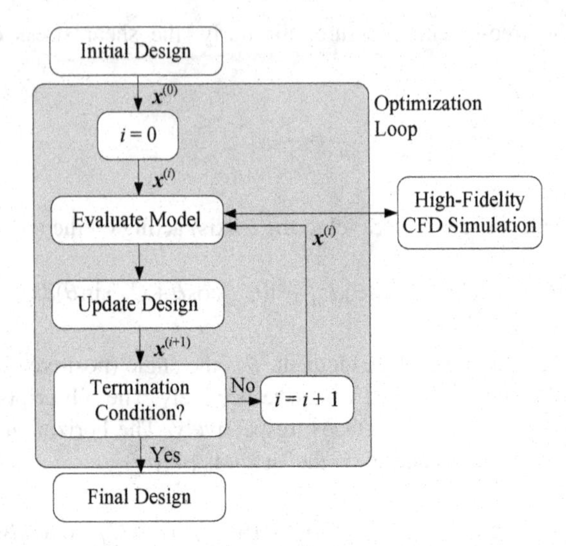

Fig. 9.10 Flowchart of the direct optimization process

9.4.1 Gradient-Based Methods

The development of numerical optimization techniques pertaining to aircraft design began when Hicks and Henne [1] coupled a gradient-based optimization algorithm with CFD codes to design airfoils and wings at both subsonic and transonic conditions. Substantial progress in gradient-based methods for aerodynamic design has been made since then. Jameson [12] introduced control theory and continuous adjoint methods to the optimal aerodynamic design for two-dimensional airfoils and three-dimensional wings, first using inviscid flow solvers [13, 14], and later using viscous flow solvers [4, 16]. The adjoint method is gradient-based, but it is very efficient since the computational expense incurred in the calculation of the gradient is effectively independent of the number of design variables.

Eyi et al. [17] apply gradient-based optimization to the design of multi-element airfoils at high-lift conditions where the necessary gradients are obtained by finite-difference methods. Nemec and Zingg apply a gradient-based Newton-Krylov algorithm to high-lift system design [18], as well as transonic wing design [19], where the gradient of the objective function is computed using the discrete adjoint approach.

Papadimitriou and Giannakoglou [10] apply continuous and discrete adjoint methods for the design optimization of a two-dimensional compressor and turbine cascades. They consider various problem formulations, such as inverse design and viscous losses minimization.

Gradient-based methods are robust for local search. However, often a large number of function evaluations are needed, and since CFD simulations can be very expensive, the overall computational cost becomes prohibitive. Furthermore, results from CFD simulations include numerical noise, which is a serious issue for gradient-based algorithms.

9.4.2 Derivative-Free Methods

Derivative-free approaches can be divided into two categories, local and global search methods. The local search methods include the pattern-search algorithm [53] and the Nelder-Mead algorithm [54]. Global search methods include Genetic Algorithms (GAs) [55], Evolutionary Algorithms (EAs) [56], Particle Swarm Optimization (PSO) [57, 58], and Differential Evolution (DE) [59], all of which are often referred to as meta-heuristic algorithms. Example applications to aerodynamic shape optimization can be found, e.g., in [20], [49] and [51]. Another global search method is Simulated Annealing (SA) [60].

The main advantages of these methods are that they do not require gradient data and they can handle noisy/discontinuous objective functions. However, the aforementioned derivative-free methods normally require a large number of function evaluations.

9.5 Surrogate-Based Optimization

In this section, we provide a brief overview of surrogate-based optimization (SBO) [32, 33]. We begin with presenting the concept of SBO. Then, we discuss the construction of the surrogate model, and, finally, we present a few popular surrogate-based optimization techniques.

9.5.1 The Concept

In many situations, the functions one wants to optimize are difficult to handle. This is particularly the case in aerodynamic design where the objective and constraint functions are typically based on CFD simulations. The major issue is computational cost of simulation which may be very high (e.g., up to several days or even weeks for high-fidelity 3D wing simulation for a single design). Another problem is numerical noise which is always present in CFD tools. Also, simulation may fail for specific sets of design variables (e.g., due to convergence issues). In order to alleviate these problems, it is often advantageous to replace—in the optimization process—the original objective function by its surrogate model. To make this replacement successful, the surrogate should be sufficiently accurate representation of the original function, yet analytically tractable (smooth), and, preferably computationally cheap (so that to reduce the overall cost of the optimization process). In practice, surrogate-based optimization if often an iterative process, where the surrogate model is re-optimized and updated using the data from the original function that is accumulated during the algorithm run.

The flow of a typical SBO algorithm is shown in Fig. 9.11. The surrogate model is optimized, in place of the high-fidelity one, to yield prediction of its minimizer. This prediction is verified by evaluating the high-fidelity model, which is typically done only once per iteration (at every new design $\mathbf{x}^{(i+1)}$). Depending on the result of this verification, the optimization process may be terminated or may continue, in which case the surrogate model is updated using the new available high-fidelity model data, and then re-optimized to obtain a new, and hopefully better approximation of the minimizer. For a well performing surrogate-based algorithm, the number of iterations is substantially smaller than for most methods optimizing the high-fidelity model directly (e.g., gradient-based one).

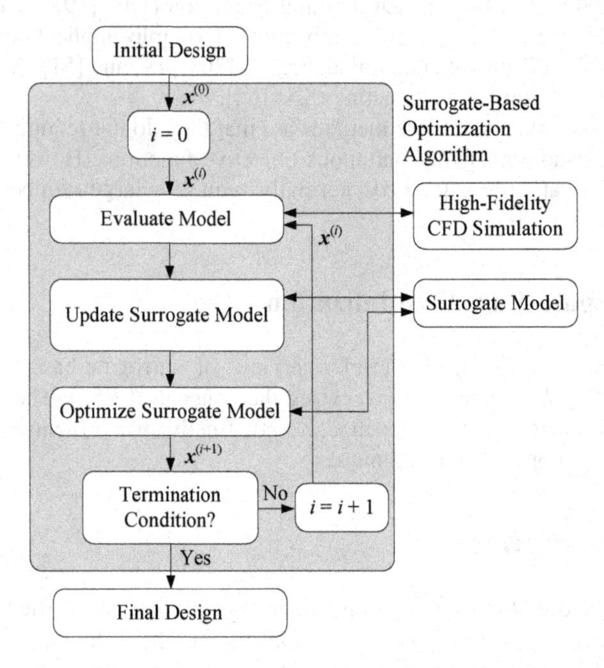

Fig. 9.11 A flowchart of a typical surrogate-based optimization algorithm

9.5.2 Surrogate Modeling

The surrogates can be created either by approximating the sampled high-fidelity model data using regression (so-called function-approximation surrogates or functional surrogates), or by correcting physics-based low-fidelity models, which are less accurate but computationally cheap representations of the high-fidelity ones [33].

9.5.2.1 Functional Surrogates

Functional surrogate models are constructed without any particular knowledge of the physical system. The construction process can be summarized as follows:

1. *Design of Experiments (DoE)*: Allocate a set of points in the design space by using a specific strategy to maximize the amount of information gained

from a limited number of samples [61]. Factorial designs, are classical DoE techniques, and these techniques typically spread the samples apart as much as possible to reduce any random error (which is important when obtaining data from physical experiments) [62]. Nowadays, space filling designs are commonly used with Latin Hypercube Sampling being probably the most popular one [61].

2. *Acquire data*: Evaluate the high-fidelity model at the points specified in step 1.
3. *Model Selection and Identification*: Choose the surrogate model and determine its parameters. A number of models are available including polynomial regression [32], radial basis functions [33], kriging [62], neural networks [62] and support vector regression [63].
4. *Model validation*: Estimate the model generalization error. The most popular method is cross-validation [32], where the data is split into k subsets, and a surrogate model is constructed k times so that k–1 subsets are used for training, and the remaining ones are used to calculate the generalization error, averaged over all k combinations of training/testing data.

The functional surrogate models are typically cheap to evaluate. However, a considerable amount of data is required to set up the surrogate model that ensures reasonable accuracy. The methodology of constructing the functional surrogates is generic, and, therefore, is applicable to a wide class of problems.

9.5.2.2 Physics-Based Surrogates

The physics-based surrogates are constructed by correcting an underlying low-fidelity model, which can be based on one of, or a combination of the following:

- *Simplified physics*: Replace the set of governing fluid flow equations by a set of simplified equations, e.g., using the Euler equations in place of the RANS equations [35]. These are often referred to as variable-fidelity physics models.
- *Coarse discretization*: Use the same fluid flow model as in the high-fidelity model, but with a coarser computational mesh discretization [36]. Often referred to as variable-resolution models.
- *Relaxed convergence criteria*: Reduce the number of maximum allowable iterations and/or reduce the convergence tolerance [63]. Sometimes referred to as variable-accuracy models.

As the low-fidelity model enjoys the same underlying physics as the high-fidelity one, it is able to predict the general behavior of the high-fidelity model. However, the low-fidelity model needs to be corrected to match the sampled data of the high-fidelity model to become a reliable and accurate predictor. Popular correction techniques include response correction [34] and space mapping [65]. One of the recent techniques is shape-preserving response prediction (SPRP) introduced in [26]. The application of this technique to the design of airfoil at high-lift and transonic conditions is given in the next chapter.

The physics-based surrogates are typically more expensive to evaluate than the functional surrogates. Furthermore, they are problem specific, i.e., reuse across different problems is rare. On the other hand, their fundamental advantage is that much less high-fidelity model data is needed to obtain a given accuracy level than in case of functional surrogates. Some SBO algorithms exploiting physics-based low-fidelity models (often referred to as variable- or multi-fidelity ones) require just a single high-fidelity model evaluation per algorithm iteration to construct the surrogate [34, 38]. One of the consequences is that the variable-fidelity SBO methods are more scalable to larger number of design variables (assuming that no derivative information is required) than SBO using functional surrogates.

9.5.3 Optimization Techniques

This section presents selected optimization techniques that employ physics-based low-fidelity surrogate models. We describe the Approximation Model Management Optimization (AMMO) algorithm [34-36] and the Surrogate Management Framework (SMF) [38]. We also briefly mention a few other techniques.

9.5.3.1 Approximation Model Management Optimization (AMMO)

The AMMO algorithm is a general approach for controlling the use of variable-fidelity models when solving a nonlinear minimization problem, such as Eq. (9.3), [34-36]. A flowchart of the AMMO algorithm is shown in Fig. 9.12. The optimizer receives the function and constraint values, as well as their sensitivities, from the low-fidelity model. The response of the low-fidelity model is corrected to satisfy at least zero- and first-order consistency conditions with the high-fidelity model, i.e., agreement between the function values and the first-order derivatives at a given iteration point. The expensive high-fidelity computations are performed outside the optimization loop and serve to re-calibrate the low-fidelity model occasionally, based on a set of systematic criteria. AMMO exploits the trust-region methodology [66], which is an adaptive move limit strategy for improving the global behavior of optimization algorithms based on local models. By combining the trust-region approach with the use of the low-fidelity model satisfying at least first-order consistency conditions, then convergence of AMMO to the optimum of the high-fidelity model can be guaranteed.

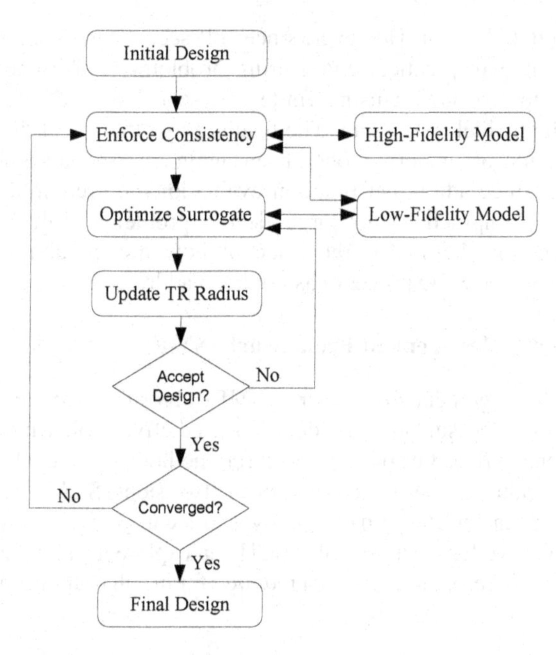

Fig. 9.12 A flowchart of the Approximation Model Management Optimization (AMMO) algorithm

The AMMO methodology has been applied to various aerodynamic design problems [34-36]. In [36], AMMO is applied to both 2D airfoil shape optimization and 3D aerodynamic wing optimization, both at transonic operating conditions. The Euler equations are used as governing fluid flow equations for the both the high- and low-fidelity models at variable grid resolution, i.e., a fine grid for the high-fidelity model and a coarse grid for the low-fidelity model. The results showed a threefold improvement in the computational cost in the 3D wing design problem, when compared to direct optimization of the high-fidelity model, and a twofold improvement for the 2D airfoil design problem. In [35], AMMO is applied to 2D airfoil design at transonic conditions using the RANS equations (representing viscous flow past the airfoil) as the high-fidelity model and the Euler equations (representing inviscid flow past the airfoil) for the low-fidelity model. The high-fidelity model is solved on a much finer mesh than the low-fidelity model. The results demonstrated a fivefold improvement when compared to direct optimization of the high-fidelity model.

First-order consistency in variable-fidelity SBO can be insufficient to achieve acceptable convergence rates, which can be similar to those achieved by first-order optimization methods, such as steepest-descent or sequential linear programming [67]. More successful optimization methods, such as sequential quadratic programming, use at least approximate second-order information to achieve super-linear or quadratic convergence rates in the neighborhood of the minimum. Eldred et al. [68] present second-order corrections methods for variable-fidelity SBO algorithms. The second-order corrections enforce consistency with the high-fidelity model Hessian. However, since full second-order information is not commonly available in practical engineering problems, consistency can also be enforced to an approximation using finite difference, quasi-Newton, or Gauss-Newton to the high-fidelity Hessian. The results show that all of these approaches outperform the first-order corrections. Then again, the second-order corrections come at a price, since additional function evaluations are required. Additionally, they can become impractical for large design problems, unless adjoint-based gradients are employed. Finally, the issue of how numerical noise affects the second-order corrected SBO process has not been addressed.

9.5.3.2 Surrogate Management Framework (SMF)

The Surrogate Management Framework (SMF) algorithm [38] is a mesh-based technique that uses the surrogate model as a predictive tool, while retaining the robust convergence properties of pattern-search methods for a local grid search.

The SMF algorithm (Fig. 9.13) consists of the two steps, SEARCH and POLL. In the SEARCH step, the surrogate model is used to identify the set of points likely to minimize the objective function. The SEARCH can explore the surrogate model globally or locally. In any case, this step is not required for the algorithm convergence.

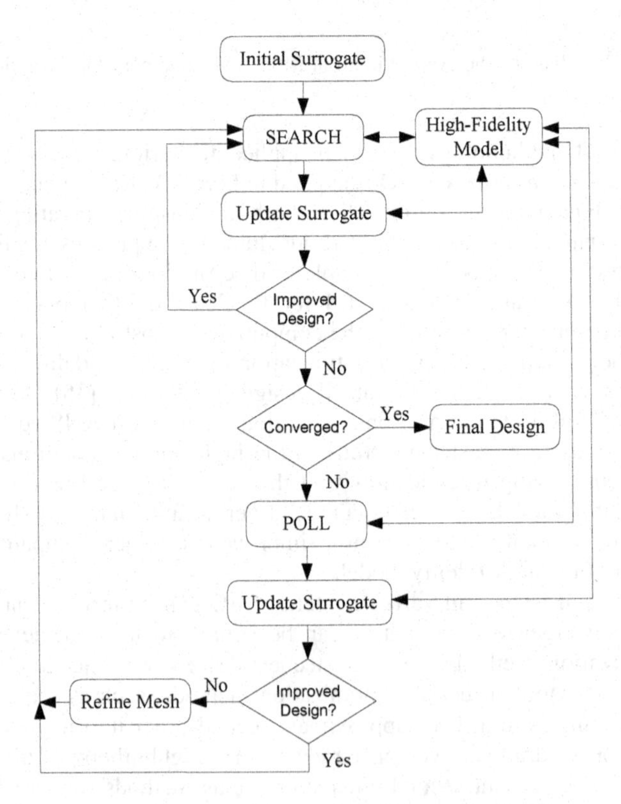

Fig. 9.13 A flowchart of the Surrogate Management Framework (SMF)

The convergence of the SMF is ensured by the POLL step, where the neighbors of the current best solution are evaluated using the high-fidelity model on the mesh in a positive spanning set of directions [69] to look for a local objective function improvement. In case the POLL step fails to improve the objective function value, the mesh is being refined and the new iteration begins starting with the SEARCH step.

The surrogate model is updated in each iteration using all accumulated high-fidelity data.

In [69], the SMF algorithm is applied to the optimal aeroacoustic shape design of an airfoil in laminar flow. The airfoil shape is designed to minimized total radiated acoustic power while constraining lift and drag. The high-fidelity model is implemented through the solution to the unsteady incompressible two-dimensional Navier-Stokes equations with a roughly 24 hour analysis time for a single CFD evaluation. The surrogate function is constructed using kriging, which is typical when using the SMF algorithm. As the acoustic noise is generated at the airfoil TE, only the upper TE of the airfoil is parameterized with a spline using five control points. Optimal shapes that minimize noise are reported. Results show a significant reduction (as much as 80%) in acoustic power with reasonable computational cost (less than 88 function evaluations).

9.5.3.3 Other Techniques

Robinson et al. [70-72] presented a provably convergent trust-region model-management (TRMM) methodology for variable-parameterization design models. This is an SBO method which uses a lower-fidelity model as a surrogate. However, the low-fidelity design space has a lower dimension than the high-fidelity design space. The design variables of the low-fidelity model can be a subset of the high-fidelity model, or they can be different from the high-fidelity model. The mathematical relationship between the design vectors is described by space mapping (SM) [65, 73-75]. Since SM does not provide provable convergence within a TRMM framework, but any surrogate that is first-order accurate does, they correct the space-mapping to be at least first-order, called corrected space-mapping.

The TRMM method has been applied to several constrained design problems. One problem involves the design of a wing planform (minimize induce drag and constrain lift for a given wingspan) using a vortex-lattice method as the high-fidelity model and a lifting-line method as the low-fidelity model. The results indicated over 76% savings in high-fidelity function calls as compared to direct optimization. Another problem involves the design of a flapping-wing where an unsteady panel-method is used as the high-fidelity model and the low-fidelity model is based on thin-airfoil theory and is assumed to be quasi-steady. Approximately 48% savings in high-fidelity function calls where demonstrated when compared to direct optimization.

Several other optimization techniques are available that exploit a surrogate constructed from a physics-based low-fidelity model. In SM, the surrogate is a composition of the low-fidelity model and simple, usually linear, transformations that re-shape the model domain (input-like SM [65]), correct the model response (output-like SM [74]) or change the overall model properties (implicit-like SM [74]) using additional variables that are not directly used in the optimization process (so-called pre-assigned parameters). Manifold mapping (MM) [76] is a special case of output-like SM that aims at enforcing first-order consistency conditions using exact or approximate high-fidelity model sensitivity data. These methods are, however, not (yet) popular in the case of aerodynamic shape optimization.

Shape-preserving response prediction (SPRP) is a relatively novel technique which was introduced in the field of microwave engineering [77], but it has been recently applied to airfoil shape optimization [26, 78]. This technique is described in detail in the next chapter.

9.5.3.4 Exploration versus Exploitation

One of the important steps of the SBO optimization process is to update the surrogate model using the high-fidelity model data accumulated during the algorithm run. In particular, the high-fidelity model is evaluated at any new design obtained from prediction provided by the surrogate model. The new points at which we evaluate the high-fidelity model are referred to as infill points [33]. Selection of these points is based on certain infill criteria. These criteria can be either exploitation- or exploration-based.

A popular exploitation-based strategy is to select the surrogate minimizer as the new infill point [33]. This strategy is able to ensure finding at least a local minimum of the high-fidelity model provided that the surrogate model satisfies zero- and first-order consistency conditions. In general, using the surrogate model optimum as a validation point corresponds to exploitation of certain region of the design space, i.e., neighborhood of a local optimum. Selecting the surrogate minimizer as the infill point is utilized by AMMO [34-36], SM [65, 73-75], MM [76], and can also be used by SMF [38].

In exploration-based strategies, the new sample points are located in between the existing ones. This allows building a surrogate model that is globally accurate. A possible infill criterion is to allocate the new samples at the points of maximum estimated error [33]. Pure exploration however, may not be a good way of updating the surrogate model in the context of optimization because the time spent on accurately modeling sub-optimal regions may be wasted if the global optimum is the only interest.

Probably the best way of performing global search is to balance exploration and exploitation of the design space. The details regarding several possible approaches can be found in [33].

9.6 Summary

Although aerodynamic shape optimization (ASO) is widely used in engineering design, there are numerous challenges involved. One of the biggest challenges is that high-fidelity computational fluid dynamic (CFD) simulations are (usually) computationally expensive. As a result, the overall computational cost of the design optimization process becomes prohibitive since, typically, a large number of simulations are required. Therefore, it is impractical to employ the high-fidelity model directly in the optimization loop.

One of the objectives of surrogate-based optimization (SBO) is to reduce the overall computational cost by replacing the high-fidelity model in the optimization loop by a cheap surrogate model. The surrogate models can be created by (9.1) approximating the sampled high-fidelity model data using regression (so-called function-approximation surrogates), or (9.2) by correcting physics-based low-fidelity models which are less accurate but computationally cheap representations of the high-fidelity models.

A variety of techniques are available to create the function-approximation surrogate model, such as polynomial regression, and kriging. Function-approximation models are versatile, however, they normally require substantial amount of data samples to ensure good accuracy. The physics-based surrogates are constructed by correcting the underlying low-fidelity models, which can be obtained through (9.1) simplified physics models, (9.2) coarser discretization, and (9.3) relaxed convergence criteria. These models are typically more expensive to evaluate than the function-approximation surrogates, but less high-fidelity model data is needed to obtain a given accuracy level.

The low-fidelity models needs to be corrected to become an accurate and reliable representation of the high-fidelity model. Popular correction methods include response correction and space mapping.

In SBO with physics-based low-fidelity models, called variable- or multi-fidelity SBO, only a single high-fidelity model evaluation is typically required per algorithm iteration. Due to this the variable-fidelity SBO method is naturally scalable to larger numbers of design variables (assuming that no derivative information is required).

The Approximation and Model Management Optimization (AMMO) is a generic SBO approach. AMMO is based on ensuring zero- and first-order consistency conditions between the high-fidelity model and the surrogate by using a suitable correction term. This requires derivative information. Another SBO technique is the Surrogate Management Framework (SMF) algorithm. SMF is a mesh-based technique that uses the surrogate model as a predictive tool, while retaining the robust convergence properties of pattern search methods for a local grid search. Typically, the surrogate model is constructed using kriging and the surrogate is updated in each iteration using all accumulated high-fidelity data. Convergence (at least to a local minimum) is ensured by the pattern search.

References

1. Hicks, R.M., Henne, P.A.: Wing Design by Numerical Optimization. Journal of Aircraft 15(7), 407–412 (1978)
2. Braembussche, R.A.: Numerical Optimization for Advanced Turbomachinery Design. In: Thevenin, D., Janiga, G. (eds.) Optimization and Computational Fluid Dynamics, pp. 147–189. Springer, Heidelberg (2008)

3. Percival, S., Hendrix, D., Noblesse, F.: Hydrodynamic Optimization of Ship Hull Forms. Applied Ocean Research 23(6), 337–355 (2001)
4. Leoviriyakit, K., Kim, S., Jameson, A.: Viscous Aerodynamic Shape Optimization of Wings including Planform Variables. In: 21st Applied Aerodynamics Conference, Orlando, Florida, June 23-26 (2003)
5. van Dam, C.P.: The aerodynamic design of multi-element high-lift systems for transport airplanes. Progress in Aerospace Sciences 8(2), 101–144 (2002)
6. Lian, Y., Shyy, W., Viieru, D., Zhang, B.: Membrane Wing Aerodynamics for Micro Air Vehicles. Progress in Aerospace Sciences 39(6), 425–465 (2003)
7. Secanell, M., Suleman, A., Gamboa, P.: Design of a Morphing Airfoil Using Aerodynamic Shape Optimization. AIAA Journal 44(7), 1550–1562 (2006)
8. Antoine, N.E., Kroo, I.A.: Optimizing Aircraft and Operations for Minimum Noise. In: AIAA Paper 2002-5868, AIAA's Aircraft Technology, Integration, and Operations (ATIO) Technical Forum, Los Angeles, California, October 1-3 (2002)
9. Hosder, S., Schetz, J.A., Grossman, B., Mason, W.H.: Airframe Noise Modeling Appropriate for Multidisciplinary Design and Optimization. In: 42nd AIAA Aerospace Sciences Meeting and Exhibit, Reno, NV, AIAA Paper 2004-0698, January 5-8 (2004)
10. Giannakoglou, K.C., Papadimitriou, D.I.: Adjoint Methods for Shape Optimization. In: Thevenin, D., Janiga, G. (eds.) Optimization and Computational Fluid Dynamics, pp. 79–108. Springer, Heidelberg (2008)
11. Celik, F., Guner, M.: Energy Saving Device of Stator for Marine Propellers. Ocean Engineering 34(5-6), 850–855 (2007)
12. Jameson, A.: Aerodynamic Design via Control Theory. Journal of Scientific Computing 3, 233–260 (1988)
13. Reuther, J., Jameson, A.: Control Theory Based Airfoil Design for Potential Flow and a Finitie Volume Discretization. In: AIAA Paper 94-0499, AIAA 32nd Aerospace Sciences Meeting and Exhibit, Reno, Nevada (January 1994)
14. Jameson, A., Reuther, J.: Control Theory Based Airfoil Design using Euler Equations. In: Proceedings of AIAA/USAF/NASA/ISSMO Symposium on Multidisciplinary Analsysis and Optimization, Panama City Beach, pp. 206–222 (September 1994)
15. Reuther, J., Jameson, A., Alonso, J.J., Rimlinger, M.J., Sauders, D.: Constrained Multipoint Aerodynamic Shape Optimization Using an Adjoint Formulation and Parallel Computing. In: 35th AIAA Aerospace Sciences Meeting & Exhibit, Reno, NV, AIAA Paper 97-0103, Januvary 6-9 (1997)
16. Kim, S., Alonso, J.J., Jameson, A.: Design Optimization of High-Lift Configurations Using a Viscous Continuous Adjoint Method. In: AIAA paper 2002-0844, AIAA 40th Aerospace Sciences Meeting & Exhibit, Reno, NV (January 2002)
17. Eyi, S., Lee, K.D., Rogers, S.E., Kwak, D.: High-lift design optimization using Navier-Stokes equations. Journal of Aircraft 33(3), 499–504 (1996)
18. Nemec, M., Zingg, D.W.: Optimization of high-lift configurations using a Newton-Krylov algorithm. In: 16th AIAA Computational Fluid Dynamics Conference, Orlando, Florida, June 23-26 (2003)
19. Nemec, M., Zingg, D.W., Pulliam, T.H.: Multi-Point and Multi-Objective Aerodynamic Shape Optimization. In: AIAA Paper 2002-5548, 9th AIAA/ISSMO Symposium on Multidisciplinary Analysis and Optimization, Altanta, Georgia, September 4-6 (2002)
20. Giannakoglou, K.C.: Design of optimal aerodynamic shapes using stochastic optimization methods and computational intelligence. Progress in Aerospace Sciences 38(2), 43–76 (2002)

21. Hicks, R.M., Murman, E.M., Vanderplaats, G.: An Assessment of Airfoil Design by Numerical Optimization. NASA TM 3092 (July 1974)
22. FLUENT, ver. 6.3.26, ANSYS Inc., Southpointe, 275 Technology Drive, Canonsburg, PA 15317 (2006)
23. GAMBIT. ver. 2.4.6, ANSYS Inc., Southpointe, 275 Technology Drive, Canonsburg, PA 15317 (2006)
24. Tannehill, J.C., Anderson, D.A., Pletcher, R.H.: Computational Fluid Mechanics and Heat Transfer, 2nd edn. Taylor & Francis, Abington (1997)
25. Hirsch, C.: Numerical Computation of Internal and External Flows – Fundamentals of Computational Fluid Dynamics, 2nd ed., Butterworth-Heinemann (2007)
26. Leifsson, L., Koziel, S.: Multi-fidelity design optimization of transonic airfoils using physics-based surrogate modeling and shape-preserving response prediction. Journal of Computational Science 1(2), 98–106 (2010)
27. Lepine, J., Guibault, F., Trepanier, J.-Y., Pepin, F.: Optimized Nonuniform Rational B-Spline Geometrical Representation for Aerodynamic Design of Wings. AIAA Journal 39(11), 2033–2041 (2001)
28. Li, W., Huyse, L., Padula, S.: Robust Airfoil Optimization to Achieve Consistent Drag Reductin Over a Mach Range. In: NASA/CR-2001-211042 (August 2001)
29. Giunta, A.A., Dudley, J.M., Narducci, R., Grossman, B., Haftka, R.T., Mason, W.H., Watson, L.T.: Noisy Aerodynamic Response and Smooth Approximations in HSCT Design. In: 5th AIAA/USAF/NASA/ISSMO Symposium on Multidisciplinary Analysis and Optimization, Panama City, FL (Septemper 1994)
30. Burman, J., Gebart, B.R.: Influence from numerical noise in the objective function for flow design optimisation. International Journal of Numerical Methods for Heat & Fluid Flow 11(1), 6–19 (2001)
31. Dudley, J.M., Huang, X., MacMillin, P.E., Grossman, B., Haftka, R.T., Mason, W.H.: Multidisciplinary Design Optimization of a High Speed Civil Transport. In: First Industry/University Symposium on High Speed Transport Vehicles, December 4-6. NC A&T University, Greensboro (1994)
32. Queipo, N.V., Haftka, R.T., Shyy, W., Goel, T., Vaidyanathan, R., Tucker, P.K.: Surrogate-Based Analysis and Optimization. Progress in Aerospace Sciences 41(1), 1–28 (2005)
33. Forrester, A.I.J., Keane, A.J.: Recent advances in surrogate-based optimization. Progress in Aerospace Sciences 45(1-3), 50–79 (2009)
34. Alexandrov, N.M., Lewis, R.M.: An overview of first-order model management for engineering optimization. Optimization and Engineering 2(4), 413–430 (2001)
35. Alexandrov, N.M., Nielsen, E.J., Lewis, R.M., Anderson, W.K.: First-Order Model Management with Variable-Fidelity Physics Applied to Multi-Element Airfoil Optimization. In: 8th AIAA/USAF/NASA/ISSMO Symposium on Multidisciplinary Design and Optimization, AIAA Paper 2000-4886, Long Beach, CA (September 2000)
36. Alexandrov, N.M., Lewis, R.M., Gumbert, C.R., Green, L.L., Newman, P.A.: Optimization with Variable-Fidelity Models Applied to Wing Design. In: 38th Aerospace Sciences Meeting & Exhibit, Reno, NV, AIAA Paper 2000-0841(January 2000)
37. Robinson, T.D., Eldred, M.S., Willcox, K.E., Haimes, R.: Surrogate-Based Optimization Using Multifidelity Models with Variable Parameterization and Corrected Space Mapping. AIAA Journal 46(11) (November 2008)
38. Booker, A.J., Dennis Jr., J.E., Frank, P.D., Serafini, D.B., Torczon, V., Trosset, M.W.: A rigorous framework for optimization of expensive functions by surrogates. Structural Optimization 17(1), 1–13 (1999)

39. Lee, D.S., Gonzalez, L.F., Srinivas, K., Periaux, J.: Multi-objective robust design optimisation using hierarchical asynchronous parallel evolutionary algorithms. In: 45th AIAA Aerospace Science Meeting and Exhibit, AIAA Paper 2007-1169, Reno, Nevada, USA, January 8-11 (2007)
40. Barrett, T.R., Bressloff, N.W., Keane, A.J.: Airfoil Design and Optimization Using Multi-Fidelity Analysis and Embedded Inverse Design. In: 47th AIAA/ASME/ASCE/AHS/ASC Structures, Structural Dynamics, and Materials Conference, Newport, AIAA Paper 2006-1820, Rhode Island, May 1-4 (2006)
41. Vanderplaats, G.N.: Numerical Optimization Techniques for Engineering Design, 3rd edn. Vanderplaats Research and Development (1999)
42. Rai, M.M.: Robust optimal design with differential evolution. In: 10th AIAA/ISSMO Multidisciplinary Analysis and Optimization Conference, AIAA Paper 2004-4588, Albany, New York, August 30 - September 1 (2004)
43. Spalart, P.R., Allmaras, S.R.: A one-equation turbulence model for aerodynamic flows. In: 30th AIAA Aerospace Sciences Meeting and Exhibit, Reno, Nevada, January 6-9 (1992)
44. Anderson, J.D.: Modern Compressible Flow – With Historical Prespective, 3rd edn. McGraw-Hill, New York (2003)
45. Katz, J., Plotkin, A.: Low-Speed Aerodynamics, 2nd edn. Cambridge University Press, Cambridge (2001)
46. Anderson, J.D.: Fundamentals of Aerodynamics, 4th edn. McGraw-Hill, New York (2007)
47. Gauger, N.R.: Efficient Deterministic Approaches for Aerodynamic Shape Optimization. In: Thevenin, D., Janiga, G. (eds.) Optimization and Computational Fluid Dynamics, pp. 111–145. Springer, Heidelberg (2008)
48. Abbott, I.H., Von Doenhoff, A.E.: Theory of Wing Sections. Dover Publications, New York (1959)
49. Peigin, S., Epstein, B.: Robust Optimization of 2D Airfoils Driven by Full Navier-Stokes Computations. Computers & Fluids 33, 1175–1200 (2004)
50. Sobieczky, H.: Parametric Airfoils and Wings. In: Fuji, K., Dulikravich, G.S. (eds.) Notes on Numerical Fluid Mechanics, vol. 68. Wiesbaden, Vieweg (1998)
51. Derksen, R.W., Rogalsky, T.: Bezier-PARSEC: An Optimized Aerofoil Parameterization for Design. Advances in Engineering Software 41, 923–930 (2010)
52. ICEM CFD. ver. 12.1, ANSYS Inc., Southpointe, 275 Technology Drive, Canonsburg, PA 15317 (2006)
53. Kolda, T.G., Lewis, R.M., Torczon, V.: Optimization by direct search: new perspectives on some classical and modern methods. SIAM Review 45(3), 385–482 (2003)
54. Nelder, J.A., Mead, R.: A simplex method for function minimization. Computer Journal 7, 308–313 (1965)
55. Goldberg, D.: Genetic Algorithms in Search, Optimization and Machine Learning. Addison-Wesley, Reading (1989)
56. Michalewicz, Z.: Genetic Algorithm + Data Structures = Evolutionary Programs, 3rd edn. Springer, Heidelberg (1996)
57. Kennedy, J., Eberhart, R.: Particle swarm optimization. In: Proceedings of IEEE International Conference on Neural Networks, pp. 1942–1948 (1995)
58. Clerc, M., Kennedy, J.: The particle swarm - explosion, stability, and convergence in a multidimensional complex space. IEEE Transactions on Evolutionary Computation 6(1), 58–73 (2002)

59. Storn, R., Price, K.: Differential evolution - a simple and efficient heuristic for global optimization over continuous spaces. Journal of Global Optimization 11, 341–359 (1997)
60. Kirkpatrick, S., Gelatt Jr, C., Vecchi, M.: Optimization by simulated annealing. Science 220(4498), 671–680 (1983)
61. Giunta, A.A., Wojtkiewicz, S.F., Eldred, M.S.: Overview of modern design of experiments methods for computational simulations. In: 41st AIAA Aerospace Sciences Meeting and Exhibit, AIAA Paper 2003-0649, Reno, NV (2003)
62. Simpson, T.W., Peplinski, J., Koch, P.N., Allen, J.K.: Metamodels for computer-based engineering design: survey and recommendations. Engineering with Computers 17(2), 129–150 (2001)
63. Gunn, S.R.: Support vector machines for classification and regression. Tech. Rep., School of Electronics and Computer Science, University of Southampton (1998)
64. Forrester, A.I.J., Bressloff, N.W., Keane, A.J.: Optimization Using Surrogate Models and Partially Converged Computationally Fluid Dynamics Simulations. Proceedings of the Royal Society A: Mathematical. Physical and Engineering Sciences 462(2071), 2177–2204 (2006)
65. Bandler, J.W., Cheng, Q.S., Dakroury, S.A., Mohamed, A.S., Bakr, M.H., Madsen, K., Søndergaard, J.: Space mapping: the state of the art. IEEE Trans. Microwave Theory Tech. 52(1), 337–361 (2004)
66. Conn, A.R., Gould, N.I.M., Toint, P.L.: Trust Region Methods. MPS-SIAM Series on Optimization (2000)
67. Rao, S.S.: Engineering Optimization: Theory and Practice, 3rd edn. Wiley, Chichester (1996)
68. Eldred, M.S., Giunta, A.A., Collis, S.S.: Second-Order Corrections for Surrogate-Based Optimizatin with Model Hierarchies. In: 10th AIAA/ISSMO Multidisciplinary Analysis and Optimization Conference, AIAA Paper 2004-4457, Albany, NY (2004)
69. Marsden, A.L., Wang, M., Dennis, J.E., Moin, P.: Optimal aeroacoustic shape design using the surrogate management framework. Optimization and Engineering 5, 235–262 (2004)
70. Robinson, T.D., Eldred, M.S., Willcox, K.E., Haimes, R.: Strategies for Multifidelity Optimization with Variable Dimensional Hierarchical Models. In: 47th AIAA/ASME/ASCE/AHS/ASC Structures, Structural Dynamics, and Materials Conference, AIAA Paper 2006-1819, Newport, Rhode Island, May 1-4 (2006)
71. Robinson, T.D., Willcox, K.E., Eldren, M.S., Haimes, R.: Multifidelity Optimization for Variable-Complexity Design. In: 11th AIAA/ISSMO Multidisciplinary Analysis and Optimization Conference, AIAA Paper 2006-7114, Portsmouth, VA (September 2006)
72. Robinson, T.D., Eldred, M.S., Willcox, K.E., Haimes, R.: Surrogate-Based Optimization Using Multifidelity Models with Variable Parameterization and Corrected Space Mapping. AIAA Journal 46(11) (November 2008)
73. Koziel, S., Bandler, J.W., Madsen, K.: A space mapping framework for engineering optimization: theory and implementation. IEEE Trans. Microwave Theory Tech. 54(10), 3721–3730 (2006)
74. Koziel, S., Cheng, Q.S., Bandler, J.W.: Space mapping. IEEE Microwave Magazine 9(6), 105–122 (2008)
75. Redhe, M., Nilsson, L.: Using space mapping and surrogate models to optimize vehicle crashworthiness design. In: 9th AIAA/ISSMO Multidisciplinary Analysis and Optimization Symp., AIAA Paper 2002-5536, Atlanta, GA, pp. 2002–5536 (Septemper 2002)

76. Echeverria, D., Hemker, P.W.: Space mapping and defect correction. CMAM Int. Mathematical Journal Computational Methods in Applied Mathematics 5(2), 107–136 (2005)
77. Koziel, S.: Efficient Optimization of Microwave Circuits Using Shape-Preserving Response Prediction. In: IEEE MTT-S Int. Microwave Symp. Dig, Boston, MA, pp. 1569–1572 (2009)
78. Koziel, S., Leifsson, L.: Multi-Fidelity High-Lift Aerodynamic Optimization of Single-Element Airfoils. In: Int. Conf. Engineering Optimization, Lisbon (September 6-9, 2010)

Chapter 10
Evolutionary Algorithms Applied to Multi-Objective Aerodynamic Shape Optimization

Alfredo Arias-Montaño, Carlos A. Coello Coello, and Efrén Mezura-Montes

Abstract. Optimization problems in many industrial applications are very hard to solve. Many examples of them can be found in the design of aeronautical systems. In this field, the designer is frequently faced with the problem of considering not only a single design objective, but several of them, i.e., the designer needs to solve a Multi-Objective Optimization Problem (MOP). In aeronautical systems design, aerodynamics plays a key role in aircraft design, as well as in the design of propulsion system components, such as turbine engines. Thus, aerodynamic shape optimization is a crucial task, and has been extensively studied and developed. Multi-Objective Evolutionary Algorithms (MOEAs) have gained popularity in recent years as optimization methods in this area, mainly because of their simplicity, their ease of use and their suitability to be coupled to specialized numerical simulation tools. In this chapter, we will review some of the most relevant research on the use of MOEAs to solve multi-objective and/or multi-disciplinary aerodynamic shape optimization problems. In this review, we will highlight some of the benefits and drawbacks of the use of MOEAs, as compared to traditional design optimization methods. In the second part of the chapter, we will present a case study on the application of MOEAs for the solution of a multi-objective aerodynamic shape optimization problem.

Alfredo Arias-Montaño · Carlos A. Coello Coello
CINVESTAV-IPN (Evolutionary Computation Group),
Departamento de Computación, Av. IPN No. 2508,
Col. San Pedro Zacatenco, México D.F. 07360, Mexico
e-mail: aarias@computacion.cs.cinvestav.mx,
ccoello@cs.cinvestav.mx

Carlos A. Coello Coello
UMI LAFMIA 3175 CNRS at CINVESTAV-IPN

Efrén Mezura-Montes
Laboratorio Nacional de Informática Avanzada (LANIA A.C.),
Rébsamen 80, Centro, Xalapa, Veracruz, 91000, Mexico
e-mail: emezura@lania.mx

S. Koziel & X.-S. Yang (Eds.): Comput. Optimization, Methods and Algorithms, SCI 356, pp. 211–240.
springerlink.com

10.1 Introduction

There are many industrial areas in which optimization processes help to find new solutions and/or to increase the performance of an existing one. Thus, in many cases a research goal can be translated into an optimization problem. Optimal design in aeronautical engineering is, by nature, a multiobjective, multidisciplinary and highly difficult problem. Aerodynamics, structures, propulsion, acoustics, manufacturing and economics, are some of the disciplines involved in this type of problems. In fact, even if a single discipline is considered, many design problems in aeronautical engineering have conflicting objectives (e.g., to optimize a wing's lift and drag or a wing's structural strength and weight). The increasing demand for optimal and robust designs, driven by economics and environmental constraints, along with the advances in computational intelligence and the increasing computing power, has improved the role of computational simulations, from being just analysis tools to becoming design optimization tools.

In spite of the fact that gradient-based numerical optimization methods have been successfully applied in a variety of aeronautical/aerospace design problems,[1] [30, 16, 42] their use is considered a challenge due to the following difficulties found in practice:

1. The design space is frequently multimodal and highly non-linear.
2. Evaluating the objective function (performance) for the design candidates is usually time consuming, due mainly to the high fidelity and high dimensionality required in the simulations.
3. By themselves, single-discipline optimizations may provide solutions which not necessarily satisfy objectives and/or constraints considered in other disciplines.
4. The complexity of the sensitivity analyses in Multidisciplinary Design Optimization (MDO[2]) increases as the number of disciplines involved becomes larger.
5. In MDO, a trade-off solution, or a set of them, are searched for.

Based on the previously indicated difficulties, designers have been motivated to use alternative optimization techniques such as Evolutionary Algorithms (EAs) [31, 20, 33]. Multi-Objective Evolutionary Algorithms (MOEAs) have gained an increasing popularity as numerical optimization tools in aeronautical and aerospace engineering during the last few years [1, 21]. These population-based methods mimic the evolution of species and the survival of the fittest, and compared to traditional optimization techniques, they present the following advantages:

(a) *Robustness:* In practice, they produce good approximations to optimal sets of solutions, even in problems with very large and complex design spaces, and are less prone to get trapped in local optima.

[1] It is worth noting that most of the applications using gradient-based methods have adopted them to find global optima or a single compromise solution for multi-objective problems.

[2] Multidisciplinary Design Optimization, by its nature, can be considered as a multi-objective optimization problem, where each discipline aims to optimize a particular performance metric.

(b) *Multiple Solutions per Run:* As MOEAs use a population of candidates, they are designed to generate multiple trade-off solutions in a single run.
(c) *Easy to Parallelize:* The design candidates in a MOEA population, at each generation, can be evaluated in parallel using diverse paradigms.
(d) *Simplicity:* MOEAs use only the objective function values for each design candidate. They do not require a substantial modification or complex interfacing for using a CFD (Computational Fluid Dynamics) or CSD/M (Computational Structural Dynamics/Mechanics) code.
(e) *Easy to hybridize:* Along with the simplicity previously stated, MOEAs also allow an easy hybridization with alternative methods, e.g., memetic algorithms, which additionally introduce specifities to the implementation, without influencing the MOEA simplicity.
(f) *Novel Solutions:* In many cases, gradient-based optimization techniques converge to designs which have little variation even if produced with very different initial setups. In contrast, the inherent explorative capabilities of MOEAs allow them to produce, some times, novel and non-intuitive designs.

An important volume of information has been published on the use of MOEAs in aeronautical engineering applications (mainly motivated by the advantages previously addressed). In this chapter, we provide a review of some representative works, dealing specifically with multi-objective aerodynamic shape optimization.

The remainder of this chapter is organized as follows: In Section 10.2, we present some basic concepts and definitions adopted in multi-objective optimization. Next, in Section 10.3, we review some of the work done in the area of multi-objective aerodynamic shape optimization. This review covers: *surrogate based optimization, hybrid MOEA optimization, robust design optimization, multidisciplinary design optimization,* and *data mining and knowledge extraction.* In Section 10.4 we present a case study and, finally, in Section 10.5. we present our conclusions and final remarks.

10.2 Basic Concepts

A Multi-Objective Optimization Problem (MOP) can be mathematically defined as follows[3]:

$$\text{minimize } \mathbf{f}(\mathbf{x}) := [f_1(\mathbf{x}), f_2(\mathbf{x}), \ldots, f_k(\mathbf{x})] \tag{10.1}$$

subject to:

$$g_i(\mathbf{x}) \leq 0 \quad i = 1, 2, \ldots, m \tag{10.2}$$

$$h_i(\mathbf{x}) = 0 \quad i = 1, 2, \ldots, p \tag{10.3}$$

[3] Without loss of generality, minimization is assumed in the following definitions, since any maximization problem can be transformed into a minimization one.

where $\mathbf{x} = [x_1, x_2, \ldots, x_n]^T$ is the vector of decision variables, which are bounded by lower (x_i^l) and upper (x_i^u) limits which define the search space $\mathscr{S}$, $f_i : \mathbf{R}^n \rightarrow \mathbf{R}$, $i = 1,...,k$ are the objective functions and $g_i, h_j : \mathbf{R}^n \rightarrow \mathbf{R}$, $i = 1,...,m$, $j = 1,...,p$ are the constraint functions of the problem.

In other words, we aim to determine from among the set $\mathscr{F} \subseteq \mathscr{S}$ ($\mathscr{F}$ is the feasible region of the search space $\mathscr{S}$) of all vectors which satisfy the constraints, those that yield the optimum values for all the k objective functions, simultaneously. The set of constraints of the problem defines $\mathscr{F}$. Any vector of variables $\mathbf{x}$ which satisfies all the constraints is considered a feasible solution. In their original version, an EA (and also a MOEA) lacks a mechanism to deal with constrained search spaces. This has motivated a considerable amount of research regarding the design and implementation of constraint-handling techniques for both EAs and MOEAs [10, 29].

10.2.1 Pareto Dominance

Pareto dominance is an important component of the notion of optimality in MOPs and is formally defined as follows:

Definition 1. A vector of decision variables $\mathbf{x} \in \mathbf{R}^n$ dominates another vector of decision variables $\mathbf{y} \in \mathbf{R}^n$, (denoted by $\mathbf{x} \preceq \mathbf{y}$) if and only if $\mathbf{x}$ is partially less than $\mathbf{y}$, i.e. $\forall i \in \{1,\ldots,k\}, f_i(\mathbf{x}) \leq f_i(\mathbf{y}) \wedge \exists i \in \{1,\ldots,k\} : f_i(\mathbf{x}) < f_i(\mathbf{y})$.

Definition 2. A vector of decision variables $\mathbf{x} \in \mathscr{X} \subset \mathbf{R}^n$ is **nondominated** with respect to $\mathscr{X}$, if there does not exist another $\mathbf{x}' \in \mathscr{X}$ such that $\mathbf{f}(\mathbf{x}') \preceq \mathbf{f}(\mathbf{x})$.

In order to say that a solution dominates another one, it needs to be strictly better in at least one objective, and not worse in any of them.

10.2.2 Pareto Optimality

The formal definition of *Pareto optimality* is provided next:

Definition 3. A vector of decision variables $\mathbf{x}^* \in \mathscr{F} \subseteq \mathscr{S} \subset \mathbf{R}^n$ is **Pareto optimal** if it is nondominated with respect to $\mathscr{F}$.

In words, this definition says that $\mathbf{x}^*$ is Pareto optimal if there exists no feasible vector $\mathbf{x}$ which would decrease some objective without causing a simultaneous increase in at least one other objective (assuming minimization). This definition does not provide us a single solution (in decision variable space), but a set of solutions which form the so-called *Pareto Optimal Set* ($\mathscr{P}^*$), whose formal definition is given by:

Definition 4. The **Pareto optimal set** $\mathscr{P}^*$ is defined by:

$$\mathscr{P}^* = \{\mathbf{x} \in \mathscr{F} | \mathbf{x} \text{ is Pareto optimal}\}$$

The vectors that correspond to the solutions included in the Pareto optimal set are said to be *nondominated*.

10.2.3 Pareto Front

When all nondominated solutions are plotted in objective function space, the nondominated vectors are collectively known as the *Pareto front* ($\mathscr{P}\mathscr{F}^*$).

Definition 5. The **Pareto front** $\mathscr{P}\mathscr{F}^*$ is defined by:

$$\mathscr{P}\mathscr{F}^* = \{\mathbf{f}(\mathbf{x}) \in \mathbf{R}^k | \mathbf{x} \in \mathscr{P}^*\}$$

The goal on a MOP consists on determining $\mathscr{P}^*$ from $\mathscr{F}$ of all the decision variable vectors that satisfy (10.2) and (10.3). Thus, when solving a MOP, we aim to find not one, but a set of solutions representing the best possible trade-offs among the objectives (the so-called Pareto optimal set).

10.3 Multi-Objective Aerodynamic Shape Optimization

10.3.1 Problem Definition

Aerodynamics is the science that deals with the interactions of fluid flows and objects. This interaction is governed by conservation laws which are mathematically expressed by means of the *Navier-Stokes* equations, which comprise a set of partial differential equations, being unsteady, nonlinear and coupled among them. Aerodynamicists are interested in the effects of this interaction, in terms of their aerodynamic forces and moments, which are the result of integrating the pressure and shear stresses distributions that the flow excerses over the object with which it is interacting. In its early days, aerodynamic designs were done by extensive use of experimental facilities. Nowadays, the use of Computational Fluid Dynamics (CFD) technology to simulate the flow of complete aircraft configurations, has made it possible to obtain very impressive results with the help of high performance computers and fast numerical algorithms. At the same time, experimental verifications are carried out in scaled flight tests, avoiding many of the inherent disadvantages and extremely high costs of wind tunnel technology. Therefore, we can consider aerodynamics as a mature engineering science.

Thus, current aerodynamic research focuses on finding new designs and/or improving current ones, by using numerical optimization techniques. In the case of multi-objective optimization, the objective functions are defined in terms of aerodynamic coefficients and/or flow conditions. Additionally, design constraints are included to render the solutions practical or realizable in terms of manufacturing and/or operating conditions. Optimization is accomplished by means of a more or less systematic variation of the design variables which parameterize the shape to be optimized. A variety of optimization algorithms, ranging from gradient-based methods to stochastic approaches with highly sophisticated schemes for the adaptation of the individual mutation step sizes, are currently available. From them, MOEAs have been found to be a powerful but easy-to-use choice. Next, we will briefly review some of the most representative works on the use of MOEAs for aerodynamic design. The review comprises the following dimensions that are identified as the most relevant, from a practical point of view, for the purposes of this chapter:

- Surrogate-based optimization,
- Hybrid MOEA optimization,
- Robust design optimization,
- Multidisciplinary design-optimization, and
- Data-mining and knowledge extraction.

10.3.2 Surrogate-Based Optimization

Evolutionary algorithms, being population-based algorithms, often require population sizes, and a number of evolution steps (generations) that might demand tremendous amounts of computing resources. Examples of these conditions are presented by Benini [4], who reported computational times of 2000 hrs. in the multi-objective re-design of a transonic turbine rotor blade, using a population with 20 design candidates, and 100 generations of evolution time, in a four-processors workstation. Thus, when expensive function evaluations are required, the required CPU time may turn prohibitive the application of MOEAs, even with today's available computing power.

For tackling the above problem, one common technique adopted in the field of aerodynamic shape optimization problems, is the use of surrogate models. These models are built to approximate computationally expensive functions. The main objective in constructing these models is to provide a reasonably accurate approximation to the real functions, while reducing by several orders of magnitude the computational cost. Surrogate models range form Response Surface Methods (RSM) based on low-order polynomial functions, Gaussian processes or Kriging, Radial Basis Funcions (RBFs), Artificial Neural Networks (ANNs), to Support Vector Machines (SVMs). A detailed description of each of these techniques is beyond the scope of this chapter, but the interested reader is referred to Jin [19] for a comprehensive review of these and other approximation techniques.

In the context of aerodynamic shape optimization problems, some researchers have used surrogates models to reduce the computational time used in the optimization process. The following is a review of some representative research that has been conducted in this area:

- Lian and Liou [26] addressed the multi-objective optimization of a three-dimensional rotor blade, namely the redesign of the NASA rotor 67 compressor blade, a transonic axial-flow fan rotor. Two objectives were considered in this case:

(i) maximization of the stage pressure rise, and (ii) minimization of the entropy generation. Constraints were imposed on the mass flow rate to have a difference less than 0.1% between the new one and the reference design. The blade geometry was constructed from airfoil shapes defined at four span stations, with a total of 32 design variables. The authors adopted a MOEA based on MOGA [14] with real numbers encoding. The optimization process was coupled to a second-order RSM, which was built with 1,024 design candidates using the Improved Hypercube Sampling (IHS) algorithm. The authors reported that the evaluation of the 1,024 sampling individuals took approximately 128 hours (5.33 days) using eight processors and a Reynolds-Averaged Navier-Stokes CFD simulation. In their experiments, 12 design solutions were selected from the RSM-Pareto front obtained, and such solutions were verified with a high fidelity CFD simulation. The objective function values slightly differed from those obtained by the approximation model, but all the selected solutions were better in both objective functions than the reference design.

- Song and Keane [46] performed the shape optimization of a civil aircraft engine nacelle. The primary goal of the study was to identify the trade-off between aerodynamic performance and noise effects associated with various geometric features for the nacelle. For this, two objective functions were defined: i) scarf angle, and ii) total pressure recovery. The nacelle geometry was modeled using 40 parameters, from which 33 were considered design variables. In their study, the authors implemented the NSGA-II [12] as the multi-objective search engine, while a commercial CFD software was used for evaluation of the three-dimensional flow characteristics. A kriging-based surrogate model was adopted in order to keep the number of designs being evaluated with the CFD tool to a minimum. In their experiments, the authors reported difficulties in obtaining a reliable Pareto front (there were large discrepancies between two consecutive Pareto front approximations). They attributed this behavior to the large number of variables in the design problem, and also to the associated difficulties to obtain an accurate kriging model for these situations. In order to alleviate this, they performed an analysis of variance (ANOVA) test to find the variables that contributed the most to the objective functions. After this test, they presented results with a reduced surrogate model, employing only 7 decision variables. The authors argued that they obtained a design similar to the previous one, but requiring a lower computational cost because of the use of a reduced number of variables in the kriging model.
- Arabnia and Ghaly [2] presented the aerodynamic shape optimization of turbine stages in three-dimensional fluid flow, so as to minimize the adverse effects of three-dimensional flow features on the turbine performance. Two objectives were considered: (i) maximization of isentropic efficiency for the stage, and (ii) minimization of the streamwise vorticity. Additionally, constraints were imposed on: (1) inlet total pressure and temperature, (2) exit pressure, (3) axial chord and spacing, (4) inlet and exit flow angles, and (5) mass flow rate. The blade geometry, both for rotor and stator blades, was based on the E/TU-3 turbine which

is used as a reference design to compare the optimization results. The multi-objective optimization consisted of finding the best distribution of 2D blade sections in the radial and circumferential directions. The authors adopted NSGA [47] as their search engine. Both objective functions were evaluated using a 3D CFD flow simulation, taking an amount of time of 10 hours per design candidate. The authors adopted an artificial neural network (ANN) based model. The ANN model with backpropagation, contained a single hidden layer with 50 nodes, and was trained and tested with 23 CFD simulations, sampling the design space using the Latin Hypercubes technique. The optimization process was undertaken by using the ANN model to estimate both the objective functions, and the constraints. Finally, the nondominated solutions obtained were evaluated with the actual CFD flow simulation. The authors indicated that they were able to obtain design solutions which were better than the reference turbine design.

10.3.2.1 Comments Regarding Surrogate-Based Optimization

The accuracy of the surrogate model relies on the number and on the distribution of samples provided in the search space, as well as on the selection of the appropriate model to represent the objective functions and constraints. One important fact is that Pareto-optimal solutions based on the computationally cheap surrogate model do not necessarily satisfy the real CFD evaluation. So, as indicated in the previous references, it is necessary to verify the whole set of Pareto-optimal solutions found from the surrogate, which can render the problem very time consuming. If discrepancies are large, this condition might atenuate the benefit of using a surrogate model. The verification process is also needed in order to update the surrogate model. This latter condition raises the question of how often in the design process it is necessary to update the surrogate model. There are no general rules for this, and many researchers rely on previous experiences and trial and error guesses.

CFD analyses rely on discretization of the flow domain and in numerical models of the flow equations. In both cases, some sort of reduced model can be used as fitness approximation methods, which can be further used to generate a surrogate model. For example, Lee et al. [24] use different grid resolutions for the CFD simulations. Coarse grids are used for global exploration, while fine grids are used for solution exploitation purposes.

Finally, many of the approaches using surrogates, build them, relating the design variables with the objective functions. However, Leifsson and Koziel [25], have recently proposed the use of physics-based surrogate models in which, they are built relating the design variables with pressure distributions (instead of objective functions). The premise behind this approach is that in aerodynamics, the objective functions are not directly related with the design variables, but with the pressure distributions. The authors have presented successful results using this new kind of surrogate model for global transonic airfoil optimization. Its extension to multiobjective aerodynamic shape optimization is straightforward and very promising.

10.3.3 Hybrid MOEA Optimization

One of the major drawbacks of MOEAs is that they are very demanding (in terms of computational time), due to the relatively high number of objective function evaluations that they typically require. This has motivated a number of approaches to improve their efficiency. One of them consists in hybridizing a MOEA with a gradient-based method. In general, gradient-based methods converge quickly for simple topologies of the objective functions but will get trapped in a local optimum if multi-modal objective functions are considered. In contrast, MOEAs can normally avoid local minima and can also cope with complex, noisy objective function topologies. The basic idea behind this hybridization is to resort to gradient-based methods, whenever the MOEA convergence is slow. Some representative works using this idea are the following:

- Lian et al. [27] deal with a multi-objective redesign of the shape blade of a single-stage centrifugal compressor. The objectives are: (i) to maximize the total head, and (ii) to minimize the input power at a design point. These objectives are conflicting with each other. In their hybrid approach, they couple a gradient-based method that uses a Sequential Quadratic Programming (SQP) scheme, with a GA-based MOEA. The SQP approach works in a confined region of the design space where a surrogate model is constructed, and optimized with gradient-based methods. In the hybrid approach of this example, the MOEA is used as a global search engine, while the SQP model is used as a local search mechanism. Both mechanisms are alternatively used under a trust-region framework until Pareto optimal solutions are obtained. By this hybridization approach, favorable characteristics of both global and local search are maintained.
- Chung et al. [9] address a multidisciplinary problem involving supersonic business jet design. The main objective of this particular problem was to obtain a trade-off design having good aerodynamic performances while minimizing the intensity of the sonic boom signature at the ground level. Multiobjective optimization was used to obtain trade-offs among the objective functions of the problem which were to minimize: (i) the aircraft drag coefficient, (ii) initial pressure rise (boom overpressure), and (iii) ground perceived noise level. In this study, the authors proposed and tested the Gradient Enhanced Multiobjective Genetic Algorithm (GEMOGA). The basic idea of this MOEA is to enhance the non-dominated solutions obtained by a genetic algorithm with a gradient-based local search procedure. One important feature of this approach was that the gradient information was obtained from the Kriging model. Therefore, the computational cost was not considerably increased.
- Ray and Tsai [38] considered a multiobjective transonic airfoil shape design optimization problem with two objectives to be minimized: (i) the ratio of the drag to lift squared coefficients, and (ii) the squared moment coefficient. Constraints were imposed on the flow Mach number and angle of attack. The MOEA used is a multi-objective particle swarm optimizer (MOPSO). This MOEA was also hybridized with a gradient-based algorithm. Contrary to standard hybridization schemes where gradient-based algorithms are used to improve the

nondominated solutions obtained (i.e., as a local search engine), in this approach the authors used the gradient information to repair solutions not satisfying the equality constraints defined in the problem. This repairing algorithm was based on the Marquardt-Levenberg algorithm. During the repairing process, a subset of the design variables was used, instead of the whole set, in order to reduce the dimensionality of the optimization problem to be solved.

10.3.3.1 Comments on Hybrid MOEA Optimization

Experience has shown that hybridizing MOEAs with gradient-based techniques can, to some extent, increase their convergence rate. However, in the examples presented above, the gradient information relies on local and/or global surrogate models. For this, one major concern is how to build a high-fidelity surrogate model with the existing designs in the current population, since, their distribution in the design space can introduce some undesired bias in the surrogate model. Additionally, there are no rules for choosing the number of points for building the surrogate model, nor for defining the number of local searches to be performed. These parameters are emprirically chosen. Another idea that has not been explored in multi-objective evolutionary optimization, is to use adjoint-based CFD solutions to obtain gradient information. Adjoint-based methods are also mature techniques currently used for single objective aerodynamic optimization [28], and gradient information with these techniques can be obtained with as much of an additional objective function evaluation.

10.3.4 Robust Design Optimization

In aerodynamic optimization, uncertainties in the environment must be taken into account. For example, the operating velocity of an aircraft may deviate from the normal condition during the flight. This change in velocity can be so high that it changes the Mach and/or Reynolds number for the flow. The variation of these parameters can substantially change the aerodynamic properties of the design. In this case, a robust optimal solution is desired, instead of the optimal solution found for ideal operating conditions. By robustness, it is meant in general that the performance of an optimal solution should be insensitive to small perturbations of the design variables or environmental parameters. In multiobjective optimization, the robustness of a solution can be an important factor for a decision maker in choosing the final solution. Search for robust solutions can be treated as a multiobjective task, i.e., to maximize the performance and the robustness simultaneously. These two tasks are very likely conflicting, and therefore, MOEAs can be employed to find a number of trade-off solutions. In the context of multi-objective aerodynamic shape optimization problems, we summarize next some work on robust design.

- Yamaguchi and Arima [51] dealt with the multi-objective optimization of a transonic compressor stator blade in which three objectives were minimized: (i) pressure loss coefficient, (ii) deviation outflow angle, and (iii) incidence toughness.

The last objective function can be considered as a robust condition for the design, since it is computed as the average of the pressure loss coefficients at two off-design incidence angles. The airfoil blade geometry was defined by twelve design variables. The authors adopted MOGA [14] with real-numbers encoding as their search engine. Aerodynamic performance evaluation for the compressor blade was done using Navier-Stokes CFD simulations. The optimization process was parallelized using 24 processors in order to reduce the computational time required.

- Rai [37] dealt with the robust optimal aerodynamical design of a turbine blade airfoil shape, taking into account the performance degradation due to manufacturing uncertainties. The objectives considered were: (i) to minimize the variance of the pressure distribution over the airfoil's surface, and (ii) to maximize the probability of constraint satisfaction. Only one constraint was considered, related to the minimum thickness of the airfoil shape. The author adopted a multi-objective version of the differential evolution algorithm and used a high-fidelity CFD simulation on a perturbed airfoil geometry in order to evaluate the aerodynamic characteristics of the airfoil generated by the MOEA. The geometry used in the simulation was perturbed, following a probability density function that is observed for manufacturing tolerances. This process had a high computational cost, which the author reduced using a neural network surrogate model.
- Shimoyama et al. [44] applied a design for multi-objective six-sigma (DFMOSS) [43] for the robust aerodynamic airfoil design of a Mars exploratory airplane. The aim is to find the trade-off between the optimality of the design and its robustness. The idea of the DFMOSS methodology was to incorporate a MOEA to simultaneously optimize the mean value of an objective function, while minimizing its standard deviation due to the uncertainties in the operating environment. The airfoil shape optimization problems considered two cases: a robust design of (a) airfoil aerodynamic efficiency (lift to drag ratio), and (b) airfoil pitching moment constraint. In both cases, only the variability in the flow Mach number was taken into account. The authors adopted MOGA [14] as their search engine. The airfoil geometry was defined with 12 design variables. The aerodynamic performance of the airfoil was evaluated by CFD simulations using the Favre-Averaged compressible thin-layer Navier-Stokes equations. The authors reported computational times of about five minutes per airfoil, and about 56 hours for the total optimization process, using a NEC SX-6 computing system with 32 processors. Eighteen robust nondominated solutions were obtained in the first test case. From this set, almost half of the population attained the 6σ condition. In the second test case, more robust nondominated solutions were found, and they satisfied a sigma level as high as 25σ.
- Lee et al. [24] presented the robust design optimization of an ONERA M6 Wing Shape. The robust optimization was based on the concept of the Taguchi method in which the optimization problem is solved considering uncertainties in the design environment, in this case, the flow Mach number. The problem had two objectives: (i) minimization of the mean value of an objective function with respect to variability of the operating conditions, and (ii) minimization of the variance

of the objective function of each candidate solution, with respect to its mean value. In the sample problems, the wing was defined by means of its planform shape (sweep angle, aspect ratio, taper ratio, etc.) and of the airfoil geometry, at three wing locations (each airfoil shape was defined with a combination of mean lines and camber distributions), using a total of 80 design variables to define the wing designs. Geometry constraints were defined by upper and lower limits of the design variables. The authors adopted the Hierarchical Asynchronous Parallel Multi-Objective Evolutionary Algorithm (HAPMOEA) algorithm [15], which is based on evolution strategies, incorporating the concept of Covariance Matrix Adaptation (CMA). The aerodynamic evaluation was done with a CFD simulation. 12 solutions were obtained in the robust design of the wing. All the nondominated solutions showed a better behavior, in terms of aerodynamic performance (lift-to-drag ratio) with a varying Mach number, as compared to the baseline design. During the evolutionary process, a total of 1100 individuals were evaluated in approximately 100 hours of CPU time.

10.3.4.1 Comments on Robust Design Optimization

As can be seen form the previous examples, robust solutions can be achieved in evolutionary optimization in different ways. One simple approach is to add perturbations to the design variables or environmental parameters before the fitness is evaluated, which is known as implicit averaging [50]. An alternative to implicit averaging is explicit averaging, which means that the fitness value of a given design is averaged over a number of designs generated by adding random perturbations to the original design. One drawback of the explicit averaging method is the number of additional quality evaluations needed, which can turn the approach impractical. In order to tackle this problem, metamodeling techniques have been considered [32].

10.3.5 Multi-Disciplinary Design Optimization

Multi-disciplinary design optimization (MDO) aims at incorporating optimization methods to solve design problems, considering not only one engineering discipline, but a set of them. The optimum of a multidisciplinary problem might be a compromise solution from the multiple disciplines involved. In this sense, multi-objective optimization is well suited for this type of problems, since it can exploit the interactions between the disciplines, and can help to find the trade-offs among them. Next, we present some work in which MOEAs have been used for aerodynamic shape optimization problems, coupled with another discipline.

- Chiba et al. [8] addressed the MDO problem of a wing shape for a transonic regional-jet aircraft. In this case, three objective functions were minimized: (i) block fuel for a required airplane's mision, (ii) maximum take-off weight, and (iii) difference in the drag coefficient between transonic and subsonic flight conditions. Additionally, five constraints were imposed, three of which were related to the wing's geometry and two more to the operating conditions in lift coefficient and to the fuel volume required for a predefined aircraft mission. The wing geometry was defined by 35 design variables. The authors adopted ARMOGA [40]. The disciplines involved included aerodynamics and structural analysis and during the optimization process, an iterative aeroelastic solution was generated in order to minimize the wing weight, with constraints on flutter and strength requirements. Also, a flight envelope analysis was done, obtaining high-fidelity Navier-Stokes solutions for various flight conditions. Although the authors used very small population sizes (eight individuals), about 880 hours of CPU time were required at each generation, since an iterative process was performed in order to optimize the wing weight, subject to aeroelastic and strength constraints. The population was reinitialized at every 5 generations for range adaptation of the design variables. In spite of the use of such a reduced population size, the authors were able to find several nondominated solutions outperforming the initial design. They also noted that during the evolution, the wing-box weight tended to increase, but this degrading effect was redeemed by an increase in aerodynamic efficiency, given a reduction in the block fuel of over one percent, which would be translated in significant savings for an airline's operational costs.

- Sasaki et al. [41] used MDO for the design of a supersonic wing shape. In this case, four objective functions were minimized: (i) drag coefficient at transonic cruise, (ii) drag coefficient at supersonic cruise, (iii) bending moment at the wing root at supersonic cruise condition, and (iv) pitching moment at supersonic cruise condition. The problem was defined by 72 design variables. Constraints were imposed on the variables ranges and on the wing section's thickness and camber, all of them being geometrical constraints. The authors adopted ARMOGA [40], and the aerodynamic evaluation of the design soutions, was done by high-fidelity Navier-Stokes CFD simulations. No aeroelastic analysis was performed, which considerably reduced the total computational cost. The objective associated with the bending moment at wing root was evaluated by numerical integration of the pressure distribution over the wing surface, as obtained by the CFD analysis. The authors indicated that among the nondominated solutions there were designs that were better in all four objectives with respect to a reference design.
- Lee et al. [23] utilized a generic Framework for MDO to explore the improvement of aerodynamic and radar cross section (RCS) characteristics of an Unmanned Combat Aerial Vehicle (UCAV). In this application, two disciplines were considered, the first concerning the aerodynamic efficiency, and the second related to the visual and radar signature of an UCAV airplane. In this case, three objective functions were minimized: (i) inverse of the lift/drag ratio at ingress condition, (ii) inverse of the lift/drag ratio at cruise condition, and (iii) frontal area. The number of design variables was of approximately 100 and only side constraints were considered in the design variables. The first two objective functions were evaluated using a Potential Flow CFD Solver (FLO22) [17] coupled to FRICTION code to obtain the viscous drag, using semi-empirical relations. The authors adopted the Hierarchical Asynchronous Parallel Multi-Objective Evolutionary Algorithm (HAPMOEA) [15]. The authors reported a processing time

of 200 hours for their approach, on a single 1.8 GHz processor. It is important to consider that HAPMOEA operates with different CFD grid levels (i.e. approximation levels): coarse, medium, and fine. In this case, the authors adopted different population sizes for each of these levels. Also, solutions were allowed to migrate from a low/high fidelity level to a higher/lower one in an island-like mechanism.

10.3.5.1 Comments on Multidisciplinary Design Optimization

The increasing complexity of engineering systems has raised the interest in multidisciplinary optimization, as can be seen from the examples presented in this section. For this task, MOEAs facilitate the integration of several disciplines, since they do not require additional information other than the evaluation of the corresponding objective functions, which is usually done by each discipline and by the use of simulations. Aditionally, an advantage of the use of MOEAs for MDO, is that they can easily manage any combination of variable types, coming from the involved disciplines i.e., from the aerodynamic discipline, the variables can be continuous, but for the structural optimization, it can happen that the variables are discrete. Kuhn et al. [22] presented an example of this condition for the multi-disciplinary design of an airship. However, one challenge in MDO is the increasing dimensionality attained in the design space, as the number of disciplines also increases.

10.3.6 Data Mining and Knowledge Extraction

Data mining tools, along with data visualization using graphical methods, can help to understand and extract information from the data contained in the Pareto optimal solutions found using any MOEA. In this sense, Multi-Objective Design Exploration (MODE), proposed by Jeong et al. [18] is a framework to extract design knowledge from the obtained Pareto optimal solutions such as trade-off information between contradicting objectives and sensitivity of each design parameter to the objectives. In the framework of MODE, Pareto-optimal solutions are obtained by a MOEA and knowledge is extracted by analyzing the design parameter values and the objective function values of the obtained Pareto-optimal solutions using data mining approaches such as Self Organizing Maps (SOMs) and analysis of variance (ANOVA). They also propose to use rough sets theory to obtain rules from the Pareto optimal solutions. MODE has been applied to a wide variety of design optimization problems as summarized next:

- Jeong et al. [18] and Chiba et al. [7, 6] explored the trade-offs among four aerodynamic objective functions in the optimization of a wing shape for a Reusable Launch Vehicle (RLV). The objective functions were: (i) The shift of the aerodynamic center between supersonic and transonic flight conditions, (ii) Pitching moment in the transonic flight condition, (iii) drag in the transonic flight condition, and (iv) lift for the subsonic flight condition. The first three objectives were minimized while the fourth was maximized. These objectives were selected for attaining control, stability, range and take-off constraints, respectively. The RLV definition comprised 71 design variables to define the wing planform, the wing position along the fuselage and the airfoil shape at prescribed wingspan stations. The authors adopted ARMOGA [40], and the aerodynamic evaluation of the RLV was done with a Reynolds-Averaged Navier-Stokes CFD simulation. A trade-off analysis was conducted with 102 nondominated individuals generated by the MOEA. Data mining with SOM was used, and some knowledge was extracted in regards to the correlation of each design variable to the objective functions in [7]; with SOM, Batch-SOM, ANOVA and rough sets in [6]; and with SOM, Batch-SOM and ANOVA in [18]. In all cases, some knowledge was extracted in regards to the correlation of each design variable to the objective functions.

- Oyama et al. [35] applied a design exploration technique to extract knowledge information from a flapping wing MAV (Micro Air Vehicle). The flapping motion of the MAV was analyzed using multi-objective design optimization techniques in order to obtain nondominated solutions. Such nondominated solutions were further analyzed with SOMs in order to extract knowledge about the effects of the flapping motion parameters on the objective functions. The conflicting objectives considered were: (i) maximization of the time-averaged lift coefficient, (ii) maximization of the time-averaged thrust coefficient, and (iii) minimization of the time-averaged required power coefficient. The problem had five design variables and the geometry of the flying wing was kept fixed. Constraints were imposed on the averaged lift and thrust coefficients so that they were positive. The authors adopted a GA-based MOEA. The objective functions were obtained by means of CFD simulations, solving the unsteady incompressible Navier-Stokes equations. Objective functions were averaged over one flapping cycle. The purpose of the study was to extract trade-off information from the objective functions and the flapping motion parameters such as plunge amplitude and frequency, pitching angle amplitude and offset.
- Tani et al. [49] solved a multiobjective rocket engine turbopump blade shape optimization design which considered three objective functions: (i) shaft power, (ii) entropy rise within the stage, and (iii) angle of attack of the next stage. The first objective was maximized while the others were minimized. The design candidates defined the turbine blade aerodynamic shape and consisted of 58 design variables. The authors adopted MOGA [14] as their search engine. The objective function values were obtained from a CFD Navier-Stokes flow simulation. The authors reported using SOMs to extract correlation information for the design variables with respect to each objective function.

10.3.6.1 Comments on Data Mining and Knowledge Extraction

When adopting the data mining techniques used in the above examples, in which analyses are done, correlating the objective functions values, with the design parameter values of the Pareto optimal solutions, some valuable information is obtained. However, in many other cases, for aerodynamic flows, the knowledge required is more related to the physics, rather than to the geometry, given by the design variables. For example, for understanding the relation between the generation of shock wave formation and aerodynamic characteristics in a transonic airfoil optimization. For this, Oyama et al. [34], have recently proposed a new approach to extract useful design information from one-dimensional, two-dimensional, and three-dimensional flow data of Pareto-optimal solutions. They use a flow data analysis by Proper Orthogonal Decomposition (POD), which is a statistical approach that can extract dominant features in the data by decomposing it into a set of optimal orthogonal base vectors of decreasing importance.

10.4 A Case Study

Here, we present a case study of evolutionary multi-objective optimization for an airfoil shape optimization problem. The test problem chosen corresponds to the airfoil shape of a standard-class glider. The optimization problem aims at obtaining optimum performance for a sailplane. In this study the trade-off among three aerodynamic objectives is evaluated using a MOEA.

10.4.1 Objective Functions

Three conflicting objective functions are defined in terms of a sailplane average weight and operating conditions [48]. They are formally defined as:

(i) Minimize C_D/C_L subject to $C_L = 0.63$, $Re = 2.04 \cdot 10^6$, $M = 0.12$

(ii) Minimize C_D/C_L subject to $C_L = 0.86$, $Re = 1.63 \cdot 10^6$, $M = 0.10$

(iii) Minimize $C_D/C_L^{3/2}$ subject to $C_L = 1.05$, $Re = 1.29 \cdot 10^6$, $M = 0.08$

In the above definitions, C_D/C_L and $C_D/C_L^{3/2}$ correspond to the inverse of the glider's gliding ratio and sink rate, respectively. Both are important performance measures for this aerodynamic optimization problem. C_D and C_L are the drag and lift coefficients. In the above objective function definitions, the aim is to maximize the gliding ratio for objectives (i) and (ii), while minimizing the sink rate in objective (iii). Each of these objectives is evaluated at different prescribed flight conditions, given in terms of Mach and Reynolds numbers.

10.4.2 Geometry Parameterization

Finding an optimum representation scheme for aerodynamic shape optimization problems is an important step for a successful aerodynamic optimization task. Several options can be used for airfoil shape parameterization.

(a) The representation used needs to be flexible to describe any general airfoil shape.

(b) The representation also needs to be efficient, in order that the parameterization can be achieved with a minimum number of parameters. Inefficient representations may result in an unnecesarily large design space which, in consequence, can reduce the search efficiency of an evolutionary algorithm.
(c) The representation should allow the use of any optimization algorithm to perform local search. This requirement is important for refining the solutions obtained by the global search engine in a more efficient way.

In the present case study, the PARSEC airfoil representation [45] is used. Fig. 10.1 illustrates the 11 basic parameters used for this representation: r_{le} leading edge radius, X_{up}/X_{lo} location of maximum thickness for upper/lower surfaces, Z_{up}/Z_{lo} maximum thickness for upper/lower surfaces, Z_{xxup}/Z_{xxlo} curvature for upper/lower surfaces, at maximum thickness locations, Z_{te} trailing edge coordinate, ΔZ_{te} trailing edge thickness, α_{te} trailing edge direction, and β_{te} trailing edge wedge angle. For the present case study, the modified PARSEC geometry representation adopted allows us to define independently the leading edge radius, both for upper and lower surfaces. Thus, 12 variables in total are used. Their allowable ranges are defined in Table 10.1.

Table 10.1 Parameter Ranges for Modified PARSEC Airfoil Representation

	r_{leup}	r_{lelo}	α_{te}	β_{te}	Z_{te}	ΔZ_{te}	X_{up}	Z_{up}	Z_{xxup}	X_{lo}	Z_{lo}	Z_{xxlo}
min	0.0085	0.002	7.0	10.0	-0.006	0.0025	0.41	0.11	-0.9	0.20	-0.023	0.05
max	0.0126	0.004	10.0	14.0	-0.003	0.0050	0.46	0.13	-0.7	0.26	-0.015	0.20

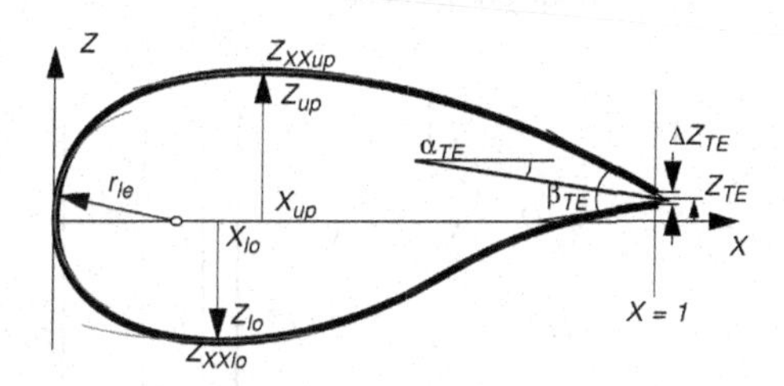

Fig. 10.1 PARSEC airfoil parameterization

The PARSEC airfoil geometry representation uses a linear combination of shape functions for defining the upper and lower surfaces. These linear combinations are given by:

$$Z_{upper} = \sum_{n=1}^{6} a_n x^{\frac{n-1}{2}} \tag{10.4}$$

$$Z_{lower} = \sum_{n=1}^{6} b_n x^{\frac{n-1}{2}} \tag{10.5}$$

In the above equations, the coefficients a_n, and b_n are determined as function of the 12 described geometric parameters, by solving the following two systems of linear equations:

Upper surface:

$$\begin{bmatrix} 1 & 1 & 1 & 1 & 1 & 1 \\ X_{up}^{1/2} & X_{up}^{3/2} & X_{up}^{5/2} & X_{up}^{7/2} & X_{up}^{9/2} & X_{up}^{11/2} \\ 1/2 & 3/2 & 5/2 & 7/2 & 9/2 & 11/2 \\ \frac{1}{2}X_{up}^{-1/2} & \frac{3}{2}X_{up}^{1/2} & \frac{5}{2}X_{up}^{3/2} & \frac{7}{2}X_{up}^{5/2} & \frac{9}{2}X_{up}^{7/2} & \frac{11}{2}X_{up}^{9/2} \\ -\frac{1}{4}X_{up}^{-3/2} & \frac{3}{4}X_{up}^{-1/2} & \frac{15}{4}X_{up}^{1/2} & \frac{35}{4}X_{up}^{3/2} & \frac{63}{4}X_{up}^{5/2} & \frac{99}{4}X_{up}^{7/2} \\ 1 & 0 & 0 & 0 & 0 & 0 \end{bmatrix} \begin{bmatrix} a_1 \\ a_2 \\ a_3 \\ a_4 \\ a_5 \\ a_6 \end{bmatrix} = \begin{bmatrix} Z_{te} + \frac{1}{2}\Delta Z_{te} \\ Z_{up} \\ tan((2\alpha_{te} - \beta_{te})/2) \\ 0 \\ Z_{xxup} \\ \sqrt{r_{leup}} \end{bmatrix} \tag{10.6}$$

It is important to note that the geometric parameters r_{leup}/r_{lelo}, X_{up}/X_{lo}, Z_{up}/Z_{lo}, Z_{xxup}/Z_{xxlo}, Z_{te}, ΔZ_{te}, α_{te}, and β_{te} are the actual design variables in the optimization process, and that the coeficients a_n, b_n serve as intermediate variables for interpolating the airfoil's coordinates, which are used by the CFD solver (we used the Xfoil CFD code [13]) for its discretization process.
Lower surface:

$$\begin{bmatrix} 1 & 1 & 1 & 1 & 1 & 1 \\ X_{lo}^{1/2} & X_{lo}^{3/2} & X_{lo}^{5/2} & X_{lo}^{7/2} & X_{lo}^{9/2} & X_{lo}^{11/2} \\ 1/2 & 3/2 & 5/2 & 7/2 & 9/2 & 11/2 \\ \frac{1}{2}X_{lo}^{-1/2} & \frac{3}{2}X_{lo}^{1/2} & \frac{5}{2}X_{lo}^{3/2} & \frac{7}{2}X_{lo}^{5/2} & \frac{9}{2}X_{lo}^{7/2} & \frac{11}{2}X_{lo}^{9/2} \\ -\frac{1}{4}X_{lo}^{-3/2} & \frac{3}{4}X_{lo}^{-1/2} & \frac{15}{4}X_{lo}^{1/2} & \frac{35}{4}X_{lo}^{3/2} & \frac{63}{4}X_{lo}^{5/2} & \frac{99}{4}X_{lo}^{7/2} \\ 1 & 0 & 0 & 0 & 0 & 0 \end{bmatrix} \begin{bmatrix} b_1 \\ b_2 \\ b_3 \\ b_4 \\ b_5 \\ b_6 \end{bmatrix} = \begin{bmatrix} Z_{te} - \frac{1}{2}\Delta Z_{te} \\ Z_{lo} \\ tan((2\alpha_{te} + \beta_{te})/2) \\ 0 \\ Z_{xxlo} \\ -\sqrt{r_{lelo}} \end{bmatrix} \tag{10.7}$$

10.4.3 Constraints

For this case study, five constraints are considered. The first three are defined in terms of flight speed for each objective function, namely the prescribed C_L values, $C_L = 0.63$ for objective (i), $C_L = 0.86$ for objective (ii), and $C_L = 1.05$ for objective (iii), enable the glider to fly at a given design speed, and to produce the necessary amount of lift to balance the gravity force for each design condition being analyzed. It is important to note that prescribing the required C_L, the corresponding angle of attack α for the airfoil is obtained as an additional variable. For this, the flow solver, given the design candidate geometry, solves the flow equations with a constraint on the C_L value, i.e., it additionally determines the operating angle of attack α. Two additional constraints are defined for the airfoil geometry. First, the maximum airfoil thickness range is defined by $13.0\% \leq t/c \leq 13.5\%$. For handling this constraint, every time a new design candidate is created by the evolutionary operators, its maximum thickness is checked and corrected before being evaluated. The correction is done by scaling accordingly the design parameters Z_{up} and Z_{lo}, which mainly define the thickness distribution in the airfoil. In this way, only feasible solutions are evaluated by the simulation process. The final constraint is the trailing edge thickness, whose range is defined by $0.25\% \leq \Delta Z_{te} \leq 0.5\%$. This constraint is directly handled in the lower and upper bounds by the corresponding ΔZ_{te} design parameter.

10.4.4 Evolutionary Algorithm

For solving the above case study, we adopted MODE-LD+SS [3] as our search algorithm. Additionaly, and for comparison purposes, we also used an implementation of the SMS-EMOA algorithm [5]. This algorithm is based on the hypervolume performance measure [53] and has also been used in the context of airfoil optimization problems.

The Multi-objective Evolutionary Algorithm MODE-LD+SS (see Algorithm 1) [3] adopts the evolutionary operators from differential evolution [36]. In the basic DE algorithm, and during the offspring creation stage, for each current vector $P_i \in \{P\}$, three parents (mutually different among them) $\mathbf{u_1}, \mathbf{u_2}, \mathbf{u_3} \in \{P\}$ ($\mathbf{u_1} \neq \mathbf{u_2} \neq \mathbf{u_3} \neq P_i$) are randomly selected for creating a mutant vector $\mathbf{v}$ using the following mutation operation:

$$\mathbf{v} \leftarrow \mathbf{u_1} + F \cdot (\mathbf{u_2} - \mathbf{u_3}) \tag{10.8}$$

$F > 0$, is a real constant *scaling factor* which controls the amplification of the difference $(\mathbf{u_2} - \mathbf{u_3})$. Using this mutant vector, a new offspring P_i' (also called trial vector in DE) is created by crossing over the mutant vector $\mathbf{v}$ and the current solution P_i, in accordance to:

$$P_j' = \begin{cases} v_j & \text{if } (rand_j(0,1) \leq CR \text{ or } j = j_{rand} \\ P_j & \text{otherwise} \end{cases} \tag{10.9}$$

Algorithm 1 MODE-LD+SS

1: **INPUT:**
$P[1,\ldots,N]$ = Population
N = Population Size
F = Scaling factor
CR = Crossover Rate
$\lambda[1,\ldots,N]$ = Weight vectors
NB = Neighborhood Size
$GMAX$ = Maximum number of generations
2: **OUTPUT:**
PF = Pareto front approximation
3: ***Begin***
4: $g \leftarrow 0$
5: Randomly create P_i^g , $i = 1,\ldots,N$
6: Evaluate P_i^g , $i = 1,\ldots,N$
7: **while** $g < GMAX$ **do**
8: $\{LND\} = \{\oslash\}$
9: **for** $i = 1$ to N **do**
10: *DetermineLocalDominance*(P_i^g, NB)
11: **if** P_i^g is locally nondominated **then**
12: $\{LND\} \leftarrow \{LND\} \cup P_i^g$
13: **end if**
14: **end for**
15: **for** $i = 1$ to N **do**
16: Randomly select $\mathbf{u_1}$, $\mathbf{u_2}$, and $\mathbf{u_3}$ from $\{LND\}$
17: $v \leftarrow CreateMutantVector(u_1, u_2, u_3)$
18: $P_i^{g+1} \leftarrow Crossover(P_i^g, v)$
19: Evaluate P_i^{g+1}
20: **end for**
21: $Q \leftarrow P^g \cup P^{g+1}$
22: Determine $z*$ for Q
23: **for** $i = 1$ to N **do**
24: $P_i^{g+1} \leftarrow MinimumTchebycheff(Q, \lambda^i, z*)$
25: $Q \leftarrow Q \backslash P_i^{g+1}$
26: **end for**
27: $PF \leftarrow \{P\}^{g+1}$
28: **end while**
29: ReturnPF
30: ***End***

In the above expression, the index j refers to the jth component of the decision variables vectors. CR is a positive constant and j_{rand} is a randomly selected integer in the range $[1,\ldots,D]$ (where D is the dimension of the solution vectors) ensuring that the offspring is different at least in one component with respect to the current solution P_i. The above DE variant is known as $Rand/1/bin$, and is the version adopted here. Additionally, the proposed algorithm incorporates two mechanisms for improving both the convergence towards the Pareto front, and the uniform distribution of nondominated solutions along the Pareto front. These mechanisms correspond to the concept of local dominance and the use of an environmental selection based on a scalar function. Below, we explain these two mechanisms in more detail.

As for the first mechanism, local dominance concept, in Algorithm 1, the solution vectors $\mathbf{u}_1, \mathbf{u}_2, \mathbf{u}_3$, required for creating the trial vector $\mathbf{v}$ (in equation (10.8)), are selected from the current population, only if they are locally nondominated in their neighborhood $\aleph$. Local dominance is defined as follows:

Definition 6. Pareto Local Dominance. Let $\mathbf{x}$ be a feasible solution, $\aleph(\mathbf{x})$ be a neighborhood structure for $\mathbf{x}$ in the decision space, and $\mathbf{f}(\mathbf{x})$ a vector of objective functions.

- We say that a solution $\mathbf{x}$ is locally nondominated with respect to $\aleph(\mathbf{x})$ if and only if there is no $\mathbf{x}'$ in the neighborhood of $\mathbf{x}$ such that $\mathbf{f}(\mathbf{x}') \prec \mathbf{f}(\mathbf{x})$

The neighborhood structure is defined as the *NB* closest individuals to a particular solution. Closeness is measured by using the Euclidean distance between solutions in the design variable space. The major aim of using the local dominance concept, as defined above, is to exploit good individuals' genetic information in creating DE trial vectors, and the associated offspring, which might help to improve the MOEA's convergence rate toward the Pareto front. From Algorithm 1, it can be noted that this mechanism has a stronger effect during the earlier generations, where the portion of nondominated individuals is low in the global population, and progressively weakens, as the number of nondominated individuals grows during the evolutionary process. This mechanism is automatically switched off, once all the individuals in the population become nondominated, and has the possibility of being switched on, as some individuals become dominated.

As for the second mechanism, *selection based on a scalar function*, it is based on the Tchebycheff scalarization function given by:

$$g(\mathbf{x}|\lambda, z^*) = \max_{1 \le i \le m} \{\lambda^i |f_i(x) - z_i^*|\} \tag{10.10}$$

In the above equation, $\lambda^i, i = 1, \ldots, N$ represents the set of weight vectors used to distribute the solutions along the entire Pareto front. In this case, this set is calculated using the procedure described in [52]. z^* corresponds to a reference point, defined in objective space and determined with the minimum objective values of the combined population Q, consistent on the actual parents and the created offspring. This reference point is updated at each generation, as the evolution progresses. The procedure *MinimumTchebycheff*$(Q, \lambda^i, z*)$ finds, from the set Q, (the combined population consistent on the actual parents and the created offspring), the solution vector that minimizes equation (10.10) for each weight vector λ^i and the reference point z^*.

The second MOEA adopted is the SMS-EMOA, which is a steady-state algorithm based on two basic characteristics: (1) non-dominated sorting is used as its ranking criterion and (2) the hypervolume[4] is applied as its selection criterion to discard that individual, which contributes the least hypervolume to the worst-ranked front.

The basic algorithm is described in Algorithm 2. Starting with an initial population of μ individuals, a new individual is generated by means of randomised variation operators. We adopted simulated binary crossover (SBX) and polynomial-based mutation as described in [11]. The new individual will become a member of the next population, if replacing another individual leads to a higher quality of the population with respect to the hypervolume.

Algorithm 2 SMS-EMOA

1: $P_o \leftarrow init()$ /* initialize random population of μ individuals */
2: $t \leftarrow 0$
3: **repeat**
4: $q_{t+1} \leftarrow generate(P_t)$ /* generate offspring by variation*/
5: $P_{t+1} \leftarrow reduce(P_t \bigcup \{q_{t+1}\})$ /* select μ best individuals */
6: **until** termination condition is fulfilled

The procedure **Reduce** used in Algorithm 2 selects the μ individuals of the subsequent population; the definition of this procedure is given in Algorithm 3. The algorithm fast-nondominated-sort used in NSGA-II [12] is applied to partition the population into v sets $\mathscr{R}_1, \ldots, \mathscr{R}_v$. The subsets are called fronts and are provided with an index representing a hierarchical order (the level of domination) whereas the solutions within each front are mutually nondominated. The first subset contains all nondominated solutions of the original set Q. The second front consists of individuals that are nondominated in the set $(Q \backslash \mathscr{R}_1)$, e.g. each member of $\mathscr{R}_2$ is dominated by at least one member of $\mathscr{R}_1$. More general, the ith front consists of individuals that are nondominated if the individuals of the fronts j with $j < i$ were removed from Q.

Algorithm 3 Reduce(Q)

1: $\{\mathscr{R}_1, \ldots, \mathscr{R}_v\} \leftarrow fast\_nondominated\_sort(Q)$ /* all v fronts of Q*/
2: $r \leftarrow argmin_{s \in \mathscr{R}_v}[\Delta_{\mathscr{S}}(s, \mathscr{R}_v)]$ /* $s \in \mathscr{R}_v$ with lowest $\Delta_{\mathscr{S}}(s, \mathscr{R}_v)$*/
3: **return** $(Q \backslash r)$

The value of $\Delta_{\mathscr{S}}(s, \mathscr{R}_v)]$ can be interpreted as the exclusive contribution of s to the hypervolume value of its appropriate front. By definition of $\Delta_{\mathscr{S}}(s, \mathscr{R}_v)]$, an individual, which dominates another is always kept and a nondominated individual is replaced by a dominated one. This measure keeps those individuals which maximize the population's S-Metric value, which implies that the covered hypervolume

[4] The **Hypervolume** (also known as the S-metric or the Lebesgue Measure) of a set of solutions measures the size of the portion of objective space that is dominated by those solutions collectively.

of a population cannot decrease by application of the **Reduce** operator. Thus, for Algorithm 2 the following invariant holds:

$$\mathscr{S}(P_t) \leq \mathscr{S}(P_{t+1}) \tag{10.11}$$

Due to the high computational effort of the hypervolume calculation, a steady state selection scheme is used. Since only one individual is created, only one has to be deleted from the population at each generation. Thus, the selection operator has to compute at most $\mu + 1$ values of the S-Metric (exactly $\mu + 1$ values in case all solutions are nondominated). These are the values of the subsets of the worst ranked front, in which one point of the front is left out, respectively. A $(\mu + \lambda)$ selection scheme would require the calculation of $\binom{\mu+\lambda}{\mu}$ possible S-Metric values to identify an optimally composed population, maximising the S-Metric net value.

The parameters used for solving the present case study, and for each algorithm were set as follows: $N = 120$ (population size) for both MOEAs, $F = 0.5$ (mutation scaling factor for MODE-LD+SS), $CR = 0.5$ (crossover rate for MODE-LD+SS), $NB = 5$ (neighborhood size for MODE-LD+SS), $\eta_m = 20$ (mutation index for SBX in SMS-EMOA), and $\eta_c = 15$ (crossover index for SBX in SMS-EMOA).

10.4.5 Results

Both, MODE-LD+SS and SMS-EMOA were run for 100 generations. The simulation process in each case took approximately 8 hrs of CPU time. Five independent runs were executed for extracting some statistics. Figs. 10.2 to 10.3 show the Pareto front approximations (of the median run) at different evolution times. For comparison purposes, in these figures the corresponding objective functions of a reference airfoil (a720o [48]) are plotted. At $t = 10$ generations (the corresponding figure is not shown due to space constraints), the number of nondominated solutions is 26 for SMS-EMOA and 27 for MODE-LD+SS. With this small number of nondominated solutions is difficult to identify the trade-off surface for this problem. However, as the number of evolution steps increases, the trade-off surface is more clearly revealed. At $t = 50$ generations (see Fig. 10.2), the number of nondominated solutions is 120 for SMS-EMOA, and 91 for MODE-LD+SS. At this point, the trade-off surface shows a steeper variation of objective (iii) toward the compromise region of the Pareto front. Also, the trade-off shows a plateau where the third objective has a small variation with respect to the other objectives. Finally, at $t = 100$ generations (see Fig. 10.3), the shape of the trade-off surface is more clearly defined, and a clear trade-off between the three objectives are evidenced. It is important to note in Fig. 10.3, that the trade-off surface shows some void regions. This condition is captured by both MOEAs and is attributed to the constraints defined in the airfoil geometry. Table 10.2 summarizes the maximum possible improvement with respect to the reference solution, that can be attained for each objective and by each MOEA.

In the context of MOEAs, it is common to compare results on the basis of some performance measures. Next, and for comparison purposes between the algorithms

Table 10.2 Maximum improvement per objective for the median run of each MOEA used

	MOEA					
	SMS-EMOA			MODE-LD+SS		
Gen	$\Delta Obj1(\%)$	$\Delta Obj2(\%)$	$\Delta Obj3(\%)$	$\Delta Obj1(\%)$	$\Delta Obj2(\%)$	$\Delta Obj3(\%)$
10	11.43	10.19	5.43	11.93	10.38	5.47
50	12.84	10.67	6.06	13.22	10.67	6.21
100	12.75	10.79	6.28	13.63	10.80	6.40

used, we present the hypervolume values attained by each MOEA, as well as the values of the two set coverage performance measure C-M(A,B) between them. Next, we present the definition for these two performance measures:

Hypervolume (Hv): Given a Pareto approximation set PF_{known}, and a reference point in objective space z_{ref}, this performance measure estimates the *Hypervolume* attained by it. Such hypervolume corresponds to the non-overlaping volume of all the hypercubes formed by the reference point (z_{ref}) and every vector in the Pareto set approximation. This is mathematically defined as:

$$HV = \{\cup_i vol_i | vec_i \in PF_{known}\} \tag{10.12}$$

vec_i is a nondominated vector from the Pareto set approximation, and vol_i is the volume for the hypercube formed by the reference point and the nondominated vector vec_i. Here, the reference point (z_{ref}) in objective space for the 3-objective MOPs was set to (0.007610 , 0.005895 , 0.005236), which corresponds to the objective values of the reference airfoil. High values of this measure indicate that the solutions are closer to the true Pareto front and that they cover a wider extension of it.

Two Set Coverage (C-Metric): This performance measure estimates the coverage proportion, in terms of percentage of dominated solutions, between two sets. Given the sets A and B, both containing only nondominated solutions, the C-Metric is mathematically defined as:

$$C(A,B) = \frac{|\{u \in B | \exists v \in A : v\, dominates\, u\}|}{|B|} \tag{10.13}$$

This performance measure indicates the portion of vectors in B being dominated by any vector in A. The sets A and B correspond to two different Pareto approximations, as obtained by two different algorithms. Therefore, the C-Metric is used for pairwise comparisons between algorithms.

For the hypervolume measure, SMS-EMOA attains a value of $Hv = 1.5617 \cdot 10^{-10}$ with a standard deviation of $\sigma = 2.4526 \cdot 10^{-12}$, while MODE-LD+SS attains a value of $Hv = 1.6043 \cdot 10^{-10}$ with a standard deviation of $\sigma = 1.2809 \cdot 10^{-12}$. These results are the average of five independent runs executed by each algorithm.

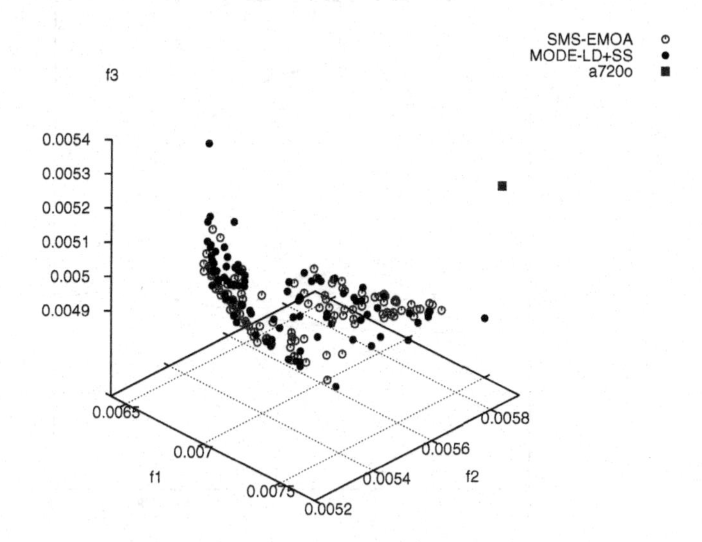

Fig. 10.2 Pareto front approximation at Gen = 50 (6000 OFEs)

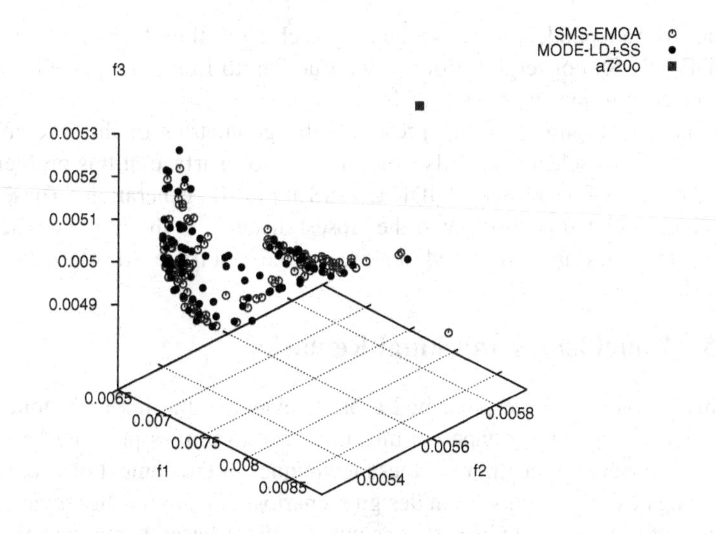

Fig. 10.3 Pareto front approximation at Gen = 100 (12000 OFEs)

As for the C-Metric, the corresponding values obtained are: $C - M(SMS - EMOA, MODE - LD + SS) = 0.07016$ with a standard deviation of $\sigma = 0.03134$, and $C - M(MODE - LD + SS, SMS - EMOA) = 0.3533$ with a standard deviation of $\sigma = 0.0510$. These latter results are the average of all the pairwise combinations

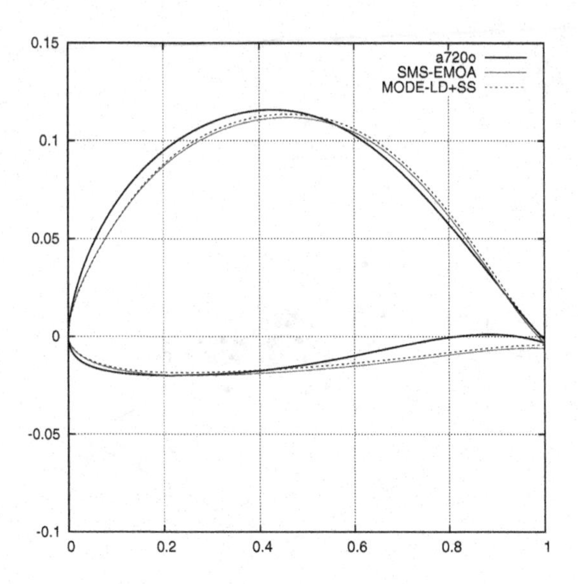

Fig. 10.4 Airfoil shape comparison

of the five independent runs executed by each algorithm. Our results indicate that MODE-LD+SS converges closer to the true Pareto front, and provides more non-dominated solutions than SMS-EMOA.

Finally, in Figure 10.4 are presented the geometries of the reference airfoil, a720o, and two selected airfoils from the trade-off surface of this problem and obtained by SMS-EMOA and MODE-LD+SS at $t = 100$ generations. These two latter airfoil are selected as those with the closest distance to the origin of the objective space, since they are considered to represent the best trade-off solutions.

10.5 Conclusions and Final Remarks

In this chapter we have presented a brief review of the research done on multi-objective aerodynamic shape optimization. The examples presented cover a wide range of current applications of these techniques in the context of aeronautical engineering design, and in several design scenarios. The approaches reviewed include the use of surrogates, hybridizations with gradient-based techniques, mechanisms to search for robust solutions, multidisciplinary approaches, and knowledge extraction techniques. It can be observed that several Pareto-based MOEAs have been successfully integrated in industrial problems. It can be anticipated that in the near future, an extended use of these techniques will be a standard practice, as the computing power available continues to increase each year. It is also worth noting that MOEAs are flexible enough as to allow their coupling to both engineering models and low-order physics-based models without major changes. They can also be easily parallelized, since MOEAs normally have low data dependency.

From an algorithmic point of view, it is clear that the use of Pareto-based MOEAs remains as a popular choice in the previous group of applications. It is also evident that, when dealing with expensive objective functions such as those of the above applications, the use of careful statistical analysis of parameters is unaffordable. Thus, the parameters of such MOEAs were simple guesses or taken from values suggested by other researchers. The use of surrogate models also appears in these costly applications. However, the use of other simpler techniques such as fitness inheritance or fitness approximation [39] seems to be uncommon in this domain and could be a good alternative when dealing with high-dimensional problems. Additionally, the authors of this group of applications have relied on very simple constraint-handling techniques, most of which discard infeasible individuals. Alternative approaches exist, which can exploit information from infeasible solutions and can make a more sophisticated exploration of the search space when dealing with constrained problems (see for example [29]) and this has not been properly studied yet. Finally, it is worth emphasizing that, in spite of the difficulty of these problems and of the evident limitations of MOEAs to deal with them, most authors report finding improved designs when using MOEAs, even when in all cases a fairly small number of fitness function evaluations was allowed. This clearly illustrates the high potential of MOEAs in this domain.

Acknowledgements. The first author acknowledges support from both CONACyT and IPN to pursue graduate studies in computer science at CINVESTAV-IPN. The second author acknowledges support from CONACyT project no. 103570. The third author acknowledges support from CONACyT project no. 79809.

References

1. Anderson, M.B.: Genetic Algorithm. In: Aerospace Design: Substantial Progress, Tremendous Potential. Technical report, Sverdrup Technology Inc./TEAS Group, 260 Eglin Air Force Base, FL 32542, USA (2002)
2. Arabnia, M., Ghaly, W.: A Strategy for Multi-Objective Shape optimization of Turbine Stages in Three-Dimensional Flow. In: 12th AIAA/ISSMO Multidisciplinary Analysis and Optimization Conference, Victoria, British Columbia Canada, September 10 –12 (2008)
3. AriasMontano, A., Coello, C.A.C., Mezura-Montes, E.: MODE-LD+SS: A Novel Differential Evolution Algorithm Incorporating Local Dominance and Scalar Selection Mechanisms for Multi-Objective Optimization. In: 2010 IEEE Congress on Evolutionary Computation (CEC 2010), Barcelona, Spain, IEEE Press, Los Alamitos (2010)
4. Benini, E.: Three-Dimensional Multi-Objecive Design optimization of a ransonic Compressor Rotor. Journal of Propulsin and Power 20(3), 559–565 (2004)
5. Beume, N., Naujoks, B., Emmerich, M.: SMS-EMOA: Multiobjective Selection-Basd on Dominated Hypervolume. European Journal of Operational Research 181, 1653–1659 (2007)

6. Chiba, K., Jeong, S., Obayashi, S., Yamamoto, K.: Knowledge Discovery in Aerodynamic Design Space for Flyback–Booster Wing Using Data Mining. In: 14th AIAA/AHI Space Planes and Hypersonic System and Technologies Conference, Canberra, Australia, November 6–9 (2006)
7. Chiba, K., Obayashi, S., Nakahashi, K.: Design Exploration of Aerodynamic Wing Shape for Reusable Launch Vehicle Flyback Booster. Journal of Aircraft 43(3), 832–836 (2006)
8. Chiba, K., Oyama, A., Obayashi, S., Nakahashi, K., Morino, H.: Multidisciplinary Design Optimization and Data Mining for Transonic Regional-Jet Wing. AIAA Journal of Aircraft 44(4), 1100–1112 (2007), doi:10.2514/1.17549
9. Chung, H.-S., Choi, S., Alonso, J.J.: Supersonic Business Jet Design using a Knowledge-Based Genetic Algorithm with an Adaptive, Unstructured Grid Methodology. In: AIAA Paper 2003-3791, 21st Applied Aerodynamics Conference, Orlando, Florida, USA, June 23-26 (2003)
10. Coello, C.A.C.: Theoretical and Numerical Constraint Handling Techniques used with Evolutionary Algorithms: A Survey of the State of the Art. Computer Methods in Applied Mechanics and Engineering 191(11-12), 1245–1287 (2002)
11. Deb, K.: Multi-Objective Optimization using Evolutionary Algorithms. John Wiley & Sons, Chichester (2001) ISBN 0-471-87339-X
12. Deb, K., Pratap, A., Agarwal, S., Meyarivan, T.: A Fast and Elitist Multiobjective Genetic Algorithm: NSGA–II. IEEE Transactions on Evolutionary Computation 6(2), 182–197 (2002)
13. Drela, M.: XFOIL: An Analysis and Design System for Low Reynolds Number Aerodynamics. In: Conference on Low Reynolds Number Aerodynamics. University of Notre Dame, IN (1989)
14. Fonseca, C.M., Fleming, P.J.: Genetic Algorithms forMultiobjective Optimization: Formulation, Discussion and Generalization. In: Stephanie, F. (ed.) Proceedings of the Fifth International Conference on Genetic Algorithms, pp. 416–423. University of Illinois at Urbana-Champaign, Morgan Kauffman Publishers, San Mateo, California (1993)
15. Gonzalez, L.F.: Robust Evolutionary Methods forMulti-objective and Multidisciplinary Design Optimization in Aeronautics. PhD thesis, School of Aerospace, Mechanical and Mechatronic Engineering, The University of Sydney, Australia (2005)
16. Hua, J., Kong, F., Liu, P.y., Zingg, D.: Optimization of Long-Endurance Airfoils. In: AIAA-2003-3500, 21st AIAA Applied Aerodynamics Conference, Orlando, FL, June 23-26 (2003)
17. Jameson, A., Caughey, D.A., Newman, P.A., Davis, R.M.: NYU Transonic Swept-Wing Computer Program - FLO22. Technical report, Langley Research Center (1975)
18. Jeong, S., Chiba, K., Obayashi, S.: DataMining for erodynamic Design Space. In: AIAA Paper 2005–5079, 23rd AIAA Applied Aerodynamic Conference, Toronto Ontario Canada, June 6–9 (2005)
19. Jin, Y.: A comprehensive survey of fitness approximation in evolutionary computation. Soft Computing 9(1), 3–12 (2005)
20. Kroo, I.: Multidisciplinary Optimization Applications in Preliminary Design – Status and Directions. In: 38th, and AIAA/ASME/AHS Adaptive Structures Forum, Kissimmee, FL, April 7-10 (1997)
21. Kroo, I.: Innovations in Aeronautics. In: 42nd AIAA Aerospace Sciences Meeting, Reno, NV, January 5-8 (2004)
22. Kuhn, T., Rösler, C., Baier, H.: Multidisciplinary Design Methods for the Hybrid Universal Ground Observing Airship (HUGO). In: AIAA Paper 2007–7781, Belfast, Northern Ireland, September 18-20 (2007)

23. Lee, D.S., Gonzalez, L.F., Srinivas, K., Auld, D.J., Wong, K.C.: Erodynamics/RCS Shape Optimisation of Unmanned Aerial Vehicles using Hierarchical Asynchronous Parallel Evolutionary Algorithms. In: AIAA Paper 2006-3331, 24th AIAA Applied Aerodynamics Conference, San Francisco, California, USA, June 5-8 (2006)
24. Lee, D.S., Gonzalez, L.F., Periaux, J., Srinivas, K.: Robust Design Optimisation Using Multi-Objective Evolutionary Algorithms. Computer & Fluids 37, 565–583 (2008)
25. Leifsson, L., Koziel, S.: Multi-fidelity design optimization of transonic airfoils using physics-based surrogate modeling and shape-preserving response prediction. Journal of Computational Science, 98–106 (2010)
26. Lian, Y., Liou, M.-S.: Multiobjective Optimization Using Coupled Response Surface Model end Evolutinary Algorithm. In: AIAA Paper 2004–4323, 10th AIAA/ISSMO Multidisciplinary Analysis and Optimization Conference, , Albany, New York, USA, August 30-September 1 (2004)
27. Lian, Y., Liou, M.-S.: Multi-Objective Optimization of Transonic Compressor Blade Using Evolutionary Algorithm. Journal of Propulsion and Power 21(6), 979–987 (2005)
28. Liao, W., Tsai, H.M.: Aerodynamic Design optimization by the Adjoint Equation Method on Overset Grids. In: AIAA Paper 2006-54, 44th AIAA Aerospace Science Meeting and Exhibit, Reno, Nevada, USA, January 9-12 (2006)
29. Mezura-Montes, E. (ed.): Constraint-Handling in Evolutionary Optimization. SCI, vol. 198. Springer, Heidelberg (2009)
30. Mialon, B., Fol, T., Bonnaud, C.: Aerodynamic Optimization Of Subsonic Flying Wing Configurations. In: AIAA-2002-2931, 20th AIAA Applied Aerodynamics Conference, St. Louis Missouri, June 24-26 (2002)
31. Obayashi, S., Tsukahara, T.: Comparison of OptimizationAlgorithms for Aerodynamic Shape Design. In: AIAA-96-2394-CP, AIAA 14th Applied Aerodynamics Conference, New Orleans, LA, USA, June 17-20 (1996)
32. Ong, Y.-S., Nair, P.B., Lum, K.Y.: Max-min surrogate-assisted evolutionary algorithm for robust design. IEEE Trans. Evolutionary Computation 10(4), 392–404 (2006)
33. Oyama, A.: Wing Design Using Evolutionary Algorithms. PhD thesis, Department of Aeronautics and Space Engineering. Tohoku University, Sendai, Japan (March 2000)
34. Oyama, A., Nonomura, T., Fujii, K.: Data Mining of Pareto-Optimal Transonic Airfoil Shapes Using Proper Orthogonal Decomposition. AIAA Journal Of Aircraft 47(5), 1756–1762 (2010)
35. Oyama, A., Okabe, Y., Shimoyama, K., Fujii, K.: Aerodynamic Multiobjective Design Exploration of a Flapping Airfoil Using a Navier-Stokes Solver. Journal Of Aerospace Computing, Information, and Communication 6(3), 256–270 (2009)
36. Price, K.V., Storn, R., Lampinen, J.A.: Differential Evolution. A Practical Approach to Global Optimization. Springer, Berlin (2005)
37. Rai, M.M.: Robust Optimal Design With Differential Evolution. In: AIAA Paper 2004-4588, 10th AIAA/ISSMO Multidisciplinary Analysis and Optimization Conference, Albany, New York, USA, August 30 - September 1 (2004)
38. Ray, T., Tsai, H.M.: A Parallel Hybrid Optimization Algorithm for Robust Airfoil Design. In: AIAA Paper 2004–905, 42nd AIAA Aerospace Science Meeting and Exhibit, Reno, Nevada, USA, January 5 -8 (2004)
39. Sierra, M.R., Coello, C.A.C.: A Study of Fitness Inheritance and ApproximationTechniques for Multi-Objective Particle Swarm Optimization. In: 2005 IEEE Congress on Evolutionary Computation (CEC 2005), vol. 1, pp. 65–72. IEEE Service Center, Edinburgh (2005)
40. Sasaki, D., Obayashi, S.: Efficient search for trade-offs by adaptive range multiobjective genetic algorithm. Journal Of Aerospace Computing, Information, and Communication 2(1), 44–64 (2005)

41. Sasaki, D., Obayashi, S., Nakahashi, K.: Navier-Stokes Optimization of Supersonic Wings with Four Objectives Using Evolutionary Algorithms. Journal Of Aircraft 39(4), 621–629 (2002)
42. Secanell, M., Suleman, A.: Numerical Evaluation of Optimization Algorithms for Low-Reynolds Number Aerodynamic Shape Optimization. AIAA Journal 10, 2262–2267 (2005)
43. Shimoyama, K., Oyama, A., Fujii, K.: A New Efficient and Useful Robust Optimization Approach –Design forMulti-objective Six Sigma. In: 2005 IEEE Congress on Evolutionary Computation (CEC 2005), vol. 1, pp. 950–957. IEEE Service Center, Edinburgh (2005)
44. Shimoyama, K., Oyama, A., Fujii, K.: Development of Multi-Objective Six-Sigma Approach for Robust Design Optimization. Journal of Aerospace Computing, Information, and Communication 5(8), 215–233 (2008)
45. Sobieczky, H.: Parametric Airfoils and Wings. In: Fuji, K., Dulikravich, G.S. (eds.) Notes on Numerical Fluid Mechanics, vol. 68, pp. 71–88. Vieweg Verlag, Wiesbaden (1998)
46. Song, W., Keane, A.J.: Surrogate-based aerodynamic shape optimization of a civil aircraft engine nacelle. AIAA Journal 45(10), 265–2574 (2007), doi:10.2514/1.30015
47. Srinivas, N., Deb, K.: Multiobjective Optimization Using Nondominated Sorting in Genetic Algorithms. Evolutionary Computation 2(3), 221–248 (1994)
48. Szöllös, A., Smíd, M., Hájek, J.: Aerodynamic optimization via multiobjective micro-genetic algorithm with range adaptation, knowledge-based reinitialization, crowding and epsilon-dominance. Advances in Engineering Software 40(6), 419–430 (2009)
49. Tani, N., Oyama, A., Okita, K., Yamanishi, N.: Feasibility study of multi objective shape optimization for rocket engine turbopump blade design. In: 44th AIAA/ASME/SAE/ASEE Joint Propulsion Conference & Exhibit, Hartford, CT, July 21 - 23 (2008)
50. Tsutsui, S., Ghosh, A.: Genetic algorithms with a robust solution searching scheme. IEEE Trans. Evolutionary Computation 1(3), 201–208 (1997)
51. Yamaguchi, Y., Arima, T.: Multi-Objective Optimization for the Transonic Compressor Stator Blade. In: AIAA Paper 2000–4909, 8th AIAA/USAF/NASA/ISSMO Symposium on Multidisciplinary Analysis and Optimization, AIAA Paper 2000–4909, 8th AIAA/USAF/NASA/ISSMO Symposium on Multidisciplinary Analysis and Optimization, September 6 - 8, Long Beach, CA, USA (2000)
52. Zhang, Q., Li, H.: MOEA/D: A Multiobjective Evolutionary Algorithm Based on Decomposition. IEEE Transactions on Evolutionary Computation 11(6), 712–731 (2007)
53. Zitzler, E., Thiele, L.: Multiobjective Evolutionary Algorithms: A Comparative Case Study and the Strength Pareto Approach. IEEE Transactions on Evolutionary Computation 3(4), 257–271 (1999)

Chapter 11
An Enhanced Support Vector Machines Model for Classification and Rule Generation

Ping-Feng Pai and Ming-Fu Hsu

Abstract. Based on statistical learning theory, support vector machines (SVM) model is an emerging machine learning technique solving classification problems with small sampling, non-linearity and high dimension. Data preprocessing, parameter selection, and rule generation influence performance of SVM models a lot. Thus, the main purpose of this chapter is to propose an enhanced support vector machines (ESVM) model which can integrate the abilities of data preprocessing, parameter selection and rule generation into a SVM model; and apply the ESVM model to solve real world problems. The structure of this chapter is organized as follows. Section 11.1 presents the purpose of classification and the basic concept of SVM models. Sections 11.2 and 11.3 introduce data preprocessing techniques, metaheuristics for selecting SVM models. Rule extraction of SVM models is addressed in Section 11.4. An enhanced SVM scheme and numerical results are illustrated in Section 11.5 and 11.6. Conclusions are made in Section 11.7.

Keywords: Support vector machines, Data preprocessing, Rule extraction, Classification.

11.1 Basic Concept of Classification and Support Vector Machines

The data mining technique observes enormous records comprising information about the target and input variables. Imagine that investors would like to classify the financial status based on characteristics of the firm, such as return on asset

Ping-Feng Pai
Department of Information Management, National Chi Nan University, Taiwan, ROC
e-mail: paipf@ncnu.edu.tw.

Ming-Fu Hsu
Department of International Business Studies, National Chi Nan University, Taiwan, ROC
e-mail: s97212903@ncnu.edu.tw

S. Koziel & X.-S. Yang (Eds.): Comput. Optimization, Methods and Algorithms, SCI 356, pp. 241–258.
springerlink.com

(ROA), quick ratio, and return on investment (ROI). This is a classification task and data mining techniques are suitable for this task. The goal of data mining is to build up a suitable model for a labeling process that approximates the original process as closely as possible. Thus, investors can adopt the well-developed model to learn the status of firm.

Support vector machines (SVM) were proposed by Vapnik [42, 43] originally for typical binary classification problems. The SVM implements the structural risk minimization (SRM) principle rather than the empirical risk minimization (ERM) principle employed by most traditional neural network models. The most important concept of SRM is the minimization of an upper bound to the generalization error instead of minimizing the training error. In addition, the SVM will be equivalent to solving a linear constrained quadratic programming (QP) problem, so that the solution for SVM is always unique and globally optimal [6, 12, 14, 41, 42, 43].

Given a training set of instance-base pairs (x_i, y_i), $i = 1,\ldots,m$, where $x_i \in R^n$ and $y_i \in \{\pm 1\}$, SVM determines an optimal separating hyperplane with the maximum margin by solving the following optimization problem:

$$\min_{w,g} \frac{1}{2} w^T w \qquad (11.1)$$
$$s.t. \quad y_i(w \cdot x_i + g) - 1 \geq 0$$

where w denotes the weight vector, and g denotes the bias term.

The Lagrange function's saddle point is the solution to the quadratic optimization problem:

$$L_h(w, g, \alpha) = \frac{1}{2} w^T \cdot w - \sum_{i=1}^{m} (\alpha_i y_i (w \cdot x_i + g) - 1) \qquad (11.2)$$

where α_i is Lagrange multipliers and $\alpha_i \geq 0$.

To identify an optimal saddle point is necessary because the L_h must be minimized with respect to the primal variable w and $g$and maximized the non-negative dual variable α_i. By discriminating w and g, and proposing the Karush Kuhn-Tucker (KKT) condition for the optimum constrained function, L_h is transformed to the dual Lagrangian $L_E(\alpha)$:

$$\max_{\alpha} \; L_E(\alpha) = \sum_{i=1}^{m} \alpha_i - \frac{1}{2} \sum_{i,j=1}^{m} \alpha_i \alpha_j y_i y_j \langle x_i, x_j \rangle \qquad (11.3)$$
$$s.t. \quad \alpha_i \geq 0, i = 1,\ldots,m \; and \; \sum_{i=1}^{m} \alpha_i y_i = 0$$

Dual Lagrangian $L_E(\alpha)$ must be maximized with respect to non-negative α_i to identify the optimal hyperplane. The parameters w^* and g^* of the optimal hyperplane were determined by the solution α_i for the dual optimization problem. Therefore, the optimal hyperplane $f(x) = sign(\langle w^* \cdot x \rangle + g^*)$ can be illustrated as:

$$f(x) = sign\left(\sum_{i=1}^{m} y_i \alpha_i^* \langle x_i, x \rangle + g^*\right) \tag{11.4}$$

In a binary classification task, only a few subsets of the Lagrange multipliers α_i usually tend to be greater than zero. These vectors are the closest to the optimal hyperplane. The respective training vectors having non-zero α_i are called support vectors, as the optimal decision hyperplane $f(x,\alpha^*,g^*)$ depends on them exclusively. Figure 11.1 illustrates the basic structure of SVM.

Very few data sets in the real world are linearly separable. What makes SVM so remarkable is that the basic linear framework is easily extended to the case where the data set is not linearly separable. The fundamental concept behind this extension is to transform the input space where the data set is not linearly separable into a higher-dimensional space, where the data are linearly separable. Figure 11.2 illustrates the mapping concept of SVM.

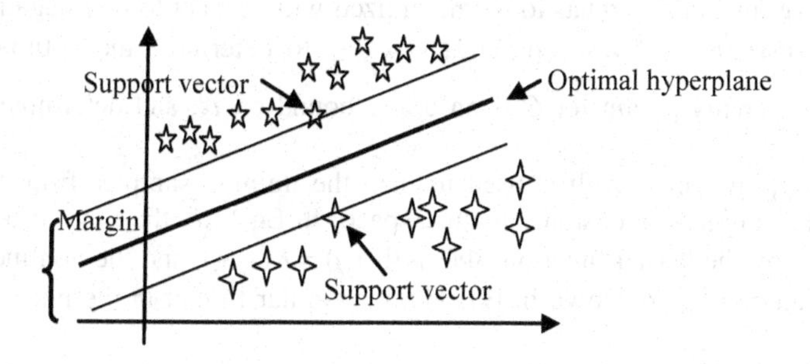

Fig. 11.1 The basic structure of the SVM [12]

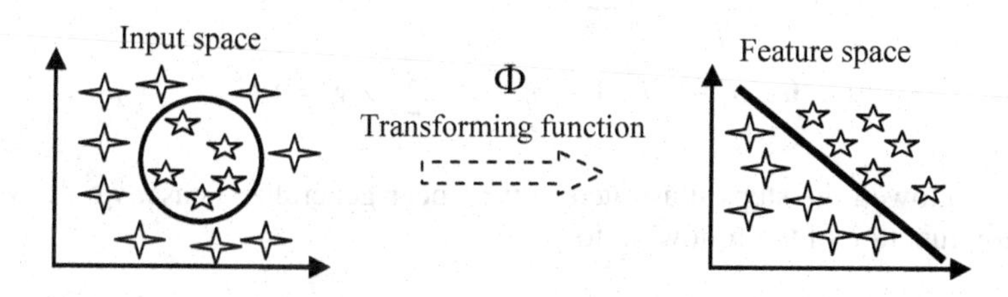

Fig. 11.2 Mapping a non-linear data set into a feature space [6]

In terms of the introduced slack variables, the problem of discovering the hyperplane with minimizing the training errors is illustrated as follows:

$$\begin{aligned} &\min_{w,g,\xi} \ \frac{1}{2} w^T \cdot w + C \sum_{i=1}^{m} \xi_i \\ &s.t. \ \ y_i(\langle w.x_i \rangle + g) + \xi - 1 \geq 0 \\ &\qquad \xi_i \geq 0 \end{aligned} \tag{11.5}$$

where C is a penalty parameter on the training error, and ξ_i is the non-negative slack variable. The constant C used to determine the trade-off between margin size and error. Observe that C is positive and cannot be zero; that is, we cannot simply ignore the slack variables by setting $C = 0$. With a large value for C, the optimization will try to discover a solution with a small number of non-zero slack variables because errors are costly [14]. Above all, it can be concluded that a large C implies a small margin, and a small C implies a large margin.

The Lagrangian method can be used to solve the optimization model, which is almost equivalent to the method for dealing with the optimization problem in the separable case. One has to maximize the dual variables Lagrangian:

$$\begin{aligned} \max_{\alpha} \ & L_E(\alpha)=\sum_{i=1}^{m}\alpha_i-\frac{1}{2}\sum_{i,j=1}^{m}\alpha_i\alpha_j y_i y_j\langle x_i\cdot x_j\rangle \\ s.t. \quad & 0\le\alpha_i\le C, i=1,\ldots,m \text{ and } \sum_{i=1}^{m}\alpha_i y_i=0 \end{aligned} \tag{11.6}$$

A dual Largrangian $L_E(\alpha)$ has to be maximized with respect to non-negative α_i under the constraints $\sum_{i=1}^{m}\alpha_i y_i=0$ and $0\le\alpha_i\le C$ to determine the optimal hyperplane. The penalty parameter C is an upper bound on α_i, and determined by the user.

The mapping function Φ is used to map the training samples from the input space into a higher-dimensional feature space. In Eq.11.6, the inner products are substituted by the kernel function $(\Phi(x_i)\cdot\Phi(y_i)) = K(x_i,x_j)$, and the nonlinear SVM dual Lagrangian $L_E(\alpha)$ shown in Eq.(11.7) is similar to that in the linear generalized case:

$$\begin{aligned} & L_E(\alpha)=\sum_{i=1}^{m}\alpha_i-\frac{1}{2}\sum_{i,j=1}^{m}\alpha_i\alpha_j y_i y_j K(x_i\cdot x_j) \\ & s.t. \ 0\le\alpha_i\le C, i=1,\ldots,m \text{ and } \sum_{i=1}^{m}\alpha_i y_i=0 \end{aligned} \tag{11.7}$$

Hence, followed the steps illustrated in the linear generalized case, we derive the decision function of the following form:

$$f(x)=sign\left(\sum_{i=1}^{m}y_i\alpha_i^*\langle\Phi(x),\Phi(x_i)\rangle+g^*\right)=sign\left(\sum_{i=1}^{m}y_i\alpha_i^*\langle K(x,x_i)\rangle+g^*\right) \tag{11.8}$$

The function K is defined as the kernel function for generating the inner products to construct machines with different types of nonlinear decision hyperplane in the input space. There are several kernel functions, depicted as follows. The determination of kernel function type depends on the problem's complexity [12].

Radial Basis Function (RBF): $K(x,x_i)=\exp\left\{-\|x-x_i\|^2/2\sigma^2\right\}$
Polynomial kernel of degree d: $K(x,x_i)=(x,x_i)^d$
Sigmoid kernel: $K(x,x_i)=\tanh(K(x,x_i)+r)$

11.2 Data Preprocessing

Data sometimes are missing, noisy and inconsistent; and irrelevant or redundant attributes of data increase the computational complexity and decrease performance of data mining models. To be useful for data mining purposes, the original data need to be preprocessed in the form of cleaning, transformation, and reduction. The data without the preprocessing procedures would cause confusion for the data mining procedure and result in unreliable output.

11.2.1 Data Cleaning

The purpose of data cleaning is to fill in missing value, eliminate the noise (outliers), and correct the inconsistencies in the data. Let us look at the following approaches for missing value [9, 21, 35, 37]:

- Ignore the missing value.
- Fill in the missing value manually.
- Apply a global constant to replace the missing value.
- Apply the mean attribute to replace the missing value.
- Apply the most probable value to fill in the missing value.

Noise data (e.g., outlier) is a random error or variance in the measured data. Even a small number of extreme values can lead to different results and impair the conclusion. There are some smoothing methods (e.g., binning, regression and clustering) to offset the effect caused by a small number of extreme values [3, 28, 37, 44]. Human error in data entry, deliberate errors and data decay are some of the reasons for inconsistent data. Missing values, noise, and inconsistent data lead to inaccurate results. Data cleaning is the first step to analyzing the original data which would lead to reliable mining result. Figure 11.3 illustrates the original data processed by the procedure of data cleaning [9, 36].

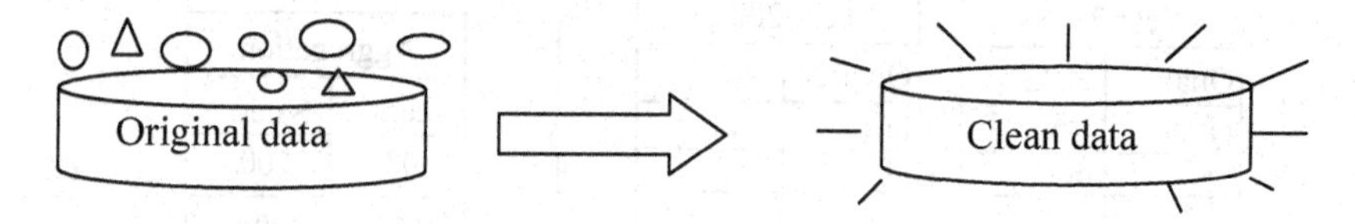

Fig. 11.3 Data cleaning [12]

11.2.2 Data Transformation

Data transformation is used to transform or consolidate data into forms suitable for the data mining process. Data transformation consists of the following processes [15, 17, 36, 38, 39]:

- Smoothing is employed to remove the noise from the data is illustrated in Fig. 11.4.
- Aggregation aggregates the data to construct the data cube for analysis.

- Generalization replaces the lower-level data with higher-level data.
- Normalization scales the attribute data to fall within a small specified range.

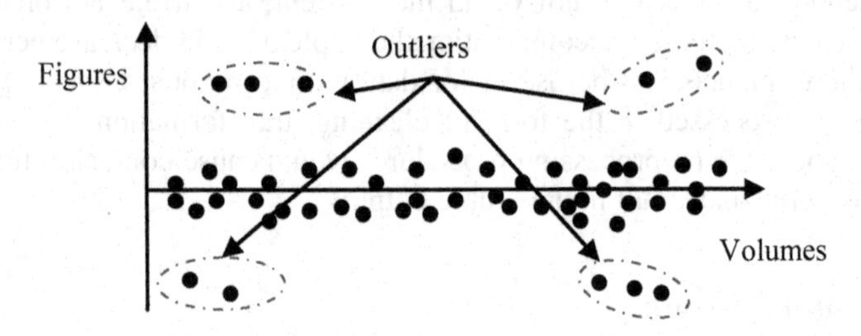

Fig. 11.4 The process of smoothing

11.2.3 Data Reduction

The purpose of the data reduction is to create a reduced representation of the dataset which is much smaller in volume yet closely sustains the integrity of the raw data. Dealing with the reduced data set enhances efficiency while producing the same analytical results. Data reduction consists of the following process [1, 2, 4, 5, 7, 18, 19, 24, 40, 45]:

- The aggregation of the data cube is employed to construct a data cube which is illustrated in Fig. 11.5.
- Attribute selection is used to remove the irrelevant, redundant or weak attributes, as shown in Fig. 11.6.
- Dimension reduction is used to reduce or compress the representation of the raw dataset. Raw data which can be reconstructed from the compressed data without losing any information is called lossless. In contrast, the approximation of the reconstructed raw data is called lossy.

Year 2008	
Quarter	Sales
Q 1	300
Q 2	400
Q 3	450
Q 4	550

Year 2009	
Quarter	Sales
Q 1	440
Q 2	410
Q 3	550
Q 4	600

Aggregation	
Years	Sales
2008	1700
2009	2000

Fig. 11.5 Aggregation of the data cube [12]

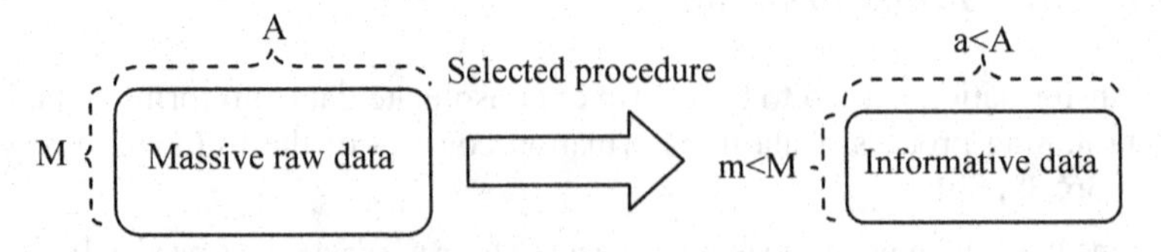

Fig. 11.6 Attribute selection [12]

11.3 Parameter Determination of Support Vector Machines by Meta-heuristics

Appropriate parameter setting can improve the performance of SVM models. Two parameters (C and σ) have to be determined in the SVM model with RBF kernel. The parameter C is the cost of penalty which influences the classification performance. If C is too large, the classification accuracy is very high in training data set, but very low in testing data set. If C is too small, the classification accuracy is inferior. The parameter σ has more influence than parameter C on classification outcome, because the value affects the partitioning outcome in the feature space. A large value for parameter σ leads to over-fitting, while a small value results in under-fitting [22]. The Grid search [24] is the most common approach to determine parameters of SVM models. Nevertheless, this approach is a local search technique, and tends to reach the local optima [20]. Furthermore, setting appropriate search intervals is an essential problem. A large search interval increases the computational complexity, while a small search interval would cause an inferior outcome. Some metaheuristics were proposed to select satisfactory parameters of SVM models [29, 30, 31, 32, 33, 34, 35]. The basic concept is to transfer the fitness functions of meta-heuristics into the forms of classification performance criteria (classification accuracy or error) of the SVM models. The fitness function of proposed metahuristics is used to measure the classification accuracy of the SVM model. Making the classification performance criteria acceptable for the metaheuristic algorithms is the most critical part of this procedure.

11.3.1 Genetic Algorithm

Holland [13] proposed the genetic algorithm (GA) to understand the adaptive processes of natural systems. Subsequently, they were employed for optimization and machine learning in the 1980's. Originally, GA was associated with the use of binary representation, but currently we can find it used with other types of representations and applied in many research domains. The basic principle is the principle of survival of the fittest. It tries to keep genetic information from generation to generation. The major merits of GA are their ability to find optimal or near optimal solutions with relatively modest computational requirements. The concept is briefly illustrated as follows and illustrated in Fig. 11.7. :

- Initialization: The initial population of chromosomes is established randomly.
- Evaluating fitness: Evaluate the fitness of each chromosome. The classification accuracy is used as the fitness function.
- Selection: Select a mating pair for reproduction.
- Crossover and mutation: Create new offspring by performing crossover and mutation operations.
- Next generation: Create a population for the next generation.
- Stop condition: If the number of generations equals a threshold, then the best chromosomes are presented as a solution; otherwise go back to step (b) [29, 31].

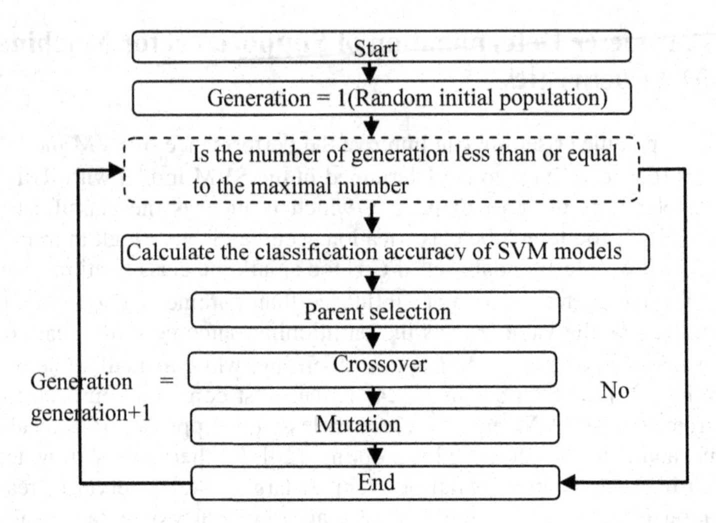

Fig. 11.7 The architecture of GA to determine parameters of SVM

11.3.2 Immune Algorithm

The immune algorithm (IA) [10] was based on the natural immune systems which efficiently distinguish all cells within the body and classify those cells as self or non-self cells. Non-self cells trigger a defense procedure which defends against foreign invaders. The antibodies are expressed by two SVM parameters. The classification error of SVM is contained in the denominator part of the affinity formula. Therefore, the reason for maximizing the affinity of IA is to minimize classification errors of the SVM model. IA search algorithm applied to determine the parameters of SVM is described as follows and illustrates in Fig. 11.8. :

- Initialization: Both the initialized antibody population and the population of the initial antibody were created randomly.
- Evaluation fitness: The classification error (CE) was treated as the fitness of IA.
- Affinity and similarity: When affinity values are high, the affinity and the similarity antibodies having higher activation levels of antigens are identified. To maintain the diversity of the antibodies stored in the memory cells, antibodies with a higher affinity value and a lower similarity value have a good likelihood of entering the memory cells. Eq. (11.9) is used to depict the affinity between the antibody and antigen:

$$Antigen = 1/1 + CE \tag{11.9}$$

A smaller CE indicates a higher affinity value. Eq. (11.10) is applied to illustrate the similarity between antibodies:

$$Antibodies = 1/1 + G_{ij} \tag{11.10}$$

where G_{ij} is the difference between the two classification errors calculated by the antibodies inside and outside the memory cells.

- Selection: Select the antibodies in the memory cells. Antibodies with higher values of *Antigen* are treated as candidates to enter the memory cell. However, the antibody candidates with $Antibodies_{ij}$ values exceeding the threshold are not qualified to enter the memory cell.
- Crossover and mutation: The antibody population is undergoing crossover and mutation. Crossover and mutation are used to generate new antibodies. When conducting the crossover operation, strings representing antibodies are paired randomly. Segments of paired strings between two predetermined break-points are swapped.
- Perform tabu search [11] on each antibody: Evaluate neighbor antibodies and adjust the tabu list. The antibody with the better classification error and not recorded on the tabu list is placed on the tabu list. If the best neighbor antibody is the same as one of the antibodies on the tabu list, then the next set of neighbor antibodies is generated and the classification error of the antibody calculated. The next set of neighbor antibodies is generated from the best neighbor antibodies in the current iteration.
- Current antibody selection by tabu search: If the best neighbor antibody is better than the current antibody, then the current antibody is replaced by the best neighbor antibody. Otherwise, the current antibody is retained.
- Next generation: From a population for the next generation.
- Stop criterion: If the number of epochs is equal to a given scale, then the best antibodies are presented as a solution; otherwise go to Step (b) [32, 33].

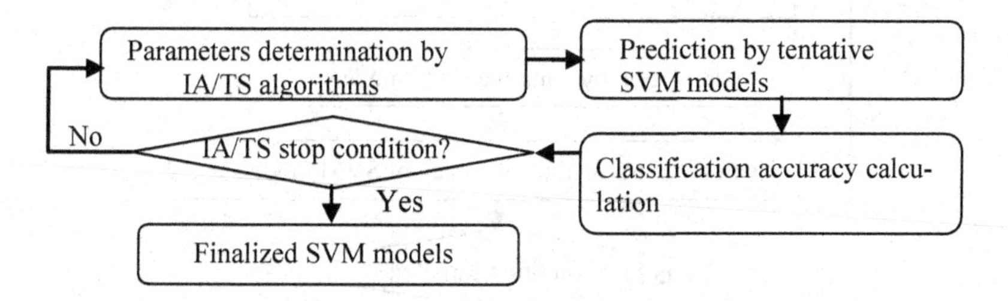

Fig. 11.8 The architecture of IA/TS to determine parameters of SVM

11.3.3 Particle Swarm Optimization

The particle swarm optimization (PSO) algorithm [16] is another population-based meta-heuristic inspired by swarm intelligence. It simulates the behavior of birds flocking to a promising position with sufficient food. A particle is considered as a point in a *G*-dimensional space and its status is characterized according to its position y_{ig} and velocity s_{ig}. The *G*-dimensional position for the particle *i* at iteration *t* is expressed as $y_i^t = \{y_{i1}^t, \ldots, y_{iG}^t\}$.

The velocity, which is also a *G*-dimensional vector, for particle *i* at iteration *t* is illustrated as $s_i^t = \{s_{i1}^t, \ldots, s_{iG}^t\}$. Let $b_i^t = \{b_{i1}^t, \ldots, b_{iG}^t\}$ be the best solution that particle *i* has obtained until iteration *t*, and $b_m^t = \{b_{m1}^t, \ldots, b_{mG}^t\}$ represents the best solution from b_i^t in the population at iteration t. To search for an optimal solution, each particle changes its velocity according to cognition and sociality. Each particle then moves to a new potential solution. The use of PSO algorithm to select SVM parameters is described as follows. First, initialize a random population of particles and velocities. Second, define the fitness of each particle. The fitness function of PSO is represented as the classification accuracy of SVM models. Each particle's velocity is expressed by Eq. (11.11). For each particle, the procedure then moves to the next position according to Eq. (11.12).

$$S_{ig}^{t} = S_{ig}^{t-1} + c_1 j_1 \left(B_{ig}^{t} - y_{ig}^{t}\right) + c_2 j_2 \left(B_{mg}^{t} - y_{mg}^{t}\right), g = 1, \ldots, G \tag{11.11}$$

where c_1 is the cognitive learning factor, c_2 is the social learning factor, and j_1 and j_2 are the random numbers uniformly distributed in $U(0,1)$.

$$Y_{ig}^{t+1} = Y_{ig}^{t} + S_{ig}^{t}, g = 1, \ldots, G \tag{11.12}$$

Finally, if the termination criterion is reached, the algorithm stops; otherwise return to the step of fitness measurement [34]. The architecture of PSO is illustrated in Fig. 11.9.

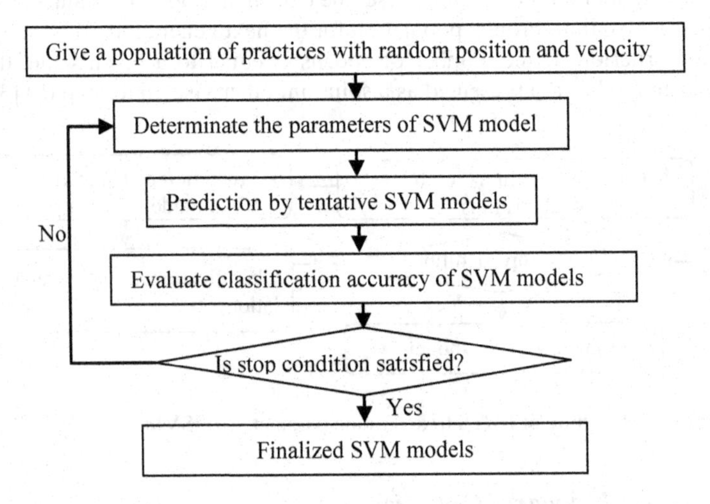

Fig. 11.9 The architecture of PSO to determine parameters of SVM

11.4 Rule Extraction Form Support Vector Machines

Support vector machines are state-of-the art data mining techniques which have proven their performance in many research domains. Unfortunately, while the models may provide a high accuracy compared to other data mining techniques, their comprehensibility is limited. In some areas, such as credit scoring, the lack of comprehensibility of a model is a main drawback causing reluctance of users to use the model [8]. Furthermore, when credit has been denied to a customer, the Equal Credit Opportunity Act of the US requires that the financial institution provide specific reasons why the application was rejected; and indefinite and vague reasons for denial are illegal [23]. Comprehensibility can be added to SVM by extracting symbolic rules from the trained model. Rule extraction techniques would be used to open up the black box of SVM and generate comprehensible decision rules with approximately the same detective power as the model itself. There are two ways to open up the black box of SVM, as shown in Fig. 11.10.

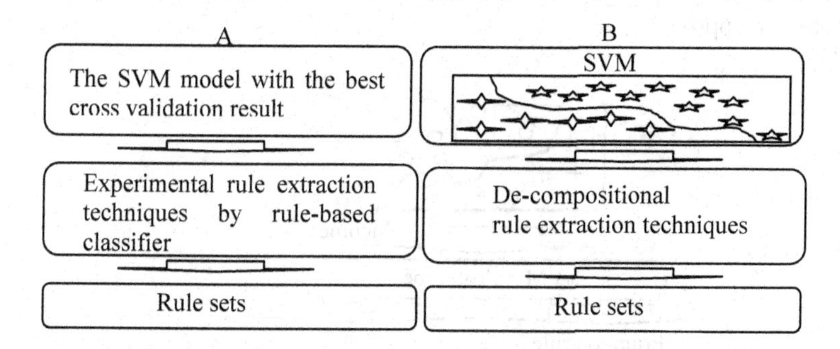

Fig. 11.10 Experimental (A) and de-compositional (B) rule extraction techniques [23]

The SVM with the best cross validation (CV) result is then fed into rule-based classifier (i.e., decision tree, rough set and so on) to derive the comprehensive decision rules for humans to understand (experimental rule extraction technique). The concept behind this procedure is the assumption that the trained model can more appropriately represent the data than can the original dataset. This is to say that the data of the best CV result is cleaner and free of curial conflicts. The CV is a re-sampling technique which adopts multiple random training and test subsamples to overcome the overfitting problem. Overfitting would lead to SVM losing its applicability, as shown in Fig. 11.11. The CV analysis would yield useful insights on the reliability of the SVM model with respect to sampling variation.

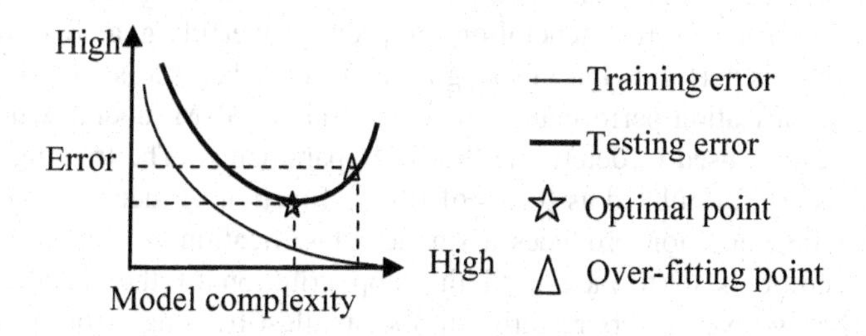

Fig. 11.11 Classification errors vs. model complexity of SVM models [12]

Decompositional rule extraction was proposed by Nunez et al. [25, 26] and proposes rule-defining regions based on the prototype and support vectors [23]. The representative of the obtained clusters is prototype vectors. The clustering task is overcome by vector quantization. There are two kinds of rules which can be proposed: equation rules and interval rules, respectively corresponding to an ellipsoid and interval region, which can be built in the following manner [18]. Applying the prototype vector as center, an ellipsoid is constructed where the axes are determined by the support vector within the partition lying the furthest from the center. The long axes of the ellipsoid are defined by the straight line connecting these two vectors. The interval regions are defined from ellipsoids parallel to the coordinate axes [23]. Figure 11.12 is used to illustrate the basic structure of SVM + Prototype approach.

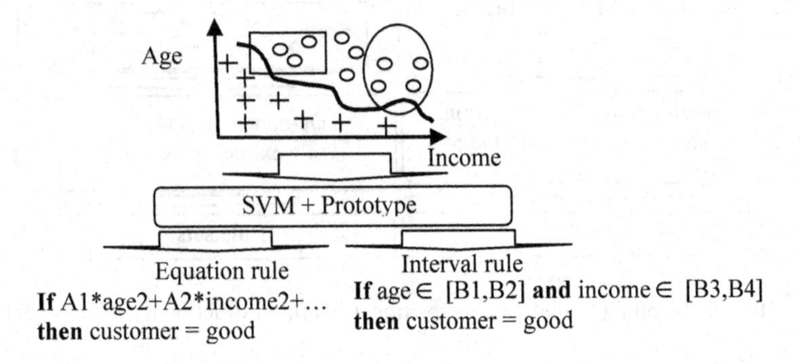

Fig. 11.12 SVM + Prototype model [25, 26]

11.5 The Proposed Enhanced SVM Model

In this section, the scheme of a proposed ESVM model is illustrated. Figure 11.13 shows the flowchart of the ESVM model, including functions of data preprocessing, parameter determination and rule generation. First, the raw data is processed by data-preprocessing techniques containing data cleaning, data transformation, feature selection, and dimension reduction. Second, the preprocessed data are divided into two sets: training and testing data sets. The training data set is used to select a data set used for rule generation. To prevent overfitting, a cross-validation (CV) procedure is performed at this stage. The testing data set is employed to examine the classification performance of a well-trained SVM model. Sequentially, metaheuristics are used to determine the SVM parameters. The training errors of SVM models are formulated as forms of fitness function of metaheuristics. Thus, each succeeding iteration produces a smaller classification error. The parameter search procedure is performed until the stop criterion of the metaheuristic is reached. The two parameters resulting in the smallest training error are then employed to undertake testing procedures and therefore testing accuracy is obtained. Finally, the CV training data set with the smallest testing error is utilized to derive decision rules by rule extraction mechanisms. Accordingly, the proposed ESVM model can provide decision rules as well as classification accuracy for decision makers.

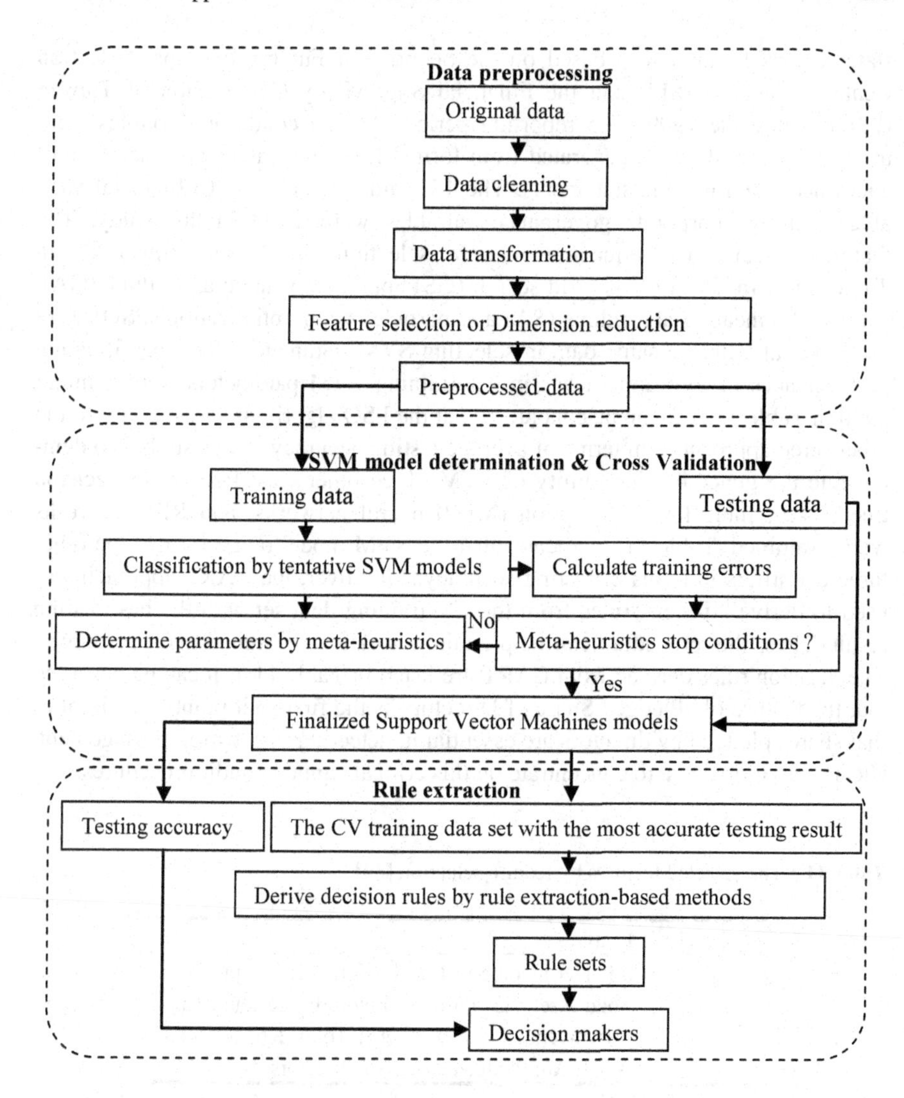

Fig. 11.13 The flowchart of the ESVM model

11.6 A Numerical Example and Empirical Results

A numerical example borrowed from Pai et al. [34] was used here to illustrate the classification and rule generation of SVM models. The original data used in this example contain 75 listed firms in Taiwan's stock market. These firms were divided into 25 fraudulent financial statement (FFS) firms and 50 non-fraudulent financial statement (non-FFS) firms. Published indication or proof of involvement in issuing FFS was found for the 25 FFS firms. The classification of a financial statement as fraudulent is based on the Security of Futures Investor Protection Center in Taiwan (SFI) and the Financial Supervisory Commission of Taiwan (FSC) during the 1999-2005 reporting period. All the condition variables were used in the sample were generated from formal financial statements, such as balance sheets and income statements. The 18 features consist of 16 financial variables and two corporate governance variables were adopt in this study. The features selected by sequential forward selection (SFS) were illustrated in Table. 11.1. In addition, the grid search (GS) approach, genetic algorithms (GA), simulated annealing algorithms (SA) and particle swam optimization (PSO) were used to deal with the same data in selecting SVM parameters. The classification performances of four approaches in determining SVM parameters were summarized in Table 11.2. It can be concluded that the PSO algorithm was superior to the other three approaches in terms of average testing accuracy in this study. To demonstrate the generalization ability of SVM, three other classifiers, C4.5 decision tree (C4.5), multi-layer perception (MLP) neural networks, and RBF networks were examined. Table 11.3 indicates that the SVM model outperformed the other three classifiers in terms of testing accuracy. Moreover, the CART approach was used to derive "if-then" rules from the CV training data set with the best testing result. Thus, this procedure can help auditors to allocate limited audit resources. The decision rules derived from CART are listed in Table 11.4. It can be observed that the feature of "Pledged Share of Directors"is the first split point. This implies that shares pledged by directors are essential in detecting FFS by top management. Clearly, auditors have to concentrate on this critical signal in audit procedures.

Table 11.1 The selected features by feature selection [34]

Method	Features
SFS	A1: Net income to Fixed asset; A2: Net profit to Total asset; A3: Earnings before Interest and Tax; A4: Inventory to Sales; A5: Total debt to Total Asset; A6: Pledged shares of Directors

Table 11.2 Classification performance of four methods in determining SVM parameters [34]

Methods	Cross-validation					Accuracy (%)
	CV-1	CV-2	CV-3	CV-4	CV-5	
Grid	86.67	80	73.33	80	80	80
GA	80	86.67	80	86.67	86.67	84
SA	80	86.67	86.67	93.33	96.67	86.67
PSO	93.33	80	93.33	93.33	93.33	92

Table 11.3 Testing accuracy of six classifiers [34]

Classifier	Cross-validation					Accuracy (%)
	CV-1	CV-2	CV-3	CV-4	CV-5	
C4.5	73.33	80	86.67	93.33	86.67	84
MLP	73.33	86.67	80	86.67	86.67	82.67
RBFNN	86.67	80	80	86.67	80	82.67
SVM	93.33	86.67	93.33	93.33	93.33	92

Table 11.4 Decision rules derived from CART [34]

(1) If "pledged shares of directors" $\geq$ 44.405 , then "FFS"

(2) If "pledged shares of directors" $<$ 44.405 and "net profit to total assets" $<$-0.3229 , then "FFS"

(3) If "pledged shares of directors" $<$44.405 , "net profit to total assets" $\geq$-0.3229 and "net income to fixed assets" $\geq$0.0497 , then "non-FFS"

(4) If "pledged shares of directors" $<$44.405 , "net profit to total assets" $\geq$-0.3229, "net income to fixed assets" $<$0.0497 and "earnings before interest and tax" $<$-42220, then "non-FFS"

(5) If "pledged shares of directors" $<$44.405, "net profit to total assets" $\geq$-0.3229, "net income to fixed assets" $<$0.0497, "earnings before interest and tax" $\geq$-42220, and "total debt to total assets " $\geq$1.48 then, "FFS"

(6) If "pledged shares of directors" $<$44.405, "net profit to total assets" $\geq$-0.3229, "net income to fixed assets" $<$0.0497, "earnings before interest and tax" $\geq$-42220, and "total debt to total assets" $<$1.48 then, "non-FFS"

11.7 Conclusion

In this chapter, the three essential issues influencing the performance of SVM models were pointed out. The three issues are: data preprocessing, parameter determination and rule extraction. Some investigations have been conducted into each issue respectively. However, this chapter is the first study proposing an enhanced SVM model which deals with three issues at the same time. Thanks to data preprocessing procedure, the computation cost decreases and the classification accuracy increases. Furthermore, the ESVM model provides rules for decision makers. Rather than the expression of complicated mathematical functions, it is easy for decision makers to realize the relation and strength between condition attributes and outcome intuitively form a set of rules. These rules can be reasoned in both forward and backward ways. For the example in Section 11.6, the forward reasoning can provide a good direction for managers to improve the current financial status; and the backward reasoning can protect the wealth of investors and sustain the stability of financial market.

Acknowledgments. The authors would like to thank the National Science Council of the Republic of China, Taiwan for financially supporting this research under Contract No. 96-2628-E-260-001-MY3 & 99-2221-E-260-006.

References

1. Agarwal, S., Agrawal, R., Deshpande, P.M., Gupta, A., Naughton, J.F., Ramakrishnan, R., Sarawagi, S.: On the computation of multidimensional aggregates. In: Proc. Int. Conf. Very Large Data Bases, pp. 506–521 (1996)
2. Barbar'a, D., DuMouchel, W., Faloutos, C., Haas, P.J., Hellerstein, J.H., Ioannidis, Y., Jagadish, H.V., Johnson, T., Ng, R., Poosala, V., Ross, K.A., Servcik, K.C.: The New Jersey data reduction report. Bull. Technical Committee on Data Engineering 20, 3–45 (1997)
3. Ballou, D.P., Tayi, G.K.: Enhancing data quality in data warehouse environments. Comm. ACM 78, 42–73 (1999)
4. Breiman, L., Friedman, J., Olshen, R., Stone, C.: Classifcation and Regression Trees, Wadsworth International Group (1984)
5. Chakrabart, S., Cox, E., Frank, E., Guiting, R.H., Han, J., Jiang, X., Kamber, M., Lightstone, S.S., Nadeau, T.P., Neapolitan, R.E., Pyle, D., Refaat, M., Schneider, M., Teorey, T.J.I., Witten, H.: Data Mining: Know It All. Morgan Kaufmann, San Francisco (2008)
6. Taylor, J.S., Cristianini, N.: Support Vector Machines and other kernel-based learning methods. Cambridge University Press, Cambridge (2000)
7. Dash, M., Liu, H.: Feature selection methods for classification. Intell. Data Anal. (1), 131–156 (1997)
8. Dwyer, D.W., Kocagil, A.E., Stein, R.M.: Moody's kmv riskcalc v3.1 model (2004)
9. English, L.: Improving Data Warehouse and Business Information Quality: Methods for Reducing Costs and Increasing. John Wiley & Sons, Chichester (1999)
10. Farmer, J.D., Packard, N.H., Perelson, A.: The immune system, adaptation, and machine learning. Physica. D 22(1–3), 187–204 (1986)
11. Glover, F., Kelly, J.P., Laguna, M.: Genetic algorithms and tabu search: hybrids for optimization. Comput. Oper. Res. 22, 111–134 (1995)
12. Hamel, L.H.: Knowledge Discovery with Support Vector Machines. Wiley, Chichester (2009)
13. Holland, J.H.: Adaptation in Natural and Artificial Systems. University of Michigan Press, Ann Arbor (1975)
14. Huang, C.L., Chen, M.C., Wang, C.J.: Credit scoring with a data mining approach based on support vector machines. Expert Systems with Applications 33(4), 847–856 (2007)
15. Kennedy, R.L., Lee, Y., Van Roy, B., Reed, C.D., Lippman, R.P.: Solving Data Mining Problems Through Pattern Recognition. Prentice-Hall, Englewood Cliffs (1998)
16. Kennedy, J., Eberhart, R.: Particle swarm optimization, In Proceedings of IEEE conference on neural network, vol. 4, pp. 1942–1948 (1995)
17. Kohavi, R., John, G.H.: Wrappers for feature subset selection. Artif. Intell. 97, 273–324 (1997)
18. Langley, P., Simon, H.A., Bradshaw, G.L., Zytkow, J.M.: Scientific Discovery: Computational Explorations of the Creative Processes. MIT Press, Cambridge (1987)

19. Liu, H., Motoda, H.: Feature Extraction, Construction, and Selection: A Data Mining Perspective. Kluwer Academic Publishers, Dordrecht (1998)
20. Lin, S.W., Shiue, Y.R., Chen, S.C., Cheng, H.M.: Applying enhanced data mining approaches in predicting bank performance: A case of Taiwanese commercial banks. Expert Syst. Appl. (36), 11543–11551 (2009)
21. Loshin, D.: Enterprise Knowledge Management: The Data Quality Approach. Morgan Kaufmann, San Francisco (2001)
22. Lopez, F.G., Torres, G.M., Batista, B.M.: Solving feature subset selection problem by parallel scatter search. Eur. J. Oper. Res. (169), 477–489 (2006)
23. Martens, D., Baesens, B., Gestel, T.V., Vanthienen, J.: Comprehensible credit scoring models using rule extraction from support vector machines. Eur. J. Oper. Res. 183(3), 1466–1476 (2007)
24. Martin, D.: Early warning of bank failure a logit regression approach. J. Bank. Financ. (1), 249–276 (1977)
25. Nunez, H., Angulo, C., Catala, A.: Rule extraction from support vector machines. In: European Symposium on Artificial Neural Networks Proceedings, pp. 107–112 (2002)
26. Nunez, H., Angulo, C., Catala, A.: Rule based learning systems from SVM and RBFNN. Tendencias de la mineria de datos en espana, Red Espaola de Minera de Datos (2004)
27. Neter, J., Kutner, M.H., Nachtsheim, C.J., Wasserman, L.: Applied Linear Statistical Models. Irwin (1996)
28. Olson, J.E.: Data Quality: The Accuracy Dimension. Morgan Kaufmann, San Francisco (2003)
29. Pai, P.F., Hong, W.C.: Forecasting regional electricity load based on recurrent support vector machines with genetic algorithms. Electr. Pow. Syst. Res. 74(3), 417–425 (2005)
30. Pai, P.F., Lin, C.S.: A hybrid ARIMA and support vector machines model in stock price forecasting. Omega 33(6), 497–505 (2005)
31. Pai, P.F.: System reliability forecasting by support vector machines with genetic algorithms. Math. Comput. Model. 433(3-4), 262–274 (2006)
32. Pai, P.F., Chen, S.Y., Huang, C.W., Chang, Y.H.: Analyzing foreign exchange rates by rough set theory and directed acyclic graph support vector machines. Expert Syst. Appl. 37(8), 5993–5998 (2010)
33. Pai, P.F., Chang, Y.H., Hsu, M.F., Fu, J.C., Chen, H.H.: A hybrid kernel principal component analysis and support vector machines model for analyzing sonographic parotid gland in Sjogren's Syndrome. International Journal of Mathematical Modelling and Numerical Optimisation (2010) (in press)
34. Pai, P.F., Hsu, M.F., Wang, M.C.: A support vector machine-based model for detecting top management fraud. Knowl.-Based Syst. 24(2), 314–321 (2011)
35. Pyle, D.: Data Preparation for Data Mining. Morgan Kaufmann, San Francisco (1999)
36. Quinlan, J.R.: Unknown attribute values in induction. In: Proc. 1989 Int. Conf. Machine Learning (ICML 1989), Ithaca, NY, pp. 164–168 (1989)
37. Redman, T.: Data Quality: Management and Technology. Bantam Books (1992)
38. Ross, K., Srivastava, D.: Fast computation of sparse datacubes. In: Proc Int. Conf. Very Large Data Bases, pp. 116–125 (1997)
39. Sarawagi, S., Stonebraker, M.: Efficient organization of large multidimensional arrays. In: Proc. Int. Conf. Data Engineering, ICDE 1994 (1994)
40. Siedlecki, W., Sklansky, J.: On automatic feature selection. Int. J. Pattern Recognition and Artificial Intelligence (2), 197–220 (1988)

41. Scholkopf, B., Smola, A.J.: Learning with Kernels: Support Vector Machines, Regularization, Optimization, and Beyond. MIT Press, Cambridge (2001)
42. Vapnik, V.: Statistical learning theory. John Wiley and Sons, New York (1998)
43. Vapnik, V., Golowich, S., Smola, A.: Support vector machine for function approximation, regression estimation, and signal processing. Advances in Neural Information processing System (9), 281–287 (1996)
44. Wang, R., Storey, V., Firth, C.: A framework for analysis of data quality research. IEEE Trans. Knowledge and Data Engineering (7), 623–640 (1995)
45. Zhao, Y., Deshpande, P.M., Naughton, J.F.: An array-based algorithm for simultaneous multi-dimensional aggregates. In: Proc. 1997 ACM-SIGMOD Int. Conf. Management of Data, pp. 159–170 (1997)

Chapter 12
Benchmark Problems in Structural Optimization

Amir Hossein Gandomi and Xin-She Yang

Abstract. Structural optimization is an important area related to both optimization and structural engineering. Structural optimization problems are often used as benchmarks to validate new optimization algorithms or to test the suitability of a chosen algorithm. In almost all structural engineering applications, it is very important to find the best possible parameters for given design objectives and constraints which are highly non-linear, involving many different design variables. The field of structural optimization is also an area undergoing rapid changes in terms of methodology and design tools. Thus, it is highly necessary to summarize some benchmark problems for structural optimization. This chapter provides an overview of structural optimization problems of both truss and non-truss cases.

12.1 Introduction to Benchmark Structural Design

New optimization algorithms are often tested and validated against a wide range of test functions so as to compare their performance. Structural optimization problems are complex and highly nonlinear, sometimes even the optimal solutions of interest do not exist. In order to see how an optimization algorithm performs, some standard structural engineering test problems are often solved. Many structural test functions exist in the literature, but there is no standard list or set of the functions one has to follow. Any new optimization algorithm should be tested using at least a subset of well-known, well-established functions with diverse

Amir Hossein Gandomi
Department of Civil Engineering, University of Akron,
Akron, OH, USA
e-mail: a.h.gandomi@gmail.com

Xin-She Yang
Mathematics and Scientific Computing,
National Physical Laboratory, Teddington, Middlesex TW11 0LW, UK
e-mail: xin-she.yang@npl.co.uk

S. Koziel & X.-S. Yang (Eds.): Comput. Optimization, Methods and Algorithms, SCI 356, pp. 259–281.
springerlink.com

properties so as to make sure whether or not the tested algorithm can solve certain types of optimization efficiently. According to the nature of the structural optimization problems, we can first divide them into two groups: truss and non-truss design problems. The selected lists of the test problems for each optimization group are listed below:

Truss design problems:

- 10-bar plane truss
- 25-space truss
- 72-bar truss
- 120-bar truss dome
- 200-bar plane truss
- 26-story truss tower

Non-truss design problems:

- Welded beam
- Reinforced concrete beam
- Compression Spring
- Pressure vessel
- Speed reducer
- Stepped cantilever beam
- Frame optimization

12.1.1 Structural Engineering Design and Optimization

Many problems in structural engineering and other disciplines involve design optimization of dozens to thousands of parameters, and the choice of these parameters can affect the performance or objectives of the system concerned. The optimization target is often measured in terms of objectives or cost functions in quantitative models. Structural engineering design and testing often require an iteration process with parameter adjustments. Optimization functions can generally be formulated as:

$$\text{Optimize: } f(X), \tag{12.1}$$

Subject to:

$$g_i(X) \geq 0, \; i = 1, 2, \ldots, N. \tag{12.2}$$

$$h_j(X) = 0, \; j = 1, 2, \ldots, M. \tag{12.3}$$

where $X = (x_1, x_2, \ldots, x_n)$, $X \text{ v } \Omega$ (parameter space).

Most design optimization problems in structural engineering involve many different design variables under complex constraints. These constraints can be written either as simple bounds such as the ranges of material properties, or as nonlinear relationships including maximum stress, maximum deflection, minimum load capacity, and geometrical configuration. Such non-linearity often results in multimodal response landscape.

The basic requirement for an efficient structural design is that the response of the structure be acceptable for given various specifications. That is, a set of parameters should at least be in a feasible design. There can be a very large number of feasible designs, but it is desirable to choose the best of these designs. The best design can be identified using minimum cost, minimum weight, maximum performance or, a combination of these [1]. Obviously, parameters may have associated uncertainties, and in this case, a robust design solution, not necessarily the best solution, is often the best choice in practice. As parameter variations are usually very large, systematic adaptive searching or optimization procedures are required. In the past several decades, researchers have developed many optimization algorithms. Examples of conventional methods are hill climbing, gradient methods, random search, simulated annealing, and heuristic methods. Examples of evolutionary or biology-inspired algorithms are genetic algorithms [2], neural network [3], particle swarm optimization [4], firefly algorithm [5], cuckoo search [6], and many others. The methods used to solve a particular structural problem depend largely on the type and characteristics of the optimization problem itself. There is no universal method that works for all structural problems, and there is generally no guarantee to find the globally optimal solution in highly nonlinear global optimization problems. In general, we can emphasize on the best estimate or suboptimal solutions under given conditions. Knowledge about a particular problem is always helpful to make the appropriate choice of the best or most efficient methods for the optimization procedure.

12.2 Classifications of Benchmarks

Generally, an optimization problem is classified according to the nature of equations with respect to design variables, the characteristics of the objectives and constraints. If the objective function and the constraints involving the design variable are linear, then the optimization is called a linear optimization problem. If even one of them is non-linear, it is classified as a non-linear optimization problem [1].

Design variables can be continuous or discrete (integer on non-integer). In structural engineering, most problems are mixed variable problems, as they contain both continuous and discrete variables. The structural optimization of bar or truss sections often includes a special set of variables which are integer multiples of certain sizes and dimensions.

According to the number of variables, constraints, and objective function(s), an optimization problem can be classified as small scale, normal scale, large scale and very large scale.

Nearly all design optimization problems in structural engineering are highly non-linear, involving many different design variables under complex, nonlinear constraints. In this study, benchmark optimization problems are classified into two groups: truss and non-truss. First, we introduce truss design problems. Truss structures are designed to carry multiple loading conditions under static constraints concerning nodal displacements, stresses in the members and critical buckling loads. This class of problems was chosen because truss structures are widely used in structural engineering [7, 8]. Also, examples of truss structure design optimization are extensively used in the literature to compare the efficiency of optimization algorithms [9-12]. Then, we introduce five examples of non-truss optimization problems under static constraints.

12.3 Design Benchmarks

12.3.1 Truss Design Problems

Truss optimization is a challenging area of structural optimization, and many researchers have tried to minimize the weight (or volume) of truss structures using different algorithms. For example, Maglaras et al. [13] compared probabilistic and deterministic optimization of trusses. They showed that probabilistic optimization provided a significant improvement. Hasancebi et al. [14] evaluated some well-known metaheuristic search techniques in the optimum design of real trusses and they found that simulated annealing and evolution strategies perform well in this area.

For most truss optimization problems, the objective function can be expressed as

$$\text{minimize } W(A) = \sum_{i=1}^{NM} \gamma_i A_i L_i \tag{12.4}$$

where W(A) is the weight of the structure; NM is the number of member in the structure; γ_i represents the material density of member *i*; L_i is the length of member i; A_i is the cross-sectional area of member i chosen between A_{min} and A_{max} (the lower bound and upper bound, respectively). Any optimal design also has to satisfy some inequality constraints that limit design variable sizes and structural responses [15].

The main issue in truss optimization is to deal with constraints because the weight of each truss structures can be simplified to an explicit formula [16]. Generally, a truss structure has one of the following three kinds of constraints:

Stress constraints: each member is under tensile or compressive strength so for each member of the structure, the positive tensile stress should be less than the allowable tensile stress (σ_{max}), while the compressive stress should be less than the allowable compressive stress (σ_{min}). In each truss optimization problem, we have 2NM stress constraints. These constraints can be formulated as follow:

$$\sigma_{i,\,min} \le \sigma_i \le \sigma_{i,\,max};\ i = 1, 2, \ldots, NM \tag{12.5}$$

Deflection constraints: The nodal deflections (displacement at each node) should be limited within the maximum deflection (δ_{max}). When a truss has NM^2=NM × NM nodes, it can be defined as:

$$\delta_j \le \delta_{j,\,max}\ j = 1, 2, \ldots, NM^2 \tag{12.6}$$

Buckling constraints: Buckling can be defined as the failure of a member due to a high compressive stress. In this case, the applied ultimate compressive stresses at the point of failure are higher than the bearing capacity of the member. When a member is in compression, the buckling status of the member is controlled according to the buckling stress (σ_b). Let NC denote the number of compression elements, and we have

$$\sigma_{k,b} \le \sigma_k \le 0;\ k = 1, 2, \ldots, NC \tag{12.7}$$

For truss optimization problems, there are only two constant mechanical properties: elastic modulus (E), and material density (γ). Structural analysis of each truss can readily be carried out using the finite element method.

12.3.1.1 10-Bar Plane Truss

This truss example is one of the most well-known structural optimization benchmarks [17]. It has been widely used by many researchers as a standard 2D benchmark for truss optimization (e.g., [16-18]). The geometry and loading of a 10-bar truss is presented in Figure 12.1. This problem has many variations and has been solved with only continuous or discrete variables. The main objective is to find minimum weight of the truss by changing the areas of elements, so it has 10 variables in total.

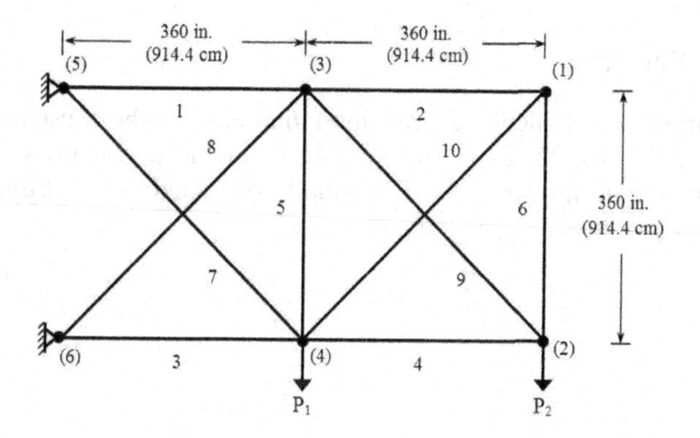

Fig. 12.1 10-bar truss structure

12.3.1.2 25-Bar Transmission Truss

This spatial truss structure has been solved by many researchers as a benchmark structural problem [19]. The topology and nodal numbers of a 25-bar spatial truss structure are shown in Figure 12.2 where 25 members are categorized into eight groups, so it has eight individual variables. This problem has been solved with various loading conditions (e.g. [10, 11, 20, 21]).

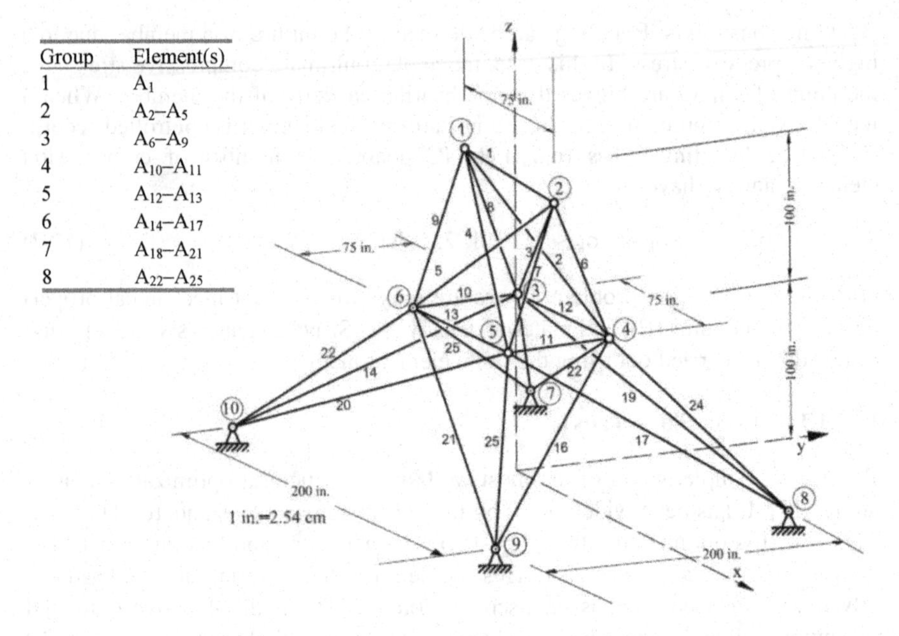

Group	Element(s)
1	A_1
2	A_2–A_5
3	A_6–A_9
4	A_{10}–A_{11}
5	A_{12}–A_{13}
6	A_{14}–A_{17}
7	A_{18}–A_{21}
8	A_{22}–A_{25}

Fig. 12.2 A twenty five-bar spatial truss [22]

12.3.1.3 72-Bar Truss

The 72-bar truss is a challenging benchmark that has also been used by many researchers (e.g., [9, 18, 22, 23]). As shown in Figure 12.3, this truss has 16 independent groups of design variables. It is usually subjected to two different loading inputs.

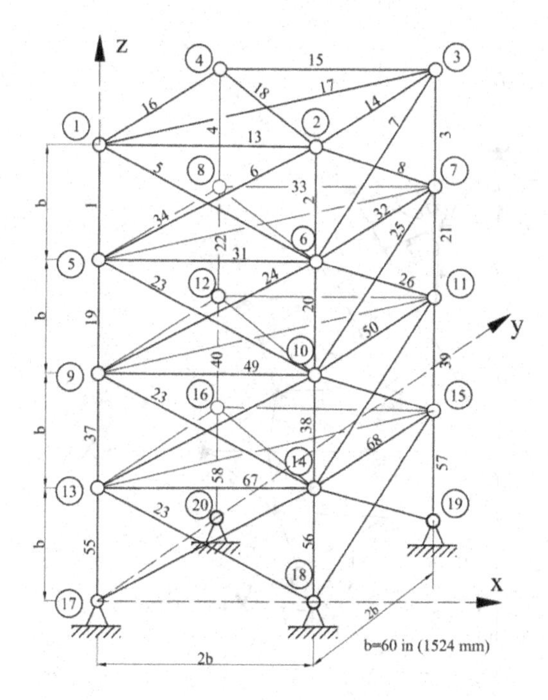

Group	Element(s)
1	A_1–A_4
2	A_5–A_{12}
3	A_{13}–A_{16}
4	A_{17}–A_{18}
5	A_{19}–A_{22}
6	A_{23}–A_{30}
7	A_{31}–A_{34}
8	A_{35}–A_{36}
9	A_{37}–A_{40}
10	A_{41}–A_{48}
11	A_{49}–A_{52}
12	A_{53}–A_{54}
13	A_{55}–A_{58}
14	A_{59}–A_{66}
15	A_{67}–A_{70}
16	A_{71}–A_{72}

Fig. 12.3 A 72-bar spatial truss [22]

12.3.1.4 120-Bar Truss Dome

The 120-bar truss dome is used as a benchmark problem in some researches (e.g., [10, 16]). This symmetrical space truss, shown in Figure 12.4, has a diameter of 31.78 m, and its 120 members are divided into 7 groups, taking the symmetry of the structure into account. Because of symmetry, the design of one-fourth of the dome is sufficient. The truss is subjected to vertical loading at all the unsupported joints. According to the American institute of steel construction (AISC) code for allowable stress design (ASD) [24] standards, the allowable tensile stress (σ_{max}) is equal to 0.6F_y (F_y is the yield stress of the steel), and the allowable compressive stress (σ_{min}) is calculated according to the slenderness.

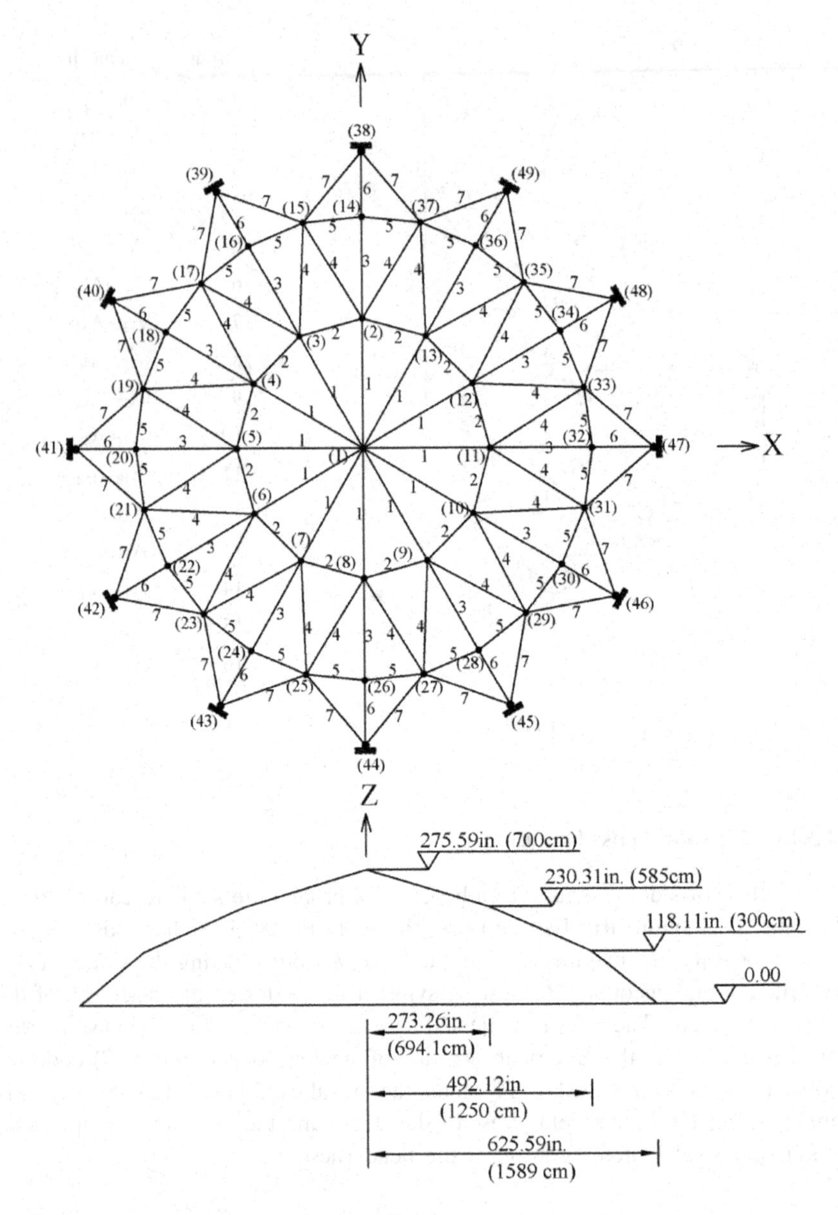

Fig. 12.4 A 120-bar dome shaped truss [10]

12.3.1.5 200-Bar Plane Truss

The benchmark 200-bar plane truss structure shown in figure 12.5 which has been solved in many papers with different number of variables. The 200 structural members of this planar truss has been categorized as 29 [11], 96 [25] or 105 [26] groups using symmetry in the literature. Some researchers have also solved it with

200 variables when each member is considered as an independent design variables [27]. This planar truss problem has been solved with three or five independent loading conditions [28].

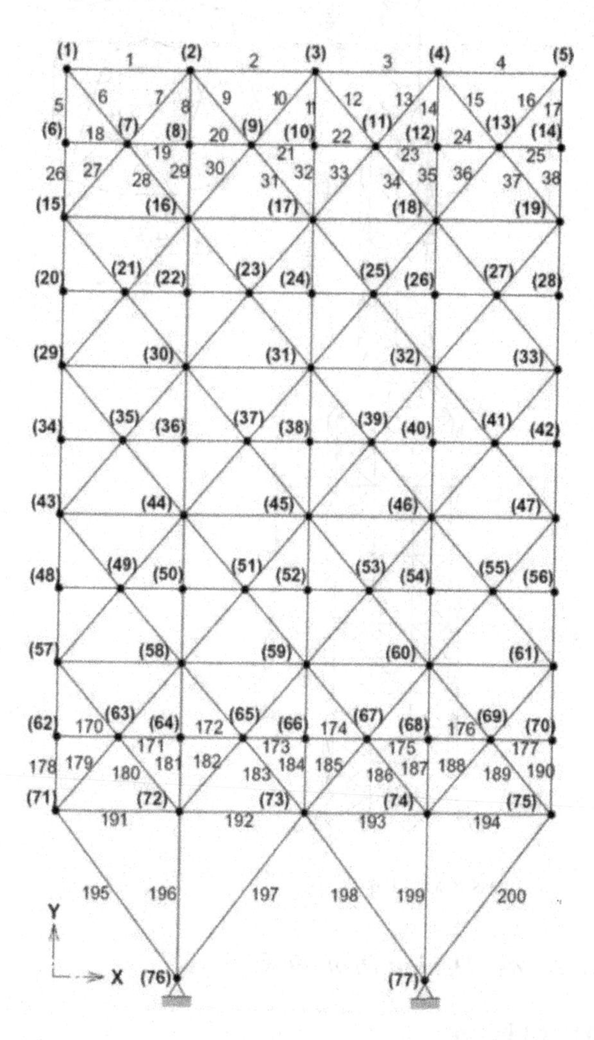

Fig. 12.5 A 200 bar plane truss

12.3.1.6 26-Story Truss Tower

Figure 12.6 shows the geometry and the element groups of the recently developed 26-story-tower space truss ([10, 11, 29-31]). This truss is a large-scale truss problem containing 244 nodes and 942 elements. In this truss structure, 59 element groups employ the symmetry of the structure. This problem has been solved as a continuous problem and as a discrete problem [30]. More details of this problem can be found in [31].

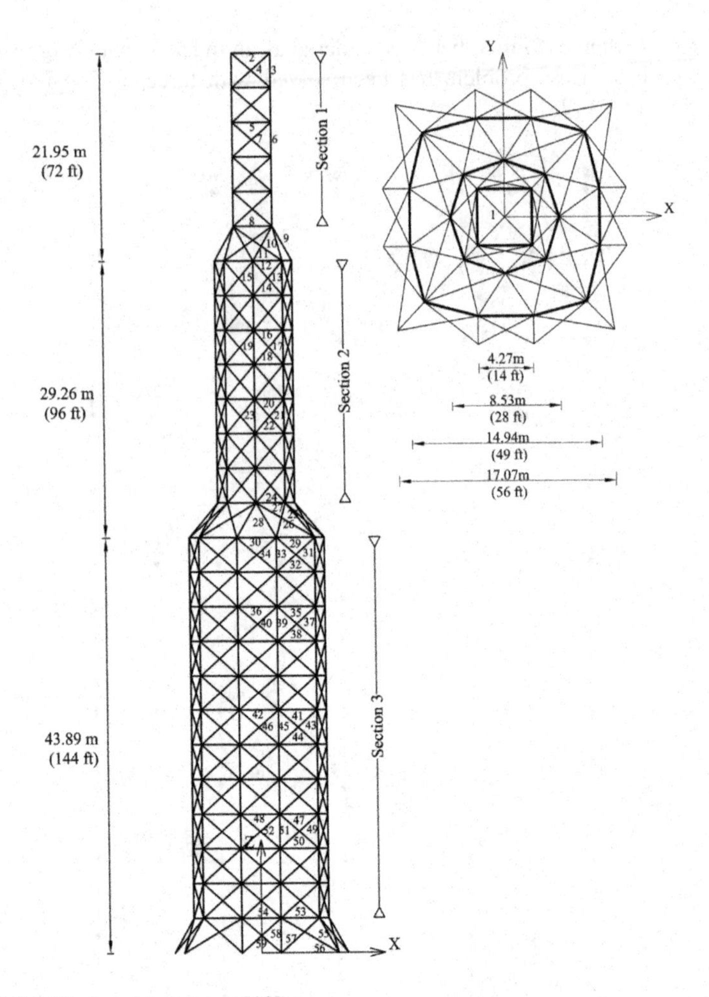

Fig. 12.6 A 26-story-truss tower [10]

12.3.1 Non-truss Design Problems

12.3.1.1 Welded Beam

The design of a welded beam which minimizes the overall cost of fabrication was introduced as a benchmark structural engineering problem by Rao [19]. Figure 12.7 shows a beam of low-carbon steel (C-1010), welded to a rigid support. The welded beam is fixed and designed to support a load (P). The thickness of the weld (h), the length of the welded joint (12.l), the width of the beam (t) and the thickness of the beam (b) are the design variables. The values of h and l can only take integer multiples of 0.0065, but many researchers consider them continuous variables [32]. The objective function of the problem is expressed as follows:

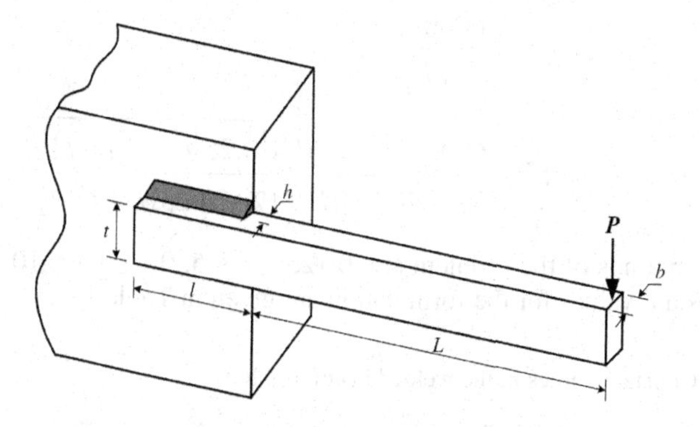

Fig. 12.7 Welded beam design problem

$$\text{Minimize: } f(h,L,t,b)=(1+C_1)h^2l+C_2tb(L+l) \tag{12.8}$$

subject to the following five constraints:
shear stress(τ)

$$g_1=\tau_d-\tau\geq 0 \tag{12.9}$$

bending stress in the beam (σ)

$$g_2=\sigma_d-\sigma\geq 0 \tag{12.10}$$

buckling load on the bar (P_c)

$$g_3=b-h\geq 0 \tag{12.11}$$

deflection of the beam (δ)

$$g_4=P_c-P\geq 0 \tag{12.12}$$

side constraints

$$g_5=0.25-\delta\geq 0 \tag{12.13}$$

where

$$\tau=\sqrt{(\tau')^2+(\tau'')^2+l\tau'\tau''/\sqrt{0.25(l^2+(h+t)^2)}} \tag{12.14}$$

$$\sigma=\frac{504000}{t^2b} \tag{12.15}$$

$$P_c=64746(1-0.0282346t)tb^3 \tag{12.16}$$

$$\delta=\frac{2.1952}{t^3b} \tag{12.17}$$

$$\tau' = \frac{6000}{\sqrt{2}hl} \quad (12.18)$$

$$\tau'' = \frac{6000\,(14+0.5l)\sqrt{0.25\left(l^2+(h+t)^2\right)}}{2\{0.707\,hl\,(l^2/12+0.25(h+t)^2)\}} \quad (12.19)$$

The simple bounds of the problem are: $0.125 \le h \le 5$, $0.1 \le l, t \le 10$ and $0.1 \le b \le 5$. The constant values for the formulation are given in Table 12.2.

Table 12.1 Constant values in the welded beam problem

Constant Item	Description	Values
C_1	cost per volume of the welded material	0.10471($/in3)
C_2	cost per volume of the bar stock	0.04811($/in3)
τ_d	design shear stress of the welded material	13600 (psi)
σ_d	design normal stress of the bar material	30000 (psi)
δ_d	design bar end deflection	0.25 (in)
E	Young's modulus of bar stock	30×10^6 (psi)
G	shear modulus of bar stock	12×10^6 (psi)
P	loading condition	6000 (lb)
L	overhang length of the beam	14 (in)

This problem has been solved by many researchers in the literature (e.g., [15, 33, 34]) here are two different solutions presented. One has an optimal function value of around 2.38 and the other one (with a difference in one of the constraints) has an optimal function value of about 1.7. Deb and Goyal [35] extended this problem to choose one of the four types of materials of the beam and two types of welded joint configurations.

12.3.1.2 Reinforced Concrete Beam

The problem of designing a reinforced concrete beam has many variations and has been solved by various researchers with different kinds of constraints (e.g., [36, 37]). A simplified optimization problem minimizing the total cost of a reinforced concrete beam, shown in Figure 12.8, was presented by Amir and Hasegawa [38]. The beam is simply supported with a span of 30 ft and subjected to a live load of 2.0 klbf and a dead load of 1.0 klbf including the weight of the beam. The concrete compressive strength (σ_c) is 5 ksi, and the yield stress of the reinforcing steel (F_y) is 50 ksi. The cost of concrete is \$0.02/in^2/linear ft and the cost of steel is \$1.0/in^2/linear ft. The aim of the design is to determine the area of the reinforcement (A_s), the width of the beam

(b) and the depth of the beam (h) such that the total cost of structure is minimized. Herein, the cross-sectional area of the reinforcing bar (A_s) is taken as a discrete type variable that must be chosen from the standard bar dimensions listed in [38]. The width of concrete beam (b) assumed to be an integer variable, and the depth (h) of the beam is a continuous variable. The effective depth is assumed to be 0.8h.

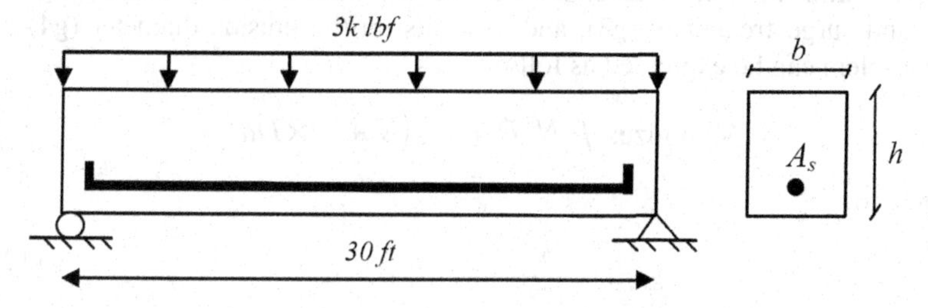

Fig. 12.8 Illustration of reinforced concrete beam

Then, the optimization problem can be expressed as:

$$\text{Minimize:} \quad f(A_s, b, h) = 2.9A_s + 0.6bh \tag{12.20}$$

The depth to width ratio of the beam is restricted to be less than, or equal, to 4, so the first constraint can be written as:

$$g_1 = \frac{h}{b} - 4 \leq 0 \tag{12.21}$$

The structure should satisfy the American concrete institute (ACI) building code 318-77 [39] with a bending strength:

$$M_u = 0.9A_sF_y(0.8h)\left(1.0 - 0.59\frac{A_sF_y}{0.8bh\,\sigma_c}\right) \geq 1.4M_d + 1.7M_l \tag{12.22}$$

where M_u, M_d and M_l are, respectively, the flexural strength, dead load and live load moments of the beam. In this case, M_d = 1350 in.kip and M_l = 2700 in.kip. This constraint can be simplified as [40]:

$$g_2 = 180 + 7.375\frac{A_s^2}{b} - A_sh \leq 0 \tag{12.23}$$

The bounds of the variables are b ϵ {28, 29, …, 40} inches, 5 ≤ h ≤ 10 inches, and A_s is a discrete variable that must be chosen from possible reinforcing bars by ACI. The best solution obtained by the existing methods so far is 359.208 with h = 34, b = 8.5 and A_s = 6.32 (15#6 or 11#7) using firefly algorithm [41].

12.3.1.3 Compression Spring

The problem of spring design has many variations and has been solved by various researchers. Sandgren [42] minimized the volume of a coil compression spring with mixed variables and Deb and Goyal [35] tried to minimize the weight of a Belleville spring. The most well-known spring problem is the design of a tension–compression spring for a minimum weight [43]. Figure 12.9 shows a tension–compression spring with three design variables: the wire diameter (d), the mean coil diameter (D), and the number of active coils (N). The weight of the spring is to be minimized, subject to constraints on the minimum deflection (g1), shear (g2), and surge frequency (g3), and to limits on the outside diameter (g4) [43]. The problem can be expressed as follows:

$$\text{Minimize: } f(N,D,d) = (N+2)\times Dd^2 \tag{12.24}$$

Subject to:

$$g_1 = 1 - \frac{D^3 N}{71785 d^4} \leq 0 \tag{12.25}$$

$$g_2 = \frac{4D^2 - Dd}{12566(Dd^3 - d^4)} + \frac{1}{5108 d^2} - 1 \leq 0 \tag{12.26}$$

$$g_3 = 1 - \frac{140.45 d}{D^2 N} \leq 0 \tag{12.27}$$

$$g_4 = \frac{D+d}{1.5} - 1 \leq 0 \tag{12.28}$$

where $0.05 \leq d \leq 1, 0.25 \leq D \leq 1.3$ *and* $2 \leq N \leq 15$.

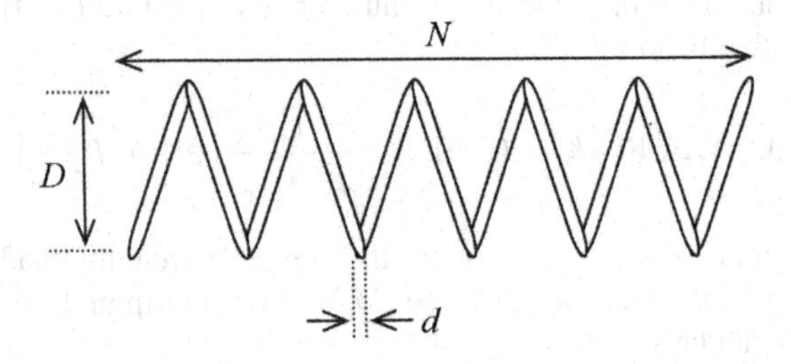

Fig. 12.9 Tension–compression spring

Many researchers have tried to solve this problem (e.g., [33, 44, 45]) and it seems the best results obtained for this problem is equal to 0.0126652 with d = 0.05169, D = 0.35673, N = 11.28846 using bat algorithm [46].

12.3.1.4 Pressure Vessel

Pressure vessel is a closed container that holds gases or liquids at a pressure, typically significantly higher than the ambient pressure. A cylindrical pressure vessel capped at both ends by hemispherical heads is presented in Figure 12.10. The pressure vessels are widely used for engineering purposes and this optimization problem was proposed by Sandgren [42]. This compressed air tank has a working pressure of 3000 psi and a minimum volume of 750 ft^3, and is designed according to the American society of mechanical engineers (ASME) boiler and pressure vessel code. The total cost, which includes a welding cost, a material cost, and a forming cost, is to be minimized. The variables are the thickness of shell (T_s), thickness of the head (T_h), the inner radius (R), and the length of the cylindrical section of the vessel (L). The thicknesses (T_s and T_h) can only take integer multiples of 0.0625 inch.

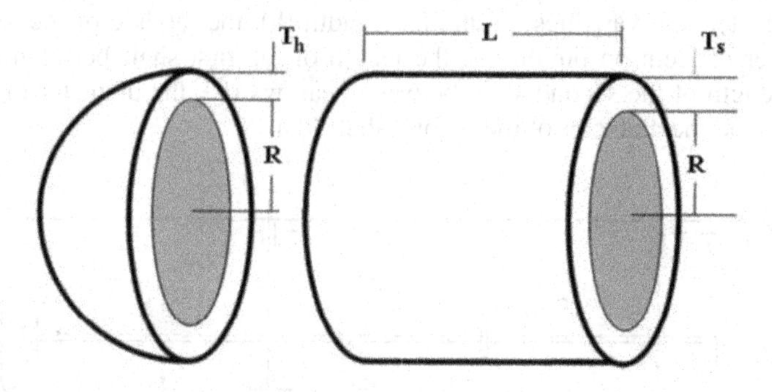

Fig. 12.10 Pressure Vessel

Then, the optimization problem can be expressed as follows:

Minimize: $f(T_s, T_h, R, L) = 0.6224T_sRL + 1.7781T_hR^2 + 3.1661T_s^2L + 19.84T_h^2L$ (12.29)

The constraints are defined in accordance with the ASME design codes where g_3 represents the constraint function of minimum volume of 750 feet^3 and others are the geometrical constraints. The constraints are as follow:

$$g_1 = -T_s + 0.0193R \leq 0 \tag{12.30}$$

$$g_2 = -T_h + 0.0095R \leq 0 \tag{12.31}$$

$$g_3 = -\pi R^2 L - \frac{4}{3}\pi R^3 + 750 \times 11728 \leq 0 \tag{12.32}$$

$$g_4 = L - 240 \leq 0 \tag{12.33}$$

where $1\times0.0625 \leq T_s, T_h \leq 99\times0.0625$, $10 \leq R$, and $L \leq 200$. The minimum cost and the statistical values of the best solution obtained in about forty different studies are reported in [47]. According to this paper, the best results are a total cost of \$6059.714. Although nearly all researchers use 200 as the upper limit of variable L, it was extended to 240 in a few studies (e.g., [41]) in order to investigate the last constrained problem region. Use this bound, the best result was decreased to about \$5850. It seems this variation may be a new challenging benchmarking problem. It should also be noted that if an approximate value for π is used in the g_3 constraint calculation, then the best result cannot be achieved (actually a smaller value will be obtain). Thus, the exact value of π should be used in this problem. From the implementation point of view, a more accurate approximation of π should be used.

12.3.1.5 Speed Reducer

A speed reducer is part of the gear box of mechanical system, and it also is used for many other types of applications. The design of a speed reducer is a more challenging benchmark [48], because it involves seven design variables As shown in Figure 12.11, these variables are the face width (b), the module of the teeth (m), the number of teeth on pinion (z), the length of the first shaft between bearings (l_1), the length of the second shaft between bearings (l_2), the diameter of the first shaft (d_1), and the diameter of the second shaft (d_2).

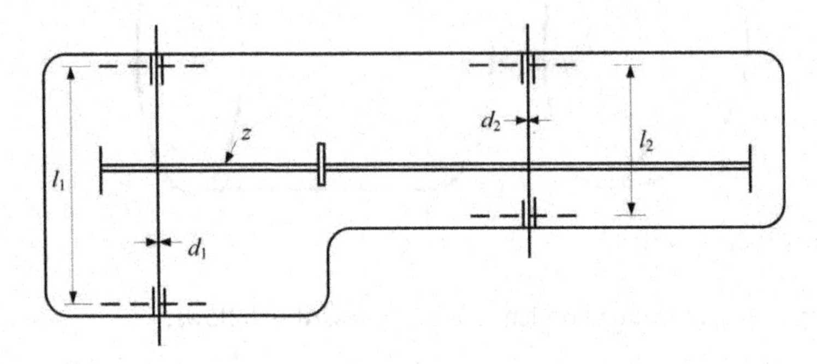

Fig. 12.11 Speed Reducer

The objective is to minimize the total weight of the speed reducer. There are nine constraints, including the limits on the bending stress of the gear teeth, surface stress, transverse deflections of shafts 1 and 2 due to transmitted force, and stresses in shafts 1 and 2.

The mathematical formulation can be summarized as follows:

$$\text{Minimize: } f(b,m,z,l_1,l_2,d_1,d_2) = 0.7854bm^2\left(3.3333z^2 + 14.9334z - 43.0934\right) - 1.508b\left(d_1{}^2 + d_2{}^2\right) + 7.477\left(d_1{}^3 + d_2{}^3\right) + 0.7854\left(l_1 d_1{}^2 + l_2 d_2{}^2\right) \quad (12.34)$$

Subject to:

$$g_1 = \frac{27}{bm^2 z} P - 1 \le 0 \quad (12.35)$$

$$g_2 = \frac{397.5}{bm^2 z^2} - 1 \le 0 \quad (12.36)$$

$$g_3 = \frac{1.93}{mzl_1{}^3 d_1{}^4} - 1 \le 0 \quad (12.37)$$

$$g_4 = \frac{1.93}{mzl_2^3 d_2^4} - 1 \le 0 \tag{12.38}$$

$$g_5 = \frac{\sqrt{\left(\frac{745 l_1}{mz}\right)^2 + 1.69 \times 10^6}}{110 d_1^3} - 1 \le 0 \tag{12.39}$$

$$g_6 = \frac{\sqrt{\left(\frac{745 l_1}{mz}\right)^2 + 157.5 \times 10^6}}{85 d_2^3} - 1 \le 0 \tag{12.40}$$

$$g_7 = \frac{mz}{40} - 1 \le 0 \tag{12.41}$$

$$g_8 = \frac{5m}{B - 1} - 1 \le 0 \tag{12.42}$$

$$g_9 = \frac{b}{12m} - 1 \le 0 \tag{12.43}$$

In addition, the design variables are also subject to the simple bounds listed in column 2 of Table 12.2. This problem has been solved by many researchers (e.g., [49, 50]) and it seems the best weight of the speed reducer is about 3000 (kg) [47, 51]. The corresponding values of this solution so far are presented in Table 12.2.

Table 12.2 Variables of the speed reducer design example

	Simple Bounds	Variables of the best solution
b	[2.6 - 3.6]	3.50000
m	[0.7 - 0.8]	0.70000
z	[17 – 28]	17.0000
l_1	[7.3 - 8.3]	7.30001
l_2	[7.3 - 8.3]	7.71532
d_1	[2.9 - 3.9]	3.35021
d_2	[5.0 - 5.5]	5.28665

12.3.1.6 Stepped Cantilever Beam

This problem is a good benchmark to verify the capability of optimization methods for solving continuous, discrete, and/or mixed variable structural design problems. This benchmark was originally presented by Thanedar and Vanderplaats [52] with ten variables, and it has been solved with continuous, discrete and mixed variables in different cases in the literature [8, 53]. Figure 12.12 illustrates a five-stepped cantilever beam with a rectangular shape. In this problem, the height and width of the beam in all five steps of the cantilever beam are the design variables, and the volume of the beam is to be minimized. The objective function is formulated as follows:

$$\text{Minimize:}\, V = D\left(b_1 h_1 l_1 + b_2 h_2 l_2 + b_3 h_3 l_3 + b_4 h_4 l_4 + b_5 h_5 l_5\right) \tag{12.44}$$

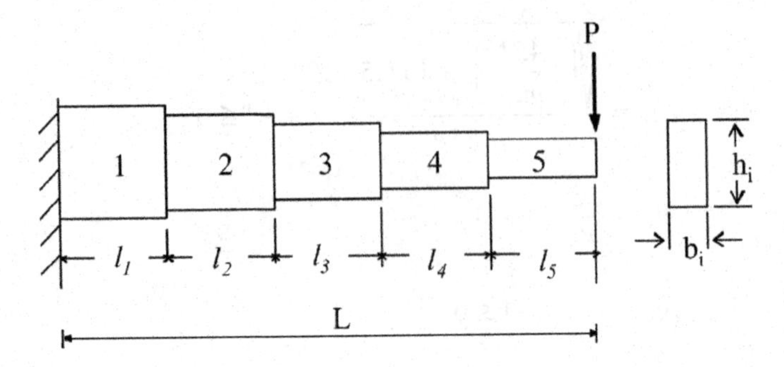

Fig. 12.12 A stepped cantilever beam

Subject to the following constraints:

- The bending stress constraint of each of the five steps of the beam are to be less than the design stress (σ_d):

$$g_1 = \frac{6Pl_s}{b_5 h_5^2} - \sigma_d \leq 0 \tag{12.45}$$

$$g_2 = \frac{6P\left(l_s + l_4\right)}{b_4 h_4^2} - \sigma_d \leq 0 \tag{12.46}$$

$$g_3 = \frac{6P\left(l_s + l_4 + l_3\right)}{b_3 h_3^2} - \sigma_d \leq 0 \tag{12.47}$$

$$g_4 = \frac{6P\left(l_s + l_4 + l_3 + l_2\right)}{b_2 h_2^2} - \sigma_d \leq 0 \tag{12.48}$$

$$g_5 = \frac{6P\left(l_s + l_4 + l_3 + l_2 + l_1\right)}{b_1 h_1^2} - \sigma_d \leq 0 \tag{12.49}$$

- One displacement constraint on the tip deflection is to be less than the allowable deflection (Δ_{max}):

$$g_6 = \frac{Pl^3}{3E}\left(\frac{1}{I_s} + \frac{7}{I_4} + \frac{19}{I_3} + \frac{37}{I_2} + \frac{61}{I_1}\right) - \Delta_{max} \leq 0 \tag{12.50}$$

- A specific aspect ratio of 20 has to be maintained between the height and width of each of the five cross sections of the beam:

$$g_7 = \frac{h_5}{b_5} - 20 \leq 0 \tag{12.51}$$

$$g_8 = \frac{h_4}{b_4} - 20 \leq 0 \tag{12.52}$$

$$g_9 = \frac{h_3}{b_3} - 20 \leq 0 \tag{12.53}$$

$$g_{10} = \frac{h_2}{b_2} - 20 \leq 0 \tag{12.54}$$

$$g_{11} = \frac{h_1}{b_1} - 20 \leq 0 \tag{12.55}$$

The initial design space for the cases with continuous, discrete and, mixed variable formulations can be found in Thanedar and Vanderplaats [52].

This problem can be used as a large-scale optimization problem if the number of segments of the beam is increased. When the beam has N segments, it has 2N+1 constrains including N stress constraints, N aspect ratio constraints and a displacement constraint. Vanderplaats [54] solved this problem as a very large structural optimization up to 25,000 segments and 50,000 variables.

12.3.1.7 Frame Structures

Frame design is one of the popular structural optimization benchmarks. Many researchers have attempted to solve frame structures as a real-world, discrete-variable problem, using different methods (e.g., [55, 56]). The design variables of frame structures are cross sections of beams and columns which have to be chosen from standardized cross sections. Recently, Hasançebi et al. [57] compared seven well-known structural design algorithms for weight minimization of some steel frames, including ant colony optimization, evolution strategies, harmony search method, simulated annealing, particle swarm optimizer, tabu search and genetic algorithms. Among these algorithms, they showed that simulated annealing and evolution strategies performed best for frame optimization.

One of the well-known frame structures was introduced by Khot et al. [58]. This problem has been solved by many researchers (e.g., [59, 60]), and now can be considered as a frame-structure benchmark. The frame has one bay, eight stories, and applied loads (see Figure 12.13). This problem has eight element groups. The values of the cross section groups are chosen from all 267 W-shapes of AISC.

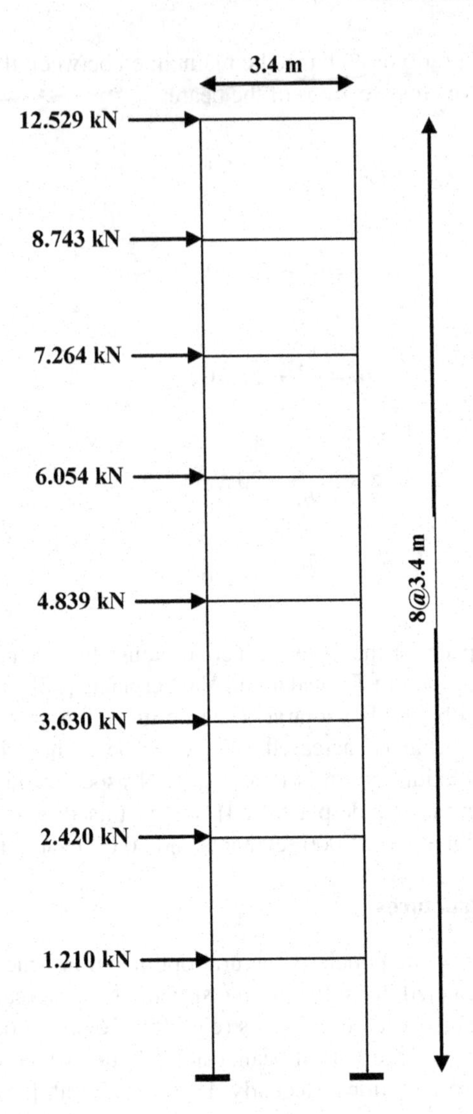

Fig. 12.13 The benchmark frame

12.4 Discussions and Further Research

A dozen benchmark problems in structural optimization are briefly introduced here, and these benchmarks are widely used in the literature. Our intention is to introduce each of these benchmarks briefly so that readers are aware of these problems and thus can refer to the cited literature for more details. The detailed description of each problem can be lengthy, here we only highlight the essence of the problems and provide enough references.

There are many other benchmark problem sets in engineering optimization, and there is no agreed upon guideline for their use. Interested readers can found more information about additional benchmarks in recent books and review articles [61, 62].

References

1. Iyengar, N.G.R.: Optimization in Structural design. DIRECTIONS, IIT Kanpur 6, 41–47 (2004)
2. Goldberg, D.E.: Genetic Algorithms in Search. In: Optimization and Machine Learning. Addison-Wesley, Reading (1989)
3. McCulloch, W.S., Pitts, W.: A logical calculus of the idea immanent in nervous activity. Bulletin of Mathematical Biophysics 5, 115–133 (1943)
4. Eberhart, R.C., Kennedy, J.: A new optimizer using particle swarm theory. In: Proceedings of the sixth international symposium on micro machine and human science, Nagoya, Japan (1995)
5. Yang, X.S.: Nature-Inspired Metaheuristic Algorithms. Luniver Press (2008)
6. Yang, X.S., Deb, S.: Cuckoo search via Levy flights. In: World Congress on Nature & Biologically Inspired Computing (NaBIC 2009), pp. 210–214. IEEE publication, Los Alamitos (2009)
7. Stolpe, M.: Global optimization of minimum weight truss topology problem with stress, displacement, and local buckling constraints using branch-and-bound. International Journal For Numerical Methods In Engineering 61, 1270–1309 (2004)
8. Lamberti, L., Pappalettere, C.: Move limits definition in structural optimization with sequential linear programming. Part II: Numerical examples, Computers and Structures 81, 215–238 (2003)
9. Kaveh, A., Talatahari, S.: Size optimization of space trusses using Big Bang–Big Crunch algorithm. Computers and Structures 87, 1129–1140 (2009)
10. Kaveh, A., Talatahari, S.: Optimal design of skeletal structures via the charged system search algorithm. Struct. Multidisc. Optim. 41, 893–911 (2010)
11. Gandomi, A.H., Yang, X.S., Talatahari, S.: Optimum Design of Steel Trusses using Cuckoo Search Algorithm (submitted for publication)
12. Floudas, C.A., Pardolos, P.M.: Encyclopedia of Optimization, 2nd edn. Springer, Heidelberg (2009)
13. Maglaras, G., Ponslet, E., Haftka, R.T., Nikolaidis, E., Sensharma, P., Cudney, H.H.: Analytical and experimental comparison of probabilistic and deterministic optimization (structural design of truss structure). AIAA Journal 34(7), 1512–1518 (1996)
14. Hasançebi, O., Çarbas, S., Dogan, E., Erdal, F., Saka, M.P.: Performance evaluation of metaheuristic search techniques in the optimum design of real size pin jointed structures. Computers and Structrures, 284–302 (2009)
15. Lee, K.S., Geem, Z.W.: A new structural optimization method based on the harmony search algorithm. Comput. Struct. 82, 781–798 (2004)
16. Iranmanesh, A., Kaveh, A.: Structural optimization by gradient-based neural networks. Int. J. Numer. Meth. Engng. 46, 297–311 (1999)
17. Kirsch, U.: Optimum Structural Design, Concepts, Methods and Applications. McGraw-Hill, New York (1981)
18. Lemonge, A.C.C., Barbosa, H.J.C.: An adaptive penalty scheme for genetic algorithms in structural optimization. Int. J. Numer. Meth. Engng. 59, 703–736 (2004)

19. Rao, S.S.: Engineering Optimization, 3rd edn. Wiley, New York (1996)
20. Patnaik, S.N., Gendys, A.S., Berke, L., Hopkins, D.A.: Modified fully utilized design (mfud) method for stress and displacement constraints. Int. J. Numer. Meth. Engng. 41, 1171–1194 (1998)
21. Kaveh, A., Rahami, H.: Analysis, design and optimization of structures using force method and genetic algorithm. Int. J. Numer. Meth. Engng. 65, 1570–1584 (2006)
22. Sonmez, M.: Discrete optimum design of truss structures using artificial bee colony algorithm. Struct. Multidisc. Optim. 43, 85–97 (2011)
23. Sedaghati, R.: Benchmark case studies in structural design optimization using the force method. International Journal of Solids and Structures 42, 5848–5871 (2005)
24. AISC-ASD Manual of steel construction-allowable stress design. 9th edn., American Institute of Steel Construction, Chicago (1989)
25. Dede, T., Bekiroğlu, S., Ayvaz, Y.: Weight minimization of trusses with genetic algorithm (2011), doi:10.1016/j.asoc.2010.10.006
26. Genovese, K., Lamberti, L., Pappalettere, C.: Improved global–local simulated annealing formulation for solving non-smooth engineering optimization problems. International Journal of Solids and Structures 42, 203–237 (2005)
27. Korycki, J., Kostreva, M.: Norm-relaxed method of feasible directions: application in structural optimization. Structural Optimization 11, 187–194 (1996)
28. Lamberti, L.: An efficient simulated annealing algorithm for design optimization of truss structures. Computers and Structures 86, 1936–1953 (2008)
29. Hasancebi, O., Erbatur, F.: On efficient use of simulated annealing in complex structural optimization problems. Acta. Mech. 157, 27–50 (2002)
30. Hasancebi, O.: Adaptive evolution strategies in structural optimization: Enhancing their computational performance with applications to large-scale structures. Comput. Struct. 86, 119–132 (2008)
31. Adeli, H., Cheng, N.T.: Concurrent genetic algorithms for optimization of large structures. J. Aerospace Eng. 7, 276–296 (1994)
32. Zhang, M., Luo, W., Wang, X.: Differential evolution with dynamic stochastic selection for constrained optimization. Information Sciences 178, 3043–3074 (2008)
33. Coello, C.A.C.: Constraint-handling using an evolutionary multiobjective optimization technique. Civil Engineering and Environmental Systems 17, 319–346 (2000)
34. Deb, K.: An efficient constraint handling method for genetic algorithms. Computer Methods in Applied Mechanics and Engineering 186, 311–338 (2000)
35. Deb, K., Goyal, M.: A combined genetic adaptive search (GeneAS) for engineering design. Comput. Sci. Informatics 26(4), 30–45 (1996)
36. Coello, C.A.C., Hernandez, F.S., Farrera, F.A.: Optimal Design of Reinforced Concrete Beams Using Genetic Algorithms. Expert systems with Applications 12(1), 101–108 (1997)
37. Saini, B., Sehgal, V.K., Gambhir, M.L.: Genetically Optimized Artificial Neural Network Based Optimum Design Of Singly And Doubly Reinforced Concrete Beams. Asian Journal of Civil Engineering (Building And Housing) 7(6), 603–619 (2006)
38. Amir, H.M., Hasegawa, T.: Nonlinear mixed-discrete structural optimization. J. Struct. Engng. 115(3), 626–645 (1989)
39. ACI 318-77, Building code requirements for reinforced concrete. American Concrete Institute, Detroit, Mich (1977)
40. Liebman, J.S., Khachaturian, N., Chanaratna, V.: Discrete structural optimization. J. Struct. Div. 107(ST11), 2177–2197 (1981)

41. Gandomi, A.H., Yang, X.S., Alavi, A.H.: Mixed Discrete Structural Optimization Using Firefly Algorithm (2010) (submitted for publication)
42. Sandgren, E.: Nonlinear Integer and Discrete Programming in Mechanical Design Optimization. J. Mech. Design 112(2), 223–229 (1990)
43. Arora, J.S.: Introduction to Optimum Design. McGraw-Hill, New York (1989)
44. Aragon, V.S., Esquivell, S.C., Coello, C.A.C.: A modified version of a T-Cell Algorithm for constrained optimization problems. Int. J. Numer. Meth. Engng. 84, 351–378 (2010)
45. Coello, C.A.C.: Self-adaptive penalties for GA based optimization. In: Proceedings of the Congress on Evolutionary Computation, vol. 1, pp. 573–580 (1999)
46. Gandomi, A.H., Yang, X.S., Alavi, A.H.: Bat Algorithm for Solving Nonlinear Constrained Engineering Optimization Tasks(2010) (submitted for publication)
47. Gandomi, A.H., Yang, X.S., Alavi, A.H.: Cuckoo Search Algorithm: A Metaheuristic Approach to Solve Structural Optimization Problems (2010) (submitted for publication)
48. Golinski, J.: An adaptive optimization system applied to machine synthesis. Mech. Mach. (1973)
49. Akhtar, S., Tai, K., Ray, T.: A socio-behavioural simulation model for engineering design optimization. Eng. Optmiz. 34(4), 341–354 (2002)
50. Kuang, J.K., Rao, S.S., Chen, L.: Taguchi-aided search method for design optimization of engineering systems. Eng. Optmiz. 30, 1–23 (1998)
51. Yang, X.S., Gandomi, A.H.: Bat Algorithm: A Novel Approach for Global Engineering Optimization (2011) (submitted for publication)
52. Thanedar, P.B., Vanderplaats, G.N.: Survey of discrete variable optimization for structural design. Journal of Structural Engineering 121(2), 301–305 (1995)
53. Huang, M.W., Arora, J.S.: Optimal Design With Discrete Variables: Some Numerical Experiments. International Journal for Numerical Methods in Engineering 40, 165–188 (1997)
54. Vanderplaats, G.N.: Very Large Scale Optimization. NASA/CR-2002 211768 (2002)
55. Degertekin, S.O.: Optimum design of steel frames using harmony search algorithm. Struct. Multidisc. Optim. 36, 393–401 (2008)
56. Greiner, D., Emperador, J.M., Winter, G.: Single and multiobjective frame optimization by evolutionary algorithms and the auto-adaptive rebirth operator. Computer Methods in Applied Mechanics and Engineering 193(33-35), 3711–3743 (2004)
57. Hasançebi, O., Çarbas, S., Dogan, E., Erdal, F., Saka, M.P.: Comparison of non-deterministic search techniques in the optimum design of real size steel frames. Computers and Structures 88, 1033–1048 (2010)
58. Khot, N.S., Venkayya, V.B., Berke, L.: Optimum structural design with stability constraints. Int. J. Numer. Methods Eng. 10, 1097114 (1976)
59. Kaveh, A., Shojaee, S.: Optimal design of skeletal structures using ant colony optimisation. Int. J. Numer. Methods Eng. 5(70), 563–581 (2007)
60. Camp, C.V., Pezeshk, S., Cao, G.: Optimized design of two dimensional structures using a genetic algorithm. J. Struct. Eng. ASCE 124(5), 551–559 (1998)
61. Alimoradi, A., Foley, C.M., Pezeshk, S.: Benchmark problems in structural design and performance optimization: past, present and future – part I. In: Senapathi, S., Hoit, C.K. (eds.) 19th ASCE Conf. Proc., State of the Art and Future Challenges in Structure. ASCE Publications (2010)
62. Yang, X.S.: Engineering Optimization: An Introduction with Metaheuristic Applications. John Wiley and Sons, Chichester (2010)

Author Index

Printed in the United States
By Bookmasters

Printed in the United States
By Bookmasters